D1662185

Smart Patches

Smart Patches

Biosensors, Graphene and Intra-Body Communications

Dominique Paret
Pierre Crégo
Pauline Solère

WILEY

First published 2023 in Great Britain and the United States by ISTE Ltd and John Wiley & Sons, Inc.

ISTE Ltd
27-37 St George's Road
London SW19 4EU
UK

www.iste.co.uk

John Wiley & Sons, Inc.
111 River Street
Hoboken, NJ 07030
USA

www.wiley.com

Library of Congress Control Number: 2023933069

British Library Cataloguing-in-Publication Data
A CIP record for this book is available from the British Library
ISBN 978-1-78630-709-5

Contents

Foreword

To begin with, we shall congratulate Dominique Paret, Pierre Crégo and Pauline Solère for taking the initiative to write a book describing, in detail, solutions relating to biosensors, graphene-based technology and Intra-Body Communications (IBC), or communications on the surface of the human body, and the numerous regulations, norms, applicative aspects, technical, technological and financial facets of the world of smart apparel.

There is relatively little technical literature on these subjects, despite how every day, we are seeing new applications emerge for health, wellbeing, sport, leisure, protection or security. In addition, France has an excellent position in the emerging markets for smart apparel, and by 2025–2030, there will be marvelous opportunities that are not to be missed. SMEs and intermediate-sized enterprises in textiles and apparel, and clients operating in application markets are gearing up to take advantage of such opportunities, as demonstrated by numerous recent initiatives (BPI, Techtera, Up-Tex, IFTH, etc.), predicting that "creative and productive fashion engineering in France will be awakened through innovation".

This highly thorough book is designed for beginner readers wishing to gain a full understanding of the complex fields of smart apparel, graphene technologies relating to biosensors or the Intra-Body Communications of today and of the future (for all applications). It is also for anyone designing such technologies, from the fundamental principles to their applications, wishing for a precise, detailed overall view.

This book has numerous laudable qualities. It re-examines the fundamental building blocks of disciplines relating to function, hardware and software, with the technological fundaments to describe possible architectures. It addresses the various communication protocols, implemented to establish connectivity, and makes designers aware of the various regulations and norms which must be adhered to. Finally, it discusses the essential subject of security at every step, including the processing of sensitive data, all by means of an excellent technical and scientific base.

Other aspects are highlighted by numerous examples, which make this book much more concrete for the readers, and will offer an understanding of the overall design of the chain of IBC solutions that are connected and secured, and their technical-financial development.

In addition, Paret and Crégo have long been renowned technical experts in numerous technologies for RFID, contactless chip cards, NFC, IoT and software development. This means that they are able to inject a high level of technicality into this book, in contrast to journalistic publications. Solère provides a fresh perspective on these various technical solutions.

We are grateful to them for sharing their expertise in this rapidly developing field: smart textiles and apparel. In 2017, the *Union des industries textiles* (Textile Industry Union) published a white paper on the subject. In addition, we applaud them for providing readers with a very thorough guided tour through this nascent industry, which sees textiles, electronics and communications technology woven together, to serve the needs of society today, and sometimes, of society tomorrow.

Buckle up, and enjoy the ride!

Florence BOST
Smart Textiles Designer
CEO of Sable chaud

Gaëlle LISSORGUES
Professor and head of department of Santé, Énergie
et Environnement (Health, Energy and the Environment)
ESIEE Paris
Université Gustave Eiffel

March 2023

Acknowledgements

There are many people to whom the authors owe a debt of gratitude – for their kindness, for lending their ears at great length, and for their constructive comments. Thus, to all of those people, who will indubitably know who they are, all three of us extend our heartfelt thanks!

In particular, we wish to thank the following people.

In the world of textiles

– Florence Bost, CEO of *Sable Chaud*, for her advice, comments and invaluable assistance in presenting the vast world of textiles.

– Laurent Houillon at the IFTH (*Institut français du textile et de l'habillement* – French Textile and Clothing Institute), who also serves as secretary to the BNITH (Bureau de normalisation des industries du textile et de l'habillement – Standardization Institute for Textile and Clothing Industries).

– Yesim Oguz-Gouillard, also from the IFTH and Group Leader on Smart Textiles for Healthcare & Medicine, for her technical expertise and unfailing good humor.

In the world of industry

We also wish to thank the following for their valuable assistance:

– Serge Gasnier and Elizabeth Patouillard, directors of the CRESITT in Orléans, France;

– Éric Devoyon, Technical Director at 3 ZA, for the many years spent working shoulder to shoulder on technical matters;

– Vincent Bouchiat, CEO and co-founder of Grapheal SAS, for his technical contributions and his friendship;

– Alain Rhélimi, Independent Technical Advisor and consultant expert in microelectronics, former expert with Schlumberger and Gemalto, for contributing his ideas;

– Emmanuelle Butaud-Stubbs, former Delegate General of the Union des industries textiles (UIT), for her undying friendship;

– Gaëlle Kermorgant, lawyer, legal consultant and lobbyist for personal data protection.

In the world of education

– Dr Gaëlle Lissorgues, Professor and thesis supervisor at the ESIEE (Université Gustave Eiffel, Marne-la-Vallée), and Head of Department for Santé, Énergie et Environnement, for her knowledge, advice and kindness.

– Dr Lionel Rousseau and Dr Magdalena Couty – respectively, the head of cleanrooms at ESIEE Paris and cleanroom engineer – for the wealth of advice they have provided us.

– Dr Ing. Delphine Bechevet, Associate Professor at HES-SO (*Haute école spécialisée de Suisse occidentale* – University of Applied Sciences and Arts of Western Switzerland), for her highly insightful reading of the draft manuscript, her advice, illuminating comments, time and friendship.

Our thanks go also to members of the panels of experts of "RGPD Associates" and "IBC Research" co-founded by the authors. In particular, we are grateful to Jean-Paul Huon, CEO of Z#BRE. He is a consultant in innovation and systems, a microelectronics engineer, a friend, and the co-author of *Secure Connected Objects* (published by ISTE in both French and English (Paret and Huon 2017)) – this book borrows a number of extracts from that one, in the interests of a complete and coherent picture. Jean-Paul is a specialist in the complete architecture of systems of communicating objects, numerous protocols and network technologies, high-level software applications and the cloud.

Finally, thanks to many, many friends, with whom we have shared good times.

Preface

To whom is this book addressed?

This book is written for anyone who, remotely or closely, needs to think about the design of fibers, cloths, textiles and apparel which come under the aegis of "smart apparel". Of course, it is also aimed at professionals and the many students in all of these fields, and/or people who are simply curious about this (relatively) new discipline. The subject covers a huge area, including multiple aspects pertaining to physics, biochemistry, technology, technicality, industry, marketing, etc. Quite deliberately, there is a fairly heady mixture of these disciplines in this book, meaning it can be useful to members of each of these various "clans" in bridging some of the gaps dividing them.

Technical level

There is no specific level of technical knowledge required to take advantage of this book – indeed, all readers are welcome. However, throughout, we seek to satisfy readers' curiosity and, in the process, raise their level of technical knowledge rather quickly.

Teaching style

Besides years of activity in the purely professional and industrial spheres, Crégo and Paret have long been teaching as well (at engineering schools) and training experts. Thus, the language and tone used are deliberately plain and accessible, but nevertheless highly precise and, in order to offer the complete picture, a very large number of examples of applications are presented. Throughout, the intention is to teach the reader because, in our minds, there is little point writing for our own sakes alone. In addition, we have provided numerous summary tables, secrets and anecdotes throughout the text. Simply put, this book is for *you*, for the pleasure of understanding and learning, for enjoyment. We remain both "technically" and "textile-ly" yours!

NOTE.– Of course, there are numerous points which have already been discussed at conferences or in our previous published works. With this in mind, certain repetitions are inevitable in this book, but unfortunately, this is the price to pay for this book to stand alone in this new domain. We therefore ask our faithful readers to forgive us for these repetitions, and to bear with us.

Preamble

To begin with, let us clarify a few points about this book:

– The topics discussed herein are by no means new. Hundreds of articles have been published, looking at each subject area individually in detail. However, there is a renewed interest in the discipline, due to numerous new applications and advances in certain technologies (components with very low energy consumption, highly integrated components/systems, energy harvesting, etc.).

– At the time of writing (in late 2021), there are still a few barriers to the concrete, industrial applications of these technologies. We describe these barriers in detail, and discuss how some of them could be overcome.

– In addition, this book is not – and is not intended to be – an encyclopedic treatment of all biosensor systems, particularly those based on graphene and Intra-Body Communications (smart and connected textiles and apparel). There are already a plethora of articles available online, addressing these subjects in varying degrees of detail. They set out wondrous and futuristic theories, discussing various markets and presenting commercial figures. For our part here, by drawing upon some of this pre-existing material, we aim to present constructive overviews of the various subdisciplines.

– Finally, to avoid needless and unproductive redundancy, we have focused solely on subjects about which there is a notable dearth of pre-existing literature – the day-to-day "nitty gritty" technical details of work in this field. Hence, this book is constructed to serve as a guide, helping readers to overlook nothing, and to avoid the pitfalls which may be encountered in designing and implementing patches, and smart, connected and secure apparel. It is all very well to speak eloquently about such matters, and to stage impressive demonstrations. However, it is another matter entirely – and a far greater achievement – to actually produce a smart, connected garment in real life for a sensible cost, and manage to sell it in large numbers at a sensible price. This is the ultimate goal, and the splash created by any project which cannot achieve it is, sadly, much ado about nothing.

March 2023

Introduction

Today's world – the world of the early 2020s – is characterized by major changes in all walks of life: our society, our environment, our way of living, the average age, the way in which we take care of our physical health, workplace legislation, the Covid-19 pandemic and all its repercussions, etc. Among the most common keywords encountered today are "Health", "Wellbeing", "Leisure", "Sport", "Working environment", "Safety", "Personal Protective Equipment" and a whole host of derivatives of these terms.

The authors have, for many years, been at the forefront of this evolution, and have published a number of books on the subject (among them, Paret and Huon (2017)). Through these earlier works, we began to observe a shift in the applications of wearables and smart textiles (Paret and Crégo 2018). Now, we wish to focus on even greater integration of electronic systems of autonomous patches into the world of smart apparel (SA), for professional or general use, including sensors and biosensors which, of course, communicate with the outside world, but also communicate with one another to make the system as a whole work. In addition, if we want these "smart garments" to be produced in large quantities on an industrial scale, then we now need to start looking at real, day-to-day applications, and large-scale production (as opposed to the production and laboratory at the Proof of Concept – "POC"). As part of this process, we need to ensure that costs are reasonable and that the products meet a range of criteria concerning their application (types of uses, soft cloth, resistant fabric, high durability in terms of number of washes, temperature, etc.). It may be that the dream is some years away from becoming a reality, but as Jules Verne said, it needs to be expressed one day… and in this case, that day is today!

For a color version of all figures in this introduction, see www.iste.co.uk/paret/smartpatches.zip.

To return to the matter, in reviewing the literature in the domain, we have found only highly specialized books, doctoral theses, major articles and/or treatises in (bio)chemistry, simplistic treatments, and popularization articles published by start-ups on a particular, specific aspect of this domain. With the exception of certain documents and books cited in the bibliography, there is a real dearth of literature in this domain. In addition, after having operated in the field for a long time, we have realized that there is a lack of knowledge of the real-world potential of electronics, and of the details of radiofrequency connectivity on the basis of the applications of such technology in the world of smart textiles, fabrics and garments, which is entirely understandable – everyone must have their own areas of expertise!

Following these observations and numerous discussions with professional colleagues and friends, we have once again taken our courage in hand – in this instance, six hands – to explore these domains and, in the hope that it fills a small part of this void, decided to write this book, which is primarily technical, designed around "(multi)-biosensors and patches built on graphene and Intra-Body Communications for and in smart apparel" for applications in healthcare, wellbeing, sport, leisure, etc. These technologies are expected to break into the mainstream market before long.

I.1. Aim of this book

The core of the authors' work is on technical applied research, which is likely to have real-world implications within four to eight years. To some, this may seem a long way off, but it is coming up fast. In that context, every day, we conduct detailed preliminary technical feasibility studies, often drilling down to the true feasibility of a project. Some years ago, after various discussions with clients, a particular project re-emerged: to make a smart garment, quite deliberately not for specialist professional use, designed for applications in the areas of health or wellbeing – a garment designed to help the wearer, using measurements taken by sensors or biosensors integrated into the fabric. The crux of the project was, of course, not new, but when looking at the subject in detail, we found that a great many matters still had to be resolved before it could become a reality.

NOTE.– In order to help readers come up with their own solutions to the questions that arise, this book considers the use of a "normal" garment: flexible, lightweight, comfortable and washable, rather than a professional, clinical or hospital garment. In that context, we set out to:

– examine the various techniques, technologies, materials, etc., which can be employed to produce patches with all sorts of biosensors and bioprocessors, which may or may not be able to be integrated into garments;

– examine the types of communication protocols which allow the patches, if necessary, to communicate with one another in a mini-network, other than by using hardwired or RF connections, which are completely inappropriate for wearer comfort;

– ensure that the whole system works, of course, without a power cell or battery which needs to be removed or replaced, etc.;

– all for a reasonable cost.

That is the point at which things become rather less simple…

Having set out the general context in which this book is situated, throughout, we will take readers through the fine details of everything behind its very long title.

I.2. How is this book constructed?

Let us take a look at the way in which this book is structured, in order to best serve its purposes.

In this field, we must be able to talk, simultaneously, about textiles, apparel, sensors, chemistry, biochemistry, biosensor technologies, electronics, communications, networks, etc., and legal language as well. It is quite a challenge to be highly skilled in all these areas at the same time. Thus, in putting together this book, we have been grateful for the cooperation of a number of experts, all specializing in one of these fields, so the book as a whole offers a coherent treatment of all these aspects. It is important to be able to see the bigger picture before we can think about designing smart apparel, be it for general public or professional use, in the fields of healthcare, wellbeing and sport.

Thus, this book aims to provide a simple, technical and accessible overview, but one which is clear and precise. We look firstly at smart apparel as a whole discipline, ranging from patches to biosensors. Secondly, we look at Body-Area Networks (BANs) in the broadest sense, and in particular, Intra-Body Communications as they relate to textiles. In addition, so that this multidisciplinary journey through techniques, economics and ergonomics in relation to the smart apparel of tomorrow is coherent and enjoyable, and so readers can easily orientate themselves, the book is structured into two main parts, followed by a conclusion.

I.3. Content and plan of the book

Over the course of the introduction, we will indicate the position of the book in the future landscape of patches for applications in health, wellbeing, sport, etc., including the topics of connected objects, hardware and software security, conventional microsensors, smart fibers/textiles/fabrics, IBC and new technologies for these applications.

– Part 1 – "Smart Apparel, Smart Patches and Biosensors"[1] – is split into two chapters. The first is dedicated to the specifics and boundaries of the field of apparel, and more specifically, so-called "smart apparel", implying garments which have integrated or on-board electronics. It should be noted, even at this early stage, that this is a vast discipline: in addition to the garment itself, we will need to look, in detail, at the numerous limitations and constraints (regulatory, health-related, etc.) that the uses of such "smart apparel" are subject to. Thus, we will examine how these sensors and patches can be worked into the industrial design. The second chapter, "Biosensors and Graphene Technology", takes an in-depth look at the most widely used techniques and biosensor technologies in healthcare, wellbeing, sport and similar applications. The technology needs to be able to withstand the strain of all sorts of clothes (being washed and ironed, etc.) that are worn on a daily basis (they need to be flexible and lightweight, etc.). This discussion will lead us down the paths of pure chemistry, biochemistry, biology, electronics, etc., and an examination of new high-performing materials such as *graphene,* which could serve as a possible generic platform for the design of numerous biosensors.

– Part 2 – "Biocontroller". At this stage, readers will know how to design and build a patch, including the bioprocessor (Chapter 3), its electronics, and a power supply through energy harvesting (of course, it must be batteryless) (Chapter 4). Then, there is one final step to be completed. The patch, the garment and the application may need to communicate with the outside world using their own communication resources, fully integrated into the garment. In addition, when multiple patches are located at different places in the garment, they may need to communicate with one another, locally, using a wireless network. At that point in the discussion, we will discover Intra-Body Communication (IBC) (Chapter 5), in which communications are sent through the human body itself, using the patches arranged in the garment.

---

1 This part was written with the kind help of Florence Bost, CEO of the textile company Sable Chaud, and Maîtres Naima Alahyane Rogeon and Isabelle Pottier, of the law firm Alain Bensoussan – Lexing.

– Conclusion – This book concludes with a detailed example: "Concrete realization of patches for smart apparel". The cost aspect will also be addressed, as will the performances and limits of IBC in such applications.

The journey of a thousand miles starts with a single step… so let us now step into the world of apparel, and get started!

I.4. Intended applications for "smart apparel" (SA): patches

Let us take a moment to remember the ultimate goal: to design a piece of smart apparel (abbreviated to *SA* throughout this book for reasons of space) for use in a wide range of applications such as healthcare, wellbeing and sport. Before proceeding any further, it is important to define how the apparel is to be made "smart" in relation to the intended applications. For this purpose, it needs to include a set of electronics, which, throughout this book, will be called a *patch,* which serves as a link between the individual's measured "bio" parameters, the garment and the services rendered or to be rendered.

I.4.1. *Functional diagram of a patch*

The functional overview of the technical content of such a patch is presented in Figure I.1. It is mainly made up of a biosensor (designed, for example, around graphene) and a (super)-biocontroller which has a range of sub-functions. Throughout this book, we will use this overview as the basis for the detailed description of the patch's functions and subfunctions.

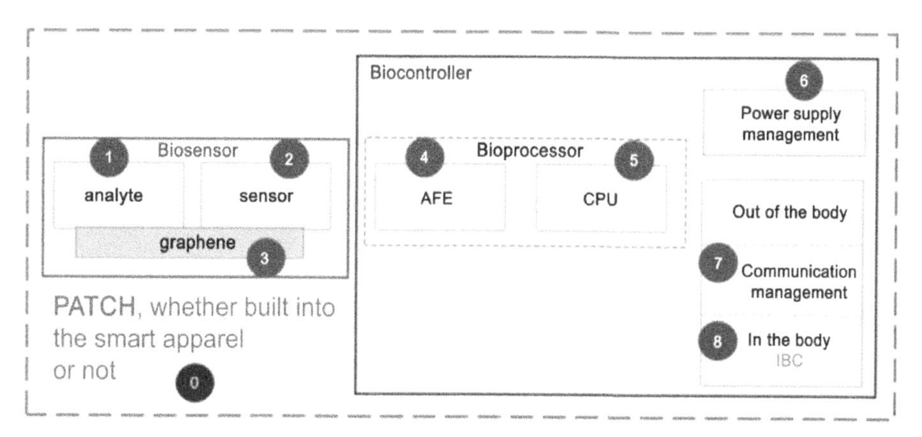

Figure I.1. *Overview of a "patch"*

We have just mentioned patches, but what are they, and what applications do they serve?

In the grand scheme of things, we hope to use patches whose edges have sensors and/or biosensors to measure and track individuals' biometric data. The patches could be stuck to the skin directly, or incorporated/integrated into the actual structure of flexible and lightweight normal clothes for applications in sport, wellbeing, senior activities, etc. They must cause no discomfort to the wearers – and thus be non-intrusive – and give the impression that the garment in question is absolutely normal (see Figure I.2).

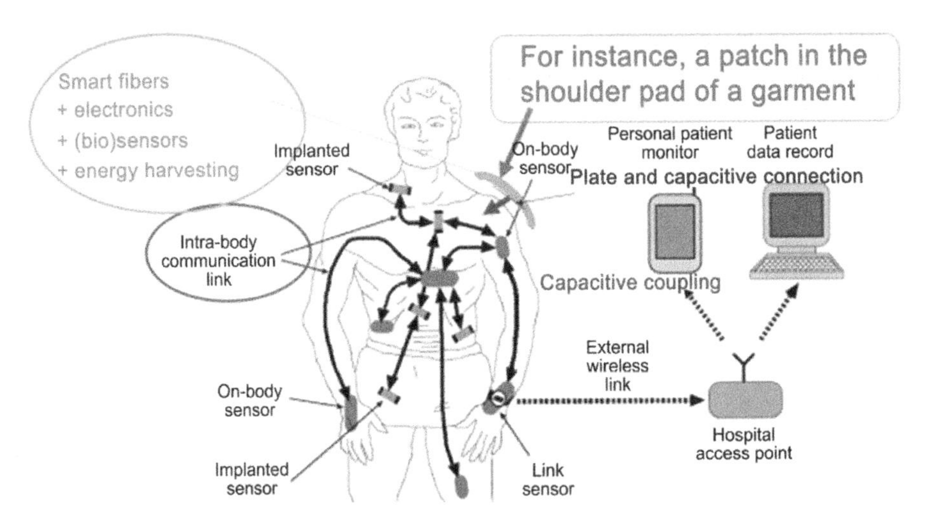

Figure I.2. *Generic example of applications with a patch directly integrated into the garment*

There are two main areas of work and applications in connection with biosensors.

I.4.2. *"Physical/mechanical" biosensors*

The production of *biosensors* (whether singly or as part of a network of multiple sensors) is termed *"physical/mechanical"*. Today, it is possible to manufacture generic fabrics which are capable of measuring data about the human body, through capacitance, resistance and bio-impedance (such as those typically used to measure heart rate). These sensors may be stuck to the skin in a pad, or integrated into the fabric directly (e.g. a piezoelectric sensor). We can take a look at a few examples.

Electroconductive thread can be used to produce a range of sensors which are flexible and do not adversely affect current apparel production processes or the quality of the end product (in terms of comfort, washability, mechanical properties, etc.). With this approach, conventional electroconductive textile threads are costly. However, thanks to vaporization and dyeing processes, it is possible to create wires which are washable, foldable and flexible. When doing this, we need to know how to formulate the most suitable material (for example, graphene) for the process in question. These wire threads are then woven together, and configured to respond to data such as pressure, temperature or sweat content, which we wish to measure and concentrate in a bracelet or a watch.

EXAMPLE I.1.– *Resistive pressure sensor*: made from two layers of cloth and a frame of parallel conductive paths incorporating a semiconductive element. Much of the technique here lies in being able to formulate the semiconductive element using modified graphene. The force of pressure alters the resistance in a repetitive, easily predictable manner (see Figure I.3).

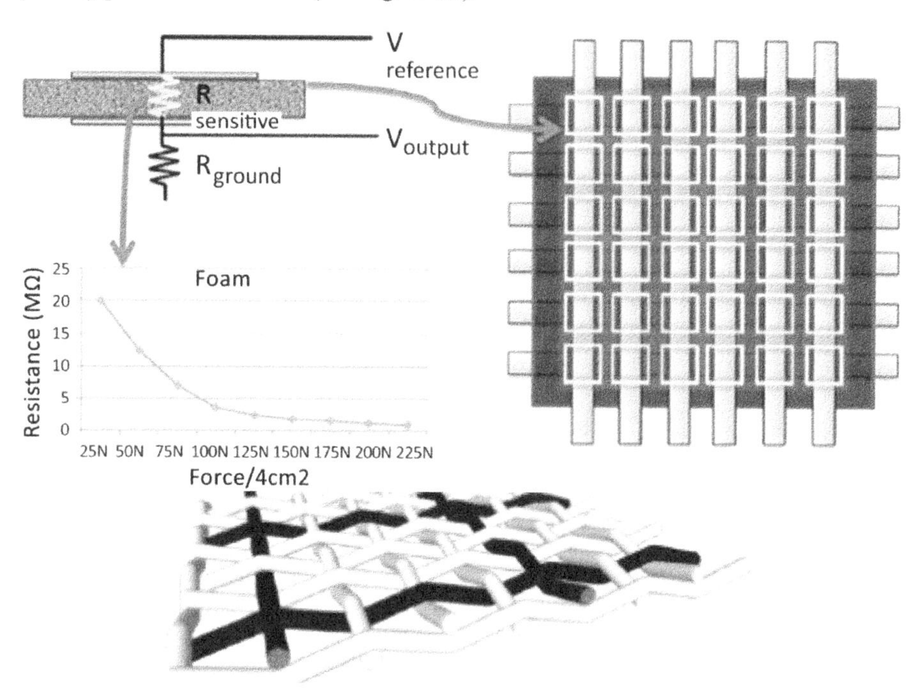

Figure I.3. *Example of a resistive pressure sensor (source: Sefar)*

EXAMPLE I.2.– *Woven temperature sensor:* connected fabrics, and devices such as bracelets or bandanas yield measurements which may vary depending on the point on the body at which they are measured, and which are therefore medically contestable. In this case, epidermic "patch" sensors can respond to medical requirements analyzed in terms of the form, portability and signal quality. Such connected fabrics can easily be put to work in other sectors, such as automobiles, transport and industry.

NOTE.– Patches made from graphene have the advantage of being biocompatible with the human body.

I.4.3. *"Biological/chemical" biosensors*

It is possible to create *biosensors* (again, singly or in a network) which are "*biological/chemical*", coated in an analyte (see Chapter 2), requiring direct contact (be it constant or intermittent) with a specific area of skin. They may or may not be connected to a piece of clothing (for example, a dressing). In both cases discussed above, the patch needs to be very small, very thin, lightweight (batteryless)… in short, purpose made.

In the interests of exhaustivity, these two areas of thinking about biosensors must be subdivided again, into so-called "isolated" patches and "local networks of patches".

I.4.4. *An "isolated" patch*

In the simplest of cases, a single, "isolated" patch may be worn by the user. It may simply serve as a dressing, for example. It will have its own power supply, and communicate "on request" with a reader – for example, by an NFC or Bluetooth connection.

I.4.5. *Patch networks*

In the case of other applications, it may be necessary or even absolutely compulsory to have multiple patches within the same smart garment. Those patches may:

– be completely isolated from one another, and each operate completely independently;

– or, from time to time, need to communicate with one another and form a mini local connection known as a body-area network (BAN).

In the latter case, we need to choose the way in which the patches are to communicate with one another. The options are to use:

– Wired connections: we can reject this solution out of hand, as it would make the garment uncomfortable to wear and the technology is now somewhat outdated.

– Radiofrequency (RF) connections: in principle, this is a suitable solution. However, whether using Bluetooth or other technologies, it will mean that additional components need to be present, which will drive up costs and therefore run counter to the economical goal of our project.

– Approaches based on:

- "Galvanic" operation: in this case, the patch must be in direct contact with the skin. This requirement may be problematic for the user, so as a matter of preference, we shall rarely use this method in the projects discussed in this book.

- "Capacitive" operation: in this case, the patch is not in direct contact with the skin, or only periodically comes into contact with the skin. The connection is made by capacitive coupling between the patch and the skin. In principle, this is far trickier to achieve, from a technical standpoint, but is much more advantageous in terms of the device being non-intrusive, and it opens up the possibility of integrating the electronics into the garment itself.

It is this latter type of operation which we shall primarily discuss in this book. The patch will be (able to be) applied directly to the skin, or integrated into the garment. Figure I.4 offers an overview of the possibilities and paradigms which can be chosen.

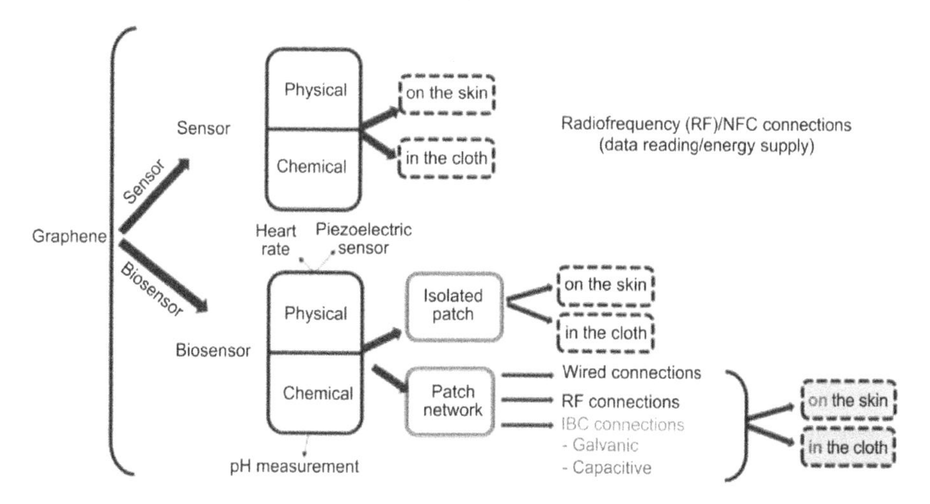

Figure I.4. *Paradigms for patch design and production*

I.5. Fields of application for patches in SA

Technical expertise is all very well, but unfortunately, we need concrete projects to live (read, generate turnover, etc.). Sooner or later, then, we have to look at the applications and commercial aspects. Thus, this book mainly focuses on areas of activity in which e-textiles and SA can be of help.

In order to define the technical and commercial context in which this book is situated, we will begin with a few general remarks about the industrial field of textiles and apparel, the related economy, market and policy. We will also discuss high-added-value "smart apparel" projects in domains such as medicine, healthcare, wellbeing, security, protection with professional protective wear, or traceability, theft prevention, etc. In addition, beyond these projects, the technology will be applied in sport, fashion, home decorating, creative arts, etc., within five years, industrially and commercially. The future is already here!

Let us briefly examine the main driving forces behind these markets.

I.5.1. *Healthcare and medicine*

There are vast ranges of possible applications for these patches in these markets, which are always growing, due to a range of factors:

– In France, one of the consequences of the demographic transition is a reduction in fertility and an increase in life expectancy. This leads to an increase in the proportion of the population aged 65 or over, which rose from 13.9% in 1990 to 18.8% in 2016. Aging is expected to continue in the European population, and by 2050, the proportion of people aged 65 and over will be 28.5%.

– Territories with a low population density have a high proportion of elderly people. Thus, the main issue facing these populations is the problem of isolation and access to services.

"Health patches" that can communicate externally, therefore, are one of the potential solutions to the isolation of the elderly, because they can be remotely connected to doctors, who can directly monitor various health parameters. In addition, for wellbeing and to address the problem of an aging population, these technologies could help lower the costs of the solutions put forward, allowing care to be provided and ensuring users/patients can stay in their own homes, in good conditions.

It is therefore important to:

– define an architecture for remote supervision of dependent people by means of a connected garment, in order to facilitate interventions by caregivers and helpers when required;

– interconnect a set of smart patches to supervise the wearer in either proximity or remote mode;

– create applications for the temporary transfer of rights and/or physical access from person to person, for private or sensitive spaces (secure rooms in pharmacies, reserved areas, etc.);

– take readings, locally and remotely, of the person's physiological data in an IAAS architecture (for example, monitoring of diabetes patients by a connected diabetic shoe);

– provide a range of sensor patches, connected sensors and continuous surveillance of at-risk patients (Covid-19 brought about a high level of teleconsultation and lockdown in elderly care homes).

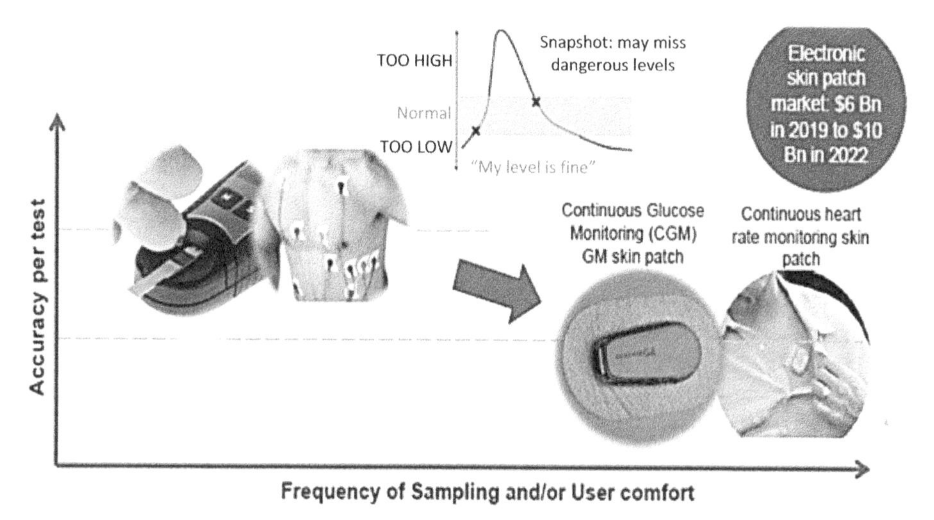

Figure I.5. *Examples of applications, from a simple sensor to a patch*

I.5.2. *Sport, wellbeing and leisure*

The market in the fields of sport, wellbeing and leisure is also highly dynamic, thanks to the rise in popularity of sports and health disciplines: running, aquabiking, fitness, yoga (especially for women), and the desire for the quantified self, which leads to the pursuit of improved personal performances. In these environments and in the context of wellbeing, it is a matter of providing reassurance and preventing risky situations: defining a range of connected "second skins" such as kneepads, bras and stretchy bands whose applications would be: remote supervision and measurement of physical efforts for training sessions and competitions.

I.5.3. *Personal Protective Equipment and working conditions*

On a professional level, the market in industrial/professional personal protective equipment (PPE) and monitoring of working conditions is regularly growing, leading to a high demand for lightweight, strong and interactive materials, in both civilian and military markets (we will come back to this point):

– employees or people exposed to high-risk situations (noise, heat, high physical intensity, such as in sports);

– vehicle access control (vehicle–driver pairing);

– drive-time monitoring;

– isolated workers;

– augmented reality;

– posturology;

– measuring physical effort;

– physical and logical access control;

– control of access to sensitive areas (Seveso, operators of vital importance, major administrative bodies under the military programming law).

Figure I.6. *Example of a PPE jacket/parka (source: DuPont)*

Note that in particular among the elements listed, which take part in or may have a role to play in the monitoring of parameters in these areas, there are two main branches of items which often include electronics: patches that are directly stuck to the skin or patches which are included in clothes, making them into "smart" apparel.

I.6. Market scope

The potential applications for such products, which have a commercial future in the fields of health, wellbeing, sport, leisure and industry are presented briefly in Figure I.7, which, for clarity's sake, gives a few numbers relating to the size of these markets.

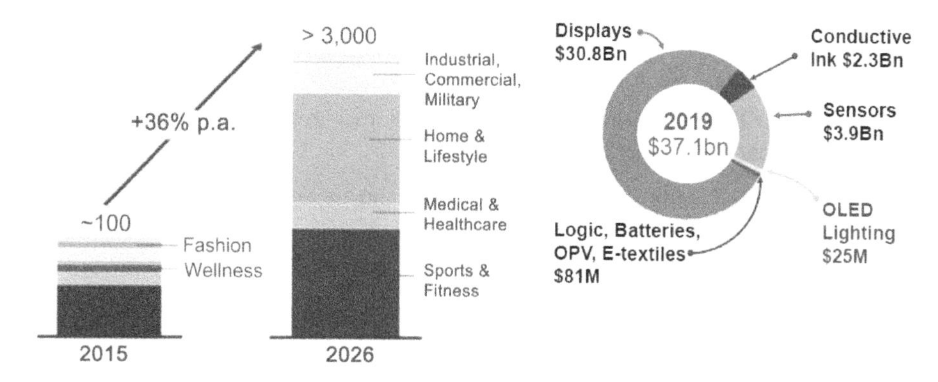

Figure I.7. *Size of the market in printed electronics, organic and flexible technologies*

I.6.1. *Defining a business model*

In parallel to and in advance of all the above, in the patch and its immediate environment, before burying ourselves in a project that may easily last months on end, it is important to have a concrete idea of its financial viability. It is therefore important to focus on and define the types of business models that can be envisaged, and then on that basis, make an overall determination of where the economic interest of such a concept lies, at what level, and for whom. For this reason, we need to estimate the final price of the patch, the price of the system, types of uses, jobs, requirements, types of potential users, sellers and buyers etc.

Let us look, for example, at the business model in Figure I.8, relating to tele-surveillance or remote assistance of a person.

A fine-grained analysis of the economic model of this value chain reveals that the price is heavily linked to the level of purchasing of the sensor material (for example, quality graphene). Also, though, the purpose of the project is to collect health data so that the users can monitor the progress of some of their vital statistics in real time. In the business model for the project, the sale of health data represents one of the major financial benefits, and in addition, the exploitation/sale of the information harvested by all the applied patches seems to offer a stable, long-lasting channel for generating profit. In time, those who win from this business model will be those who are able to:

– produce quality graphene (exfoliated, liquid, etc.) at reasonable prices;

– produce integrated sensors (multisensors) on an industrial scale and at low cost;

– produce IoT applications, including sensors that are widely connected and autonomous in terms of energy supply;

– provide cloud architectures for data processing with a significant "artificial intelligence" element in order to predict human behavior, the evolution of human health, and that of plant health.

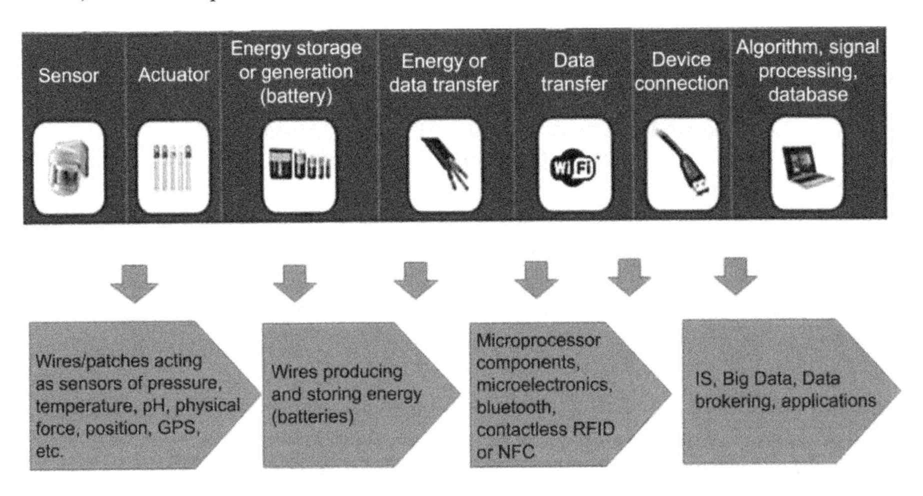

Figure I.8. *Value chain of a telesurveillance solution*

I.7. A few remarks

In view of the dearth of patch manufacturers working with graphene, there is a great deal of added value in the design and manufacture. There is also a lot of added value in patches with energy autonomy. At present, these sensors are more often at the laboratory development stage rather than mass production. However, there is a plethora of studies and prototypes on the market.

With respect to microelectronic components, the market does offer mature components to which improvements could be made in terms of power consumption and software handling, because there are not many API-type software applications in this sector.

In terms of processing Big Data and organizing the sale of such data, the studies to define the different business models for the possible applications must also take the issues of cybersecurity, and the constraints of the GDPR in each sector (medical, wellbeing, etc.), amongst other things, into account.

Having now thoroughly outlined the situation as it stands, let us begin by examining the constraints that a patch must meet in the apparel sector.

PART 1

Smart Apparel, Smart Patches and Biosensors

1

Smart Apparel, Smart Patches and the Related Constraints

1.1. Reminders and definitions

We have already detailed a large number of points relating to smart, communicating fibers, textiles, fabrics and apparel in an earlier work (Paret and Crégo 2018). However, for those who are new to the field, let us begin here with a brief description of the main families of textiles[1].

1.1.1. *Main families of textiles*

Generally speaking, the world of textiles can be subdivided into a variety of categories. Two very broad ones are "ordinary textiles" and "technical textiles" (TTs).

1.1.1.1. *Ordinary textiles and technical textiles (TTs)*

These two large families are generally defined as follows:

– normal textiles cover "clothing and homeware";

– TTs are "textiles for technical and professional uses".

The names reflect the purposes for which they were originally designed. The same is true of ordinary textiles. A more detailed view of what these categories contain is given in Figure 1.1.

---

1 This first chapter has been written with the invaluable help of Florence Bost, and documents that she uses at conferences and training sessions organised by her company, *Sable Chaud* (Hot Sand).

	Names	Applications
Technical textiles (TTs)	Agrotech	Medtech, Medtech, Sporttech, Clothtech, Mobiltech, Mobiltech, Agrotech, Indutech, Geotech, Protech, Buildtech, Hometech
	Buildtech	
	Mobiltech	
	Geotech	
	Protech	
	Oekotech	
	Sporttech	
	Medtech	
	Hometech	
	Clothtech	
	Packtech	
	Indutech	
Ordinary textiles (clothing, household linen)	Outdoor activities and sports	
	Wellbeing	
	Decoration	
	Clothing	

Figure 1.1. *Ordinary textiles and technical textiles. For a color version of this figure, see www.iste.co.uk/paret/smartpatches.zip*

Often, in the eyes of the general public, there is an overlap between some of these categories, because sometimes, the boundaries between them are rather fuzzy or flimsy, or poorly defined. In reality, despite the seemingly close fields of application, as shown in Figure 1.2, there are often separate worlds which must not be confused. Examples include:

– Medtech and wellbeing;

– Clothtech and clothing;

– Sporttech and outdoor sports;

– Hometech and decoration.

Technical textiles		Ordinary textiles (clothing, household linen)	
Agrotech	Sporttech	Outdoor activity and sport	–
Buildtech	Medtech	Wellbeing	–
Mobiltech	Hometech	Decoration	–
Geotech	Clothtech	Clothing	–
Protech	Packtech	–	–
Oekotech	Indutech	–	–

Figure 1.2. *Areas of confusion between ordinary textiles and technical textiles. For a color version of this figure, see www.iste.co.uk/paret/smartpatches.zip*

Let us now briefly examine some textile systems.

1.1.1.2. *Smart textile systems and types thereof*

As ever, we shall begin with a few definitions.

1.1.1.2.1. Smart textile material

A "smart textile material" is a functional textile material (a technical textile) which actively interacts with its environment – i.e. which responds or adapts to changes occurring in its surroundings.

1.1.1.2.2. Textile system

A "textile system" is a set of components, both textile and non-textile in nature, integrated within a product which retains the properties of the textiles (for example, clothing, a carpet or a mattress).

1.1.1.2.3. Smart textile system

A "smart textile system" is a textile system displaying a pre-determined, exploitable response either to changes in its environment or to an external signal.

1.1.1.2.4. Types of textile systems

According to the European standardization body, CEN, smart textile materials and systems, which have the ability to interact with their environment, are characterized by two functions: firstly their energy function, and secondly that of external communication. This gives us the definition of four categories of terms, presented in the matrix in Figure 1.3.

		Energy function	
		Without	**With**
Communication function	**Without**	NoE-NoCom	E-NoCom
	With	NoE-Com	E-Com

Figure 1.3. *Definition of "smart textile systems" (source: CEN)*

This figure clearly shows that there may be smart textile systems that do not have an energy function or communication ability, but are able to interact with their environment by means of two main functions:

– External communication by means of actuators, sensors and an information-management device. An example would be a garment made with shape-memory material.

– An energy function, using optical fibers, conductive wires, thermal heating, and fluorescent textiles, which make use of specific properties provided by the material, its composition, construction or finish.

Note that the majority of smart textile systems are able to fulfill at least one of these functions – for example, communication (respiration sensors) or energy (a backpack or pocket harvesting photovoltaic energy), or even both (a heat detector in a firefighter's jacket which triggers a warning light).

1.1.1.3. *Levels of integration of electronics into textiles*

The normative work of CEN Working Group WG31 distinguishes between four levels of integration of electronic components, which add functions to textiles. Each level brings different constraints in terms of development and technological possibility. Different legislation also applies to each category. They have differing effects on the human body (which must be taken into account when conducting risk assessments) and different implications for product safety.

Let us briefly examine each of these four levels of integration.

1.1.1.3.1. Level-1 integration

"The integrated electronic component can be removed without damaging the product." The components can be treated as separate units, so there is no need for separate standards for products such as these.

> EXAMPLE.– The textile and the electronics are side by side – The electronic component is attached to the textile by external pieces, and remains structurally sound in its own right. Examples include:
>
> – a jacket with channels designed to contain headphone cables, for entertainment use;
>
> – a jacket with a screen which can be removed, in order to clean, wash and iron the garment.

1.1.1.3.2. Level-2 integration

"The electronic component is attached to the textile in such a way that it is impossible to remove it without destroying the product."

> EXAMPLE.– This is a hybrid solution – The electronics are more closely coupled with the textile. In principle, they are made flexible and washable so that the garment can be cleaned. An example is a jacket with headphones built into the hood. In this case, the components can no longer be treated separately, and the system must be addressed as a whole.

1.1.1.3.3. Level-3 integration

"One or more components are textiles or have a textile-like finish. They are combined with electronic components which are connected, permanently or semi-permanently, to the textile matrix."

> EXAMPLE.– This is an integrated solution – The electronics are integrated into the textiles, and even woven into the threads. An example is a light-emitting diode (LED) connected to a woven conductive fabric. Depending on whether or not the components are removable, the consequences of levels 1 and/or 2 of integration need to be considered. The limitations of textile-based electronic components at level 3 must also be taken into account.

1.1.1.3.4. Level-4 integration

"All the components of the electronic device are textiles or have textile-like finishing (a fully textile solution)." The limitations inherent to textile-based electronic components need to be taken into account. In the majority of cases, it is necessary to develop dedicated standards for these types of systems or components.

EXAMPLE.– An intrinsic solution – The electronics are made of textiles.

1.1.1.3.5. In summary

Figure 1.4 summarizes the normative definitions of the four possible levels of integration of electronics in connected apparel.

CEN WG 31 nomenclature	**Level-1 integration**	**Level-2 integration**	**Level-3 integration**	**Level-4 integration**
Technology	**Electronics connected to the fabric**		**Electronics woven into the fabric**	
	Removable	Non-removable.	Integrated into the fabric.	Fabric/wire is integral.
	The electronic component can be removed from the smart textile without destroying the product.	The electronic component is attached to the textile, so it is impossible to remove it without destroying the product.	One or more components are textiles or have a textile-like finish. They are combined with electronic components which are connected, permanently or semi-permanently, to the textile matrix.	All components in the electronic devices are textiles or have a textile-like finish (a completely textile solution).
	The electronics are removable so that the components can be cleaned, washed and ironed.	The electronics are attached more closely to the textile and, in principle, are made flexible and washable like the textile itself.	The electronics are integrated into the textile and even into the threads.	The electronics are part of the construction of the textile itself.
	Parrot jacket: the screen is removable.	Google jacket: Bluetooth tag in the button.	–	–

Figure 1.4. *Levels of integration of electronics in connected garments*

1.1.1.4. *The three families of smart textiles*

Let us now look at so-called *smart textiles*. From a technological standpoint, this branch is divided into three categories:

– "active textiles";

– "e-textiles", which are a combination of textiles and electronics; this book discusses this family only, and e-clothing;

– "ecotechno-textiles".

1.1.1.4.1. Active function textiles

In principle, we can define three categories of "active function textiles":

– Active textiles: these fabrics have the peculiarity of emitting or diffusing molecules, light or heat, and have the ability to transition from one state to another. They have applications in the fields of cosmetics, paramedicine and safety. The technologies associated with active textiles include micro-encapsulation, luminescence, shape-memory polymers, thermochromism, etc.

– E-textiles: these fabrics need the passage of electrical currents in order to work. Usually, these currents serve to supply an electronic device, but may also serve directly to generate heat in the case of resistive wires.

– Ecotechno-textiles: this category covers all textiles whose creation, manufacture and application obeys a sustainable approach. The direct goal, however, is not usually one of eco-friendliness.

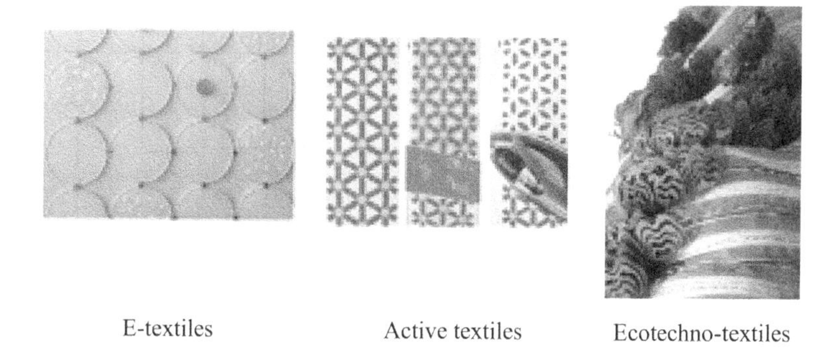

Figure 1.5. *Examples of "active function textiles". For a color version of this figure, see www.iste.co.uk/paret/smartpatches.zip*

Having set out this brief inventory, looked at some of the usual vocabulary and introduced our subject, touching on the near future, we shall now turn to the shape of this vast world, and the markets which it involves.

1.1.1.5. *Main areas of use of smart textiles*

The main fields of application for e-textiles (all types) are those which are highlighted in Figure 1.6.

Technical textiles		Ordinary textiles (clothing, household linen)	
Agrotech	Sporttech	Outdoor activity and sport	–
Buildtech	Medtech	Wellbeing	–
Mobiltech	Hometech	Decoration	–
Geotech	Clothtech	Clothing	–
Protech	Packtech	–	–
Oekotech	Indutech	–	–

Figure 1.6. *Predominant fields of use of e-textiles. For a color version of this figure, see www.iste.co.uk/paret/smartpatches.zip*

1.1.1.6. *Predominant fields of use of smart apparel*

Following this foray into e-textiles, we turn our attention to smart apparel, with onboard electronics, whose main applications are those which are highlighted in Figure 1.7.

Technical textiles		Ordinary textiles (clothing, household linen)	
Agrotech	Sporttech	Outdoor activity and sport	–
Buildtech	Medtech	Wellbeing	–
Mobiltech	Hometech	Decoration	–
Geotech	Clothtech	Clothing	–
Protech	Packtech	–	–
Oekotech	Indutech	–	–

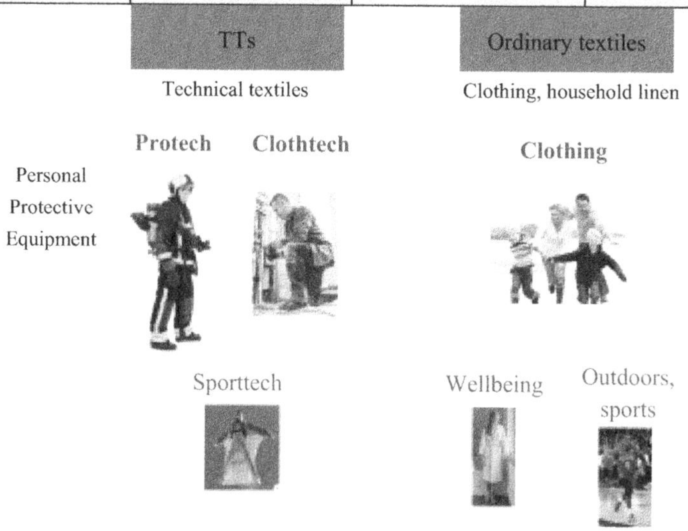

TTs
Technical textiles
Ordinary textiles
Clothing, household linen
Protech
Clothtech
Clothing
Personal
Protective
Equipment
Sporttech
Wellbeing
Outdoors,
sports

Figure 1.7. *Fields of application of smart apparel. For a color version of this figure, see www.iste.co.uk/paret/smartpatches.zip*

1.1.2. *Apparel*

"Apparel" covers a range of areas. Let us focus on two particular points.

1.1.2.1. *Conventional clothing*

Whatever the electronic component that we add to a garment, it will follow the standard value chain for clothing, the architecture of which is represented in Figure 1.8. After the manufacture of the fabrics, the next step is "finishing".

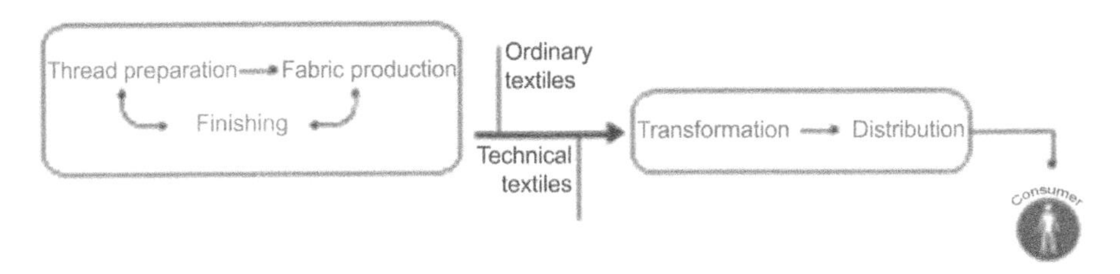

Figure 1.8. *The standard chain of activity for clothing. For a color version of this figure, see www.iste.co.uk/paret/smartpatches.zip*

1.1.2.1.1. Finishing

"Finishing" is the generic name for the various stages of decorative final touches, and the techniques which, during and/or after its manufacture, lend added value to the fabric by altering the way it feels to the touch, its appearance or its properties. These operations are carried out no matter what the family of textiles (warp and weft, mesh, non-woven, etc.). There are three main areas within the finishing of materials and textiles, taking place:

– During *thread preparation* (reeling), over the course of all operations to transform textile materials into thread.

– During *coloring* (textile dyeing and printing). Dyeing is a process whereby a color is applied to a medium. The purpose of this operation is to permanently apply a colorant into the material being handled, by means of penetration, and so to color the whole fabric.

– During *finishing* (mechanical and chemical priming). Printing is the stage of finishing which is carried out on the so-called "loom-finished" fabric. Methods other than printing are also available to create specific types of fabrics. Priming refers to a series of operations that textiles undergo at the end of the manufacture process.

Hence, the patches must be able to stand up to the finishing processes wanted or needed for the applications of the textiles we wish to use.

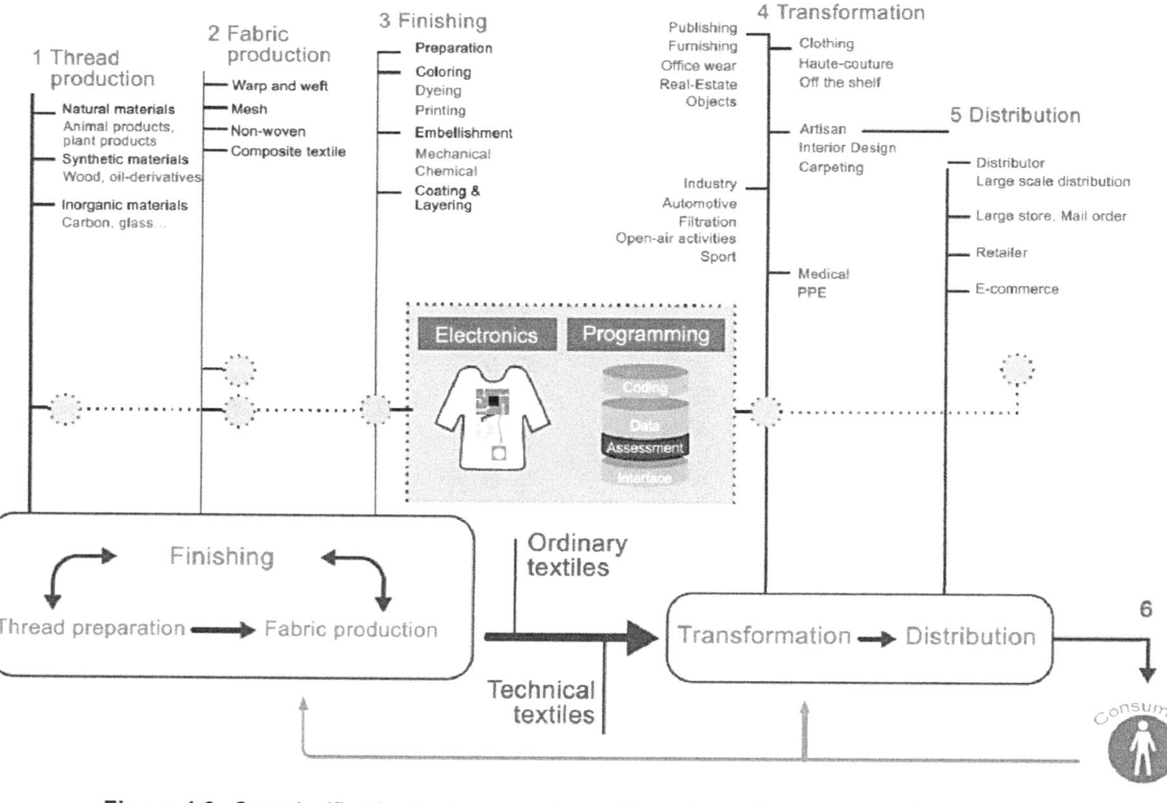

Figure 1.9. *Complexification in the case of e-textiles: introduction of electronic components. For a color version of this figure, see www.iste.co.uk/paret/smartpatches.zip*

1.1.2.2. *Smart apparel*

For the purposes of this book, we need to look a little more closely at the process of complexification shown in Figure 1.9, in the case of e-textiles into which electronic components are inserted. In this figure, we can see three possible levels of integration of the electronics into the garment:

– during the thread reeling;

– during the production of the material;

– during the finishing process.

1.2. The smart textile market from a consumer's point of view

There are four important aspects that shape the views expressed by end users – that is, ordinary people – in relation to the purchasing of smart apparel. Which levers can be used to persuade consumers to buy smart apparel? Which constituent elements most often present a barrier to the purchase of a smart garment? Which types of solutions can increase consumer confidence? Which are the most relevant devices which would facilitate the buying of a smart garment?

1.2.1. *Purchase levers*

The levers which can persuade consumers to buy, which smart apparel must have in order to be attractive, are their unique nature, the performances the product delivers, the innovative aspect of the product, and an attractive price point.

1.2.2. *Barriers to the purchase of smart apparel*

Below are the main barriers that exist and/or are perceived/cited by consumers and the general public when buying:

– price, lack of perceived utility, complexity in use, and lack of comfort in the solutions;

– data exchanges and data security;

– fear of breakdowns, and lack of after-sales services;

– general health risks, sensitivity to electromagnetic waves, and the effects of such waves on the brain or reproductive organs.

These are crucial points in relation to smart apparel, whether connected or otherwise, and a few remarks must be made about these points.

Lack of perceived utility, lack of comfort in the products, price, complexity in use, and the absence of after-sales services are problems that stem from a choice. They need to be addressed in the strategy employed by the company making the garments; in their design of a product line in relation to the market; their choice of marketing strategy; their choice of short-, medium- and long-term pricing policies; and the implementation of an after-sales service policy and strategy; etc.

1.2.2.1. *Personal data security*

Readers are invited to refer to the sections on the GDPR relating to social obligations for the protection of personal data and security. Succinctly put, it is easy to implement data security, and any good designer of SA should know what they need to do in this area. Therefore, it merely becomes a question of the company's sales direction and marketing strategy, which need to take data security into account. The obligation to comply with the GDPR is a reality, and it is not something which is particularly difficult to do.

1.2.2.2. *Concern about breakdowns*

As is the case in any system used by the general public (and/or a professional system, of course), there is a most extensive range of usage conditions which must be satisfied. Consequently, product design and development, which genuinely take account of the product's range of potential applications in the field, and well-written documentation that clearly and understandably sets out its concrete applications, will help to prevent a large proportion of concerns about breakdowns, which then become matters of misuse by the consumers, who have been properly informed before using the products.

1.2.2.3. *Health risks*

The health risks which are sometimes cited in this field often relate to electromagnetic sensitivity and the effects of electromagnetic waves on the brain or the reproductive organs. In this domain, the information carried by the general press ("I heard it through the grapevine") is often tendentious or improperly informed, throwing up certain barriers to the purchase of smart apparel and many other types of products. Badly needed responses to these concerns are included in the section on RF and health, which examines the regulations on RF pollution and health, and the effects of EM waves on the human body in the event of prolonged exposure to such radiation. That discussion will lay many readers' fears to rest. With that said, every professional in the world of smart apparel is supposed to be well informed about the law. In addition, they must comply with the technical regulations (issued by the ETSI, for example) and health regulations (issued by such bodies as the ANES, ICNIRP, the WHO, etc.) in place, and have measurements of the health effects taken by an independent authority in order to reassure consumers. Furthermore, the electrical and electronic components must not be irritant, allergenic, or otherwise dangerous to the users (dangers may include the risk of cutting, burning, fire, electrocution or explosion).

1.2.3. *Solutions to instill confidence*

What are the solutions that can be implemented to instill confidence in consumers? Which mechanisms are most effective in persuading consumers to buy a connected textile? For information, according to a study carried out in early 2017 (already five years ago, at the time of writing), the following factors have an impact, in order of decreasing importance:

– a formal study on the impact on individuals' health;

– a quality label borne by the product;

– direct feedback from other consumers;

– ease of use;

– positive reviews on the Internet;

– favorable reviews in the press;

– a good after-sales service.

1.2.4. *The hype curve for innovations*

Before embarking on a project to create smart apparel, we may legitimately ask whether the market for smart or connected apparel is sustainable. In order to answer this question, let us look at the "Hype cycle" published by the Gartner Group (Figures 1.10 and 1.11). This curve, whilst not always absolutely accurate, is usually not too far from the truth either. It offers a projected view of the development of emerging technologies. This hype cycle (or curve) comprises five key phases in terms of a technology's visibility and maturity.

– *Phase 1*: the *Technology Trigger*. At the very outset of a new branch of activity, there are a great many innovative ideas circulating – some good, some less good, some idealistic but not particularly constructive, etc. This creates buzz and interest in the media. Some of the most enterprising will begin creating their future startups in their garages. At this stage, only mock-ups/prototypes (otherwise known as *Proof of Concept*, POC) are available, and the products' commercial viability has not been proven.

– *Phase 2*: the *Peak of Inflated Expectations*. Publicity has led others to follow in the footsteps of the original innovators. There are a great many new players entering the field, and numerous startups, micro-enterprises, SMEs and SMIs are being founded. At this point, we are beginning to hear *success stories* about businesses flourishing, but there is also some *bad buzz* from other quarters. This is the time to swing into production and make the product available, because the public harbor great expectations at this point.

– *Phase 3*: the *Trough of Disillusionment*. There is almost always a period of slowdown, linked to the fact that new products are not always available in time and do not completely live up to expectations, there are too many different products available, prices are still somewhat too high, there is a lack of standardization in the newly emerging market, there are too many proprietary protocols and standards, with little or no interoperability between the different systems, etc. In short, public interest dips, and companies need to decide whether they want/are able to invest to bring the product into line with what *early adopters* in the market really want. During this phase, a great many startups fail and fall by the wayside, due to lack of liquidity, funding, capital increases, assistance packages, and a solid financial footing. We therefore see many "crashes", and the broken corpses of companies litter the field.

– *Phase 4*: the *Slope of Enlightenment*. Now, the project is reaching its final development phase. The surviving companies are gaining a greater understanding of the market which is actually available to them. During this phase, we see the formation of joint ventures, common interest groups, and buyouts of the best startups by larger companies or groups. This is the time when second or third generations of the product are issued.

– *Phase 5*: the *Plateau of Productivity*. Finally, the market as it truly is becomes clear; the use of the technology becomes more widespread, and it is finally adopted by an "early majority". The viability criteria are more clearly defined, the relevance of the innovation is more convincingly proven, and finally, the companies begin making profit. And about time, too!

NOTE.– The durations and extents of these five phases vary depending on the technologies and the markets in which they emerge. Certain products may reach the plateau of productivity within two years; with others, it may take a decade; and others may find their technology obsolete before ever reaching that point.

Drawing upon real-world experience, Gartner succeeded in defining some 100 reference curves, spread across different technological sectors: e-commerce, telemedicine, transport, software, wearables, smart apparel, etc.

Interest in smart apparel for the general public should arise once the technology becomes truly mature – that is, in three to five years, so between 2024–2026. When it reaches the plateau of productivity, people's interest will experience another upswing.

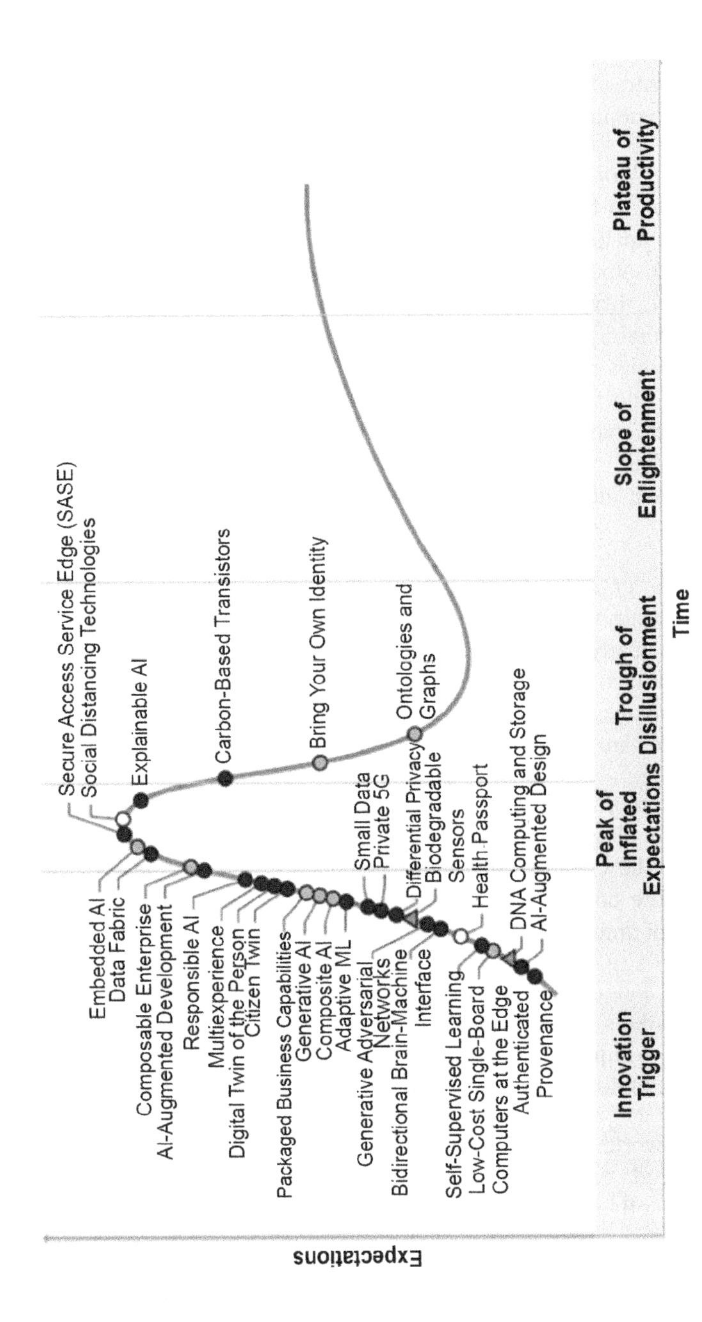

Figure 1.10. *Hype cycle for technologies (source: Gartner, August 2020). For a color version of this figure, see www.iste.co.uk/paret/smartpatches.zip*

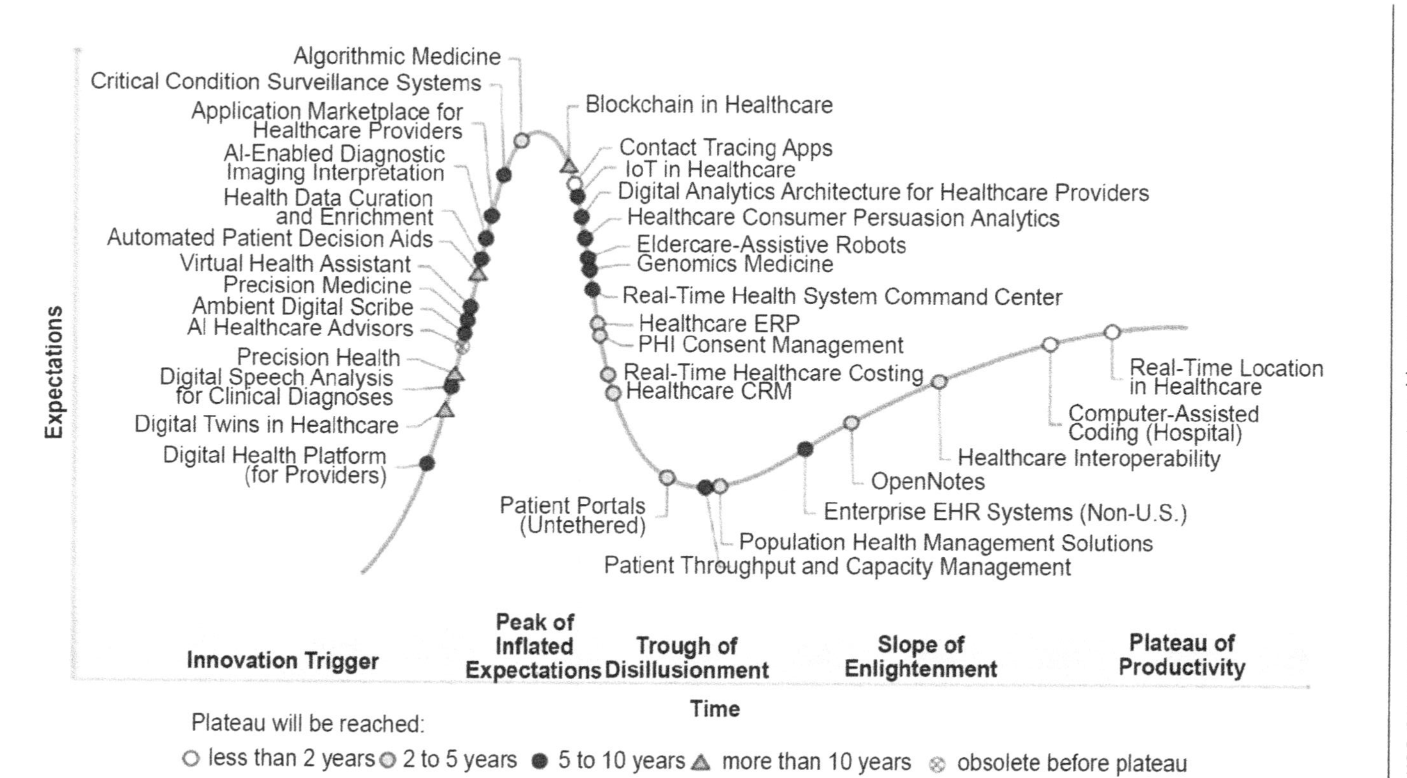

Figure 1.11. *Hype cycle for healthcare technologies (source: Gartner, August 2020). For a color version of this figure, see www.iste.co.uk/paret/smartpatches.zip*

As indicated previously, this book will describe the basic steps that must be followed and observed for a project to avoid falling into the trough of disillusionment. These steps can also ensure that the project proceeds directly from the initial innovation to healthy production, or allow the project to escape from the immaterial, virtual world and become a real-world reality. With this goal in mind, the following sections guide readers through how to comply with the constraints of the regulations and norms – technical, financial, ergonomic, etc. The journey may seem a little long, and somewhat daunting, but it holds no surprises. You have been warned!

1.3. Constraints surrounding an SA project

Before embarking on a project, whatever it may be, it is always wise to know where to tread, otherwise we risk severe handicaps. The numerous sections which follow set out the majority of concerns and "questions that must be addressed before starting" with even the smallest piece of hardware or software in the field of smart apparel, and in particular, that of patches for applications in health, wellbeing and sport, designed for mass production.

Before designing or producing patches and smart and/or connected apparel (with or without Internet connection), the following aspects should be considered, which are relevant in one way or another to the concrete production of these technologies on an industrial scale. These are major waymarkers which are absolutely necessary whatever the project, and the multiple constraints which must be recognized, resolved or managed – financial, marketing, technical, industrial, regulatory and normative, pertaining to security or costs, etc. In addition, each of these aspects tends to be subdivided into a range of sub-aspects which are mandatory to address, with no other option available. Thus, close attention must be paid to whether a project is industrially viable. If any doubt remains in any of these areas before you embark on your project, you are not ready to start.

Now that we have defined the overarching context of this lengthy discussion, which is absolutely crucial to understand before proceeding further, we can now take a look at each aspect in a certain amount of depth.

1.3.1. *Financial and marketing aspects*

The financial and marketing category includes aspects relating to commercial strategy, cost, financial aspects, the financial resources at the users' disposal, etc., marketing aspects, the quest to determine the true usefulness of the product, ergonomic aspects, definition of users' needs and wants, and so forth.

These aspects, of course, are of prime importance, and must be studied in detail before investing any technical efforts in a project. After all, if we are developing a product, the goal is ultimately to make money. Indeed, it is all very well to make a product with a view to selling it, but then we need customers to buy it! This statement may seem so obvious as to be trivial, but unfortunately, many people forget about it. It is here that we begin to see the major difference between a "sellable" product and a "buyable" product. In this book, we will speak only of products which are designed to be buyable. What is it, though, that makes a product buyable? To find out, we need to precisely define the fields of application for the prospective product, and which potential markets we are developing it for (also see section I.6.1).

– If there is already a product on the market which serves the same application, we must think carefully about the so called "substitution value", and how it will be weighted as we attempt to break into the market and thereby earn money. If the substitution value is insufficient, then frankly the project is not viable and we should not embark upon it.

– If the application is innovative and/or represents a technological breakthrough, then the key points to consider are: how much are people willing to spend in order to buy it? In parallel, we must consider whether the product represents a slice of the pie (market), or whether, in fact, it represents only a crumb of the mini-pie of sales?

1.3.1.1. *ROI (Return on Investment)*

Return on investment is usually a sticking point for smart apparel. How can we estimate the return on investment (ROI) for a project that achieves industrial rollout, and how long is that ROI going to take to materialize? What will the product's estimated lifespan be in the existing market? Above all, it is a matter of balancing the "usage value", which the users perceive with the proposed "production value", which the industrial players attach to the product. In conclusion, it is worth remembering that in order for a solution to be attractive for everyone involved in the chain of its production, everyone needs to make some money out of it. Otherwise, the project will fall flat!

1.3.2. *Ergonomic aspects*

Ergonomics is another important point which must be given due consideration: "Poor ergonomics can see an otherwise excellent product dead and buried." The shape, design, material, comfort, functional ergonomics[2], etc., of a garment must be

2 Stemming from the Greek: *érgon* (work) and *νόμος/nómos* (law/study). Thus, ergonomics refers not to the design or visual attractiveness of a product, but the degree of comfort when working with it.

studied in minute detail. For example, we need to look at the choice of forms, esthetic attractiveness, cost, production time, etc. Beyond the design, the true ergonomics and flexibility of use must be given careful consideration, and of course, from the outset, we must take the needs of elderly, disabled or reduced-mobility users into account, so that they too can easily interact with the SA.

1.3.3. *Technical aspects*

From a purely technical standpoint (or, more accurately, an electronic standpoint), the following essential technical industrial aspects must be carefully considered when embarking on a project to create smart apparel for industrial purposes:

– technical aspects, technical specifications, etc.;

– energy-related aspects, such as power consumption, lifespan of power reserves, etc.;

– industrial aspects including prototypes, early-series models, full production, costs, etc.

This set of considerations marks a considerable difference from the POC (Proof of Concept) with which many people are familiar. Ultimately, the POC is only the first tiny step towards industrial production and rollout.

1.3.3.1. *Life cycle of a new product*

It is necessary to define a life cycle for a project, which includes the phases of innovation, definition of user requirements, and so forth.

1.3.3.2. *Techno-economic assessment*

During the techno-economic assessment (TEA) phase, we must look carefully at the technical and functional specifications for the proposed product. This may include a POC.

Similarly, we need to begin looking at the matters of sourcing, accreditation, choice of suppliers/partners, and also calculate budgets for R&D, industrialization, equipment, and cost price of the final product. Thus, the project can be given the green light, shelved (amber light) or abandoned (red light) – otherwise referred to as “GO or NO GO”.

1.3.3.3. *Design*

Next comes the active phase of hardware and software development, choices must be made, normative constraints must be observed, the design-to-cost (DTC) approach should be implemented, prototypes must be produced and assessed on technical criteria, and the necessary certifications must be sought.

1.3.3.4. *Industrialization, manufacturing process and quality*

This phase includes the validation of the design, the production of prototypes, processes to move the various subcomponents of the product to industrial readiness, the building and testing of systems, the accreditations/certifications, and finally the launch of industrial pre-series and full production of a series, all stages of accreditation of suppliers, prototypes, and the pre-series process.

1.3.4. *Energy-related aspects*

Among the crucial technical aspects of a smart (and connected) garment, we must remember how its intelligent electronic components will be supplied with energy, what its intrinsic power consumption will be, and all aspects relating to the battery life. This is often one of the fundamentally important points when considering smart apparel.

In this context, depending on the technical and technological design of the SA and the intended applications, the means of energy supply may need to be different. We will examine this matter from various angles in Chapter 4.

1.3.4.1. *Battery life and product lifespan*

When designing and selling smart apparel, the battery life and lifespan when performing the desired function must be part of the first discussions had with the representatives of the intended end user (buyers, retailers, etc.). It is essential to reach an agreement on these matters at the outset, or to inform the end user of the battery life and lifespan of the designed garment, and if possible, avoid making false promises for the sake of commercialization and marketing. These discussions may be lengthy, and failing to hold them may cost the company dearly in terms of corporate image.

1.3.5. *Industrial aspects*

As previously stated, the goal of this book is to lead to the concrete production of SA for commercial purposes. Thus, the stages of "tweaks", "workarounds", reference designs, Proofs of Concept, laboratory prototypes – made from small commercial modules which all appear to work on the drawing board, but are still a long way from being industrial products – must be completed.

One of the first questions that we must ask ourselves is: "Do I have the technical and financial resources at my disposal to bring an industrial project to fruition on my own… or do I need a partner by my side, and if so, for which parts of the project?" The answers to these questions have an enormous impact in terms of time, workload, financial support and final viability of the project.

At the end of this book (see Conclusion), we present an example[3] of a dashboard whose purpose is to successfully plan for each stage of a project.

1.3.6. *Regulatory aspects and recommendations*

In the specific context of smart apparel, the following sections detail the regulations (i.e. mandatory requirements), recommendations and norms (i.e. non-mandatory but highly recommended) that need to be respected, at the national, European and global levels.

Briefly put, a "regulation" consists of a series of official documents/rules issued by an organization attached to a "state" or a "community of states" (such as the EU), which must be complied with in light of prescriptions, rules and regulations, laws, decrees and/or other legal texts which govern a corporate activity. At present, the regulatory constraints applying to smart apparel and the associated worlds are generally applicable at five fundamental levels which we will now examine: levels of radiation and radiofrequency pollution; health risks from human exposure to non-ionizing electromagnetic fields; privacy; regulations on medical devices; the Public Health Code and Labor Code; and waste management. This is already a lot to deal with!

1.3.6.1. *Radiofrequency regulations*

The "smart" part of smart technology, with radiofrequency (RF) connectivity often raises a number of questions, on the part of the general public, consumer associations and other media; however, it is not terribly difficult to set these various concerns to rest. At a technical level, all we need to do is obtain the relevant documentation, primarily from:

3 The example of an industrial dashboard presented in the Conclusion takes quantities, stages of production, subcontracting of hardware and software, mechanical processes (molds, etc.), SE security, Cloud Computing, etc. into accout.

– in France, the AnFr, ARCEP and ANSES, with its Santé-Travail (occupational health) subdivision;

– in Europe, for emissions, ERC 7003 documents issued by the European Radiocommunications Office (ERO), and for electromagnetic pollution levels, the ETSI;

– at the global level, as for human exposure to radiation, the ICNIRP (International Commission on Non-Ionizing Radiation Protection).

1.3.6.2. *Regulations pertaining to radiation and pollution*

All SA designed for the general public (jackets, sports gear, etc.) or for professional purposes (PPE, medical equipment, etc.) including RF connectivity will ultimately have antennas that emit RF waves. The regulations (enacted by the ERC, FCC, ETSI, etc.) give indications of permitted/authorized values of frequencies emitted, bandwidths, authorized radiated powers, specific patterns, occupation time, constraints and restrictions (radiation, pollution, susceptibility, etc.), governing equipment for contactless applications in the broadest sense (RFID, NFC, IoT, geolocation, etc.), and of course, connected smart apparel. One of the fundamental constraints, for obvious reasons, is that they must respect the levels of radiation and RF pollution set by national or international regulations, and indeed with global regulations on RF.

There are a number of organizations which govern and preside over the designing of parameters which directly impact communicating RF systems (including smart apparel), be they SRDs (Short-Range Devices) or LRDs (Long-Range Devices). These are recommendations which have then been passed into legally binding regulations:

– At global level, the ITU (International Telecommunications Union).

– In the United States, under the egis of the ANSI (American National Standards Institute), the FCC (Federal Communications Commission) issued the reference document "US Code of Federal Regulations (CFR) Title 47, Chapter I, Part 15. 'Radio Frequency Devices,".

– In Europe, the recommendation CEPT/ERC/REC 70-03 "Relating to the use of Short Range Devices (SRD)" published by the ERO (European Regulation Organization) is the binding standard. In addition, the measuring and testing methods of the ETSI, "Electromagnetic compatibility and Radio spectrum Matters (ERM); – EN 300 – xxx – frequencies from 9 kHz to x GHz", conform to the recommendations of the ERO.

– In France, the ANFR (*Agence nationale des fréquences*) and the ARCEP (*Autorité de régulation des communications électroniques et des postes*) are in charge of assigning and distributing frequencies that can be used for applications in the IoT, which includes smart apparel, and how they are used. These organizations refer to the European recommendations and produce documents serving as the basis for the development of France's own norms and regulations pertaining to *Non-Specific Short-Range Devices* (smart apparel applications fall within this category).

1.3.6.3. *Health recommendations*

Certain entities, associations, etc., do not have the legal standing to directly impose new regulations on states, but are so expert in their particular fields that they can recommend that states adopt certain values and criteria. After that point, the states themselves may accept, recommend, adopt, impose (etc.) such values by means of laws or decrees. The examples that follow will make this clearer.

1.3.6.3.1. Human exposure to electromagnetic fields

In the applications that we are dealing with here, sources of EM and RF fields represent sources of concern in terms of health impacts. These technologies, which are likely to increase the exposure of their users, and indeed of the general population, to EM fields give rise to new behaviors, and a great many questions (about the biological and clinical effects of such waves, epidemiology, regulation, use, metering, etc.) and concerns. Therefore, it is necessary to address the health consequences of human exposure to electromagnetic fields. The ICNIRP (International Commission on Non-Ionizing Radiation Protection), which is the global organization in charge of such matters, is represented in France by the ANES (*Agence nationale de sécurité sanitaire de l'alimentation, de l'environnement et du travail*), and is regularly in attendance at the meetings of the World Health Organization (WHO).

1.3.6.3.2. ICNIRP and ANSES

The ICNIRP issues "recommendations" on values which should not be exceeded – in particular, for the Specific Absorption Rate (SAR)[4].

The ANSES provides the French government, and the national and European bodies, with the information needed to make decisions on preventing risks to the general public or to professionals in occupational health, and back up the principles of public policy in the matter. The ANSES contributes to knowledge of emerging occupational hazards (relating to magnetic fields, electromagnetic waves, RF wireless communication technologies (Bluetooth, Wi-Fi, RFID, mobile telephony, etc.), electromagnetic hypersensitivity, etc.), and contributes to research and the imposition of reference values to protect workers and users.

---

4 SAR of 2 W/kg for 10 g of human tissue for signals of up to 300 GHz.

1.3.6.4. *Health-related regulations*

In marketing materials, a great many parameters and quality documents relating to smart apparel refer, not quite accurately, to "health", often mixing up a device that is for athletic performance, comfort and/or wellbeing, etc., and one which is a true "medical device".

To avoid certain application pitfalls, the following sections attempt to put certain misconceptions to rest, so as to start right, and eliminate some of the reigning confusion. In order to do this, let us come back to the sources.

1.3.6.4.1. Regulation on medical devices

In France, the *Code de la santé publique* (CSP – Public Health Code) fully includes the Medical Deontology Code, which healthcare professionals must abide by when administering healthcare. For example, there is a legal obligation to keep a formal medical record for every patient treated – this applies both to freelance doctors and to healthcare institutions[5]. These medical files represent a dataset which, often, in view of the volume involved, need to be processed automatically and stored by a specialist service-provider. The sensitive nature of these data, which pertain to patients' private lives, means they need a higher level of protection, not only because of their intimate nature, but also because of the automated processing to which they are subjected[6]. In addition, the procedure for accreditation of personal health data hosters (informally known as the "Kouchner law", after the Health Minister at the time in 2002) is intended to safeguard the security, confidentiality and availability of personal health data, when they are hosted externally by certified hosters.

> APPLICATION TO SMART APPAREL.–
>
> Smart apparel applications for "health" and/or "medical" purposes, which feed back information of this type to be centralized in some unknown location or cloud server, must comply with these regulations and constraints.

---

5 Indeed, all hospitals and similar institutions are required to maintain an up-to-date file. Article R 4127-45 requires self-employed doctors to keep a medical file for each of their patients.

6 Reminder: article 2 of France's law on information technology and freedoms thus reads: "Processing of personal data is any operation or set of operations carried out on those data, irrespective of the process used – in particular, the collection, recording, organization, conservation, adaptation or modification, extraction, viewing, use, communication by transmission, dissemination or any other form of making the data available, compilation or interconnection, in addition to locking, erasure or destruction."

Although the smart apparel for the general public to which this book refers does not fall into the category of medical devices, which is very tightly regulated (that is, true medical devices, not fitness aids), it should be noted that the EU MDR (Medical Device Regulation) came into force on 25 May 2020, and since that date, the pompous terminology can no longer be used for fitness clothes that measure EKGs, cardiac arrhythmia, etc. There is no longer any excuse: the use of certain commercial words and expressions surrounding the term "medical device" is punishable with heavy sanctions.

1.3.6.5. *Regulations on "individual and societal freedoms"*[7]

Smart apparel feeds out a huge amount of information/data specific to the wearer (biometric, behavioral data, etc.), and such "personal data" must be treated with all necessary care. Let us briefly examine the history of individual freedoms, computing and privacy.

1.3.6.5.1. France's Loi informatique et libertés

In France, since 1978, article 1 of Law no. 78-17 (and its successive amendments) on information technology, files and freedoms states that: "Information technology must serve the needs of all citizens. It must be developed in a context of international cooperation. It must not infringe on human identity, human rights, privacy, or individual or public freedoms." That is all that needs to be said in terms of the application!

In France, the *Commission nationale de l'informatique et des libertés* (CNIL – National Commission on Information Technologies and Freedoms) was set up with the aim of guiding professionals in complying with the law, and assisting private citizens to maintain control of their personal data and exercise their rights. The CNIL also analyzes the impact of technological innovations and emerging usages on privacy and freedoms, and works closely with its European and international counterparts to produce harmonized regulations. In 2008, the European Commission published "Mandate 436", detailing all aspects of the problems raised by matters of privacy, individual freedoms and societal aspects of RFID, of which NFC, the IoT and SA are simply derivatives or specific branches.

1.3.6.6. *General Data Protection Regulation (GDPR)*

Since then, the European Union has conducted an in-depth examination of all types of data that may be affected, and in 2016, published the General Data Protection

7 For further information, please refer to Dominique Paret's earlier works: Paret and Huon (2017); Paret and Crégo (2018).

Regulation[8] (GDPR). This text, which came into force on 25 May 2018, is the authoritative frame of reference in personal data protection. It strengthens and unifies data protection for individuals across the European Union. The main purposes of the GDPR are to improve the protection for "data subjects" (people whose data are processed) and ensure those who do the processing act responsibly. In order to do so, it gives a broad definition of standard data and health data, and strengthens the obligation of transparency and informing the data subjects. Whenever personal data are collected, the medium being used (e.g. a form, a questionnaire, etc.) must include informational statements, specifying why the data are being collected, which authority will process the data, who has access to such data, how long the data will be kept for, how the data subjects can exercise their rights in this regard, and whether the data are to be transferred outside the EU. Data subjects have rights (the right of access, the right to rectify information held about them, the right to object to the processing of their data, the right to erasure, the right to data portability, and the right to limit their processing), which are shored up by the GDPR. The GDPR does not draw upon specific norms, because in itself, it defines the requirements that must be satisfied in order to process data properly, and also the rules that apply when the data are breached, including the applicable fines, at a very high level. Thus, the GDPR is applicable without the need for national legislation – it is a shared data-protection law to which all EU Member States have been subject since May 2018.

Before exploring the salient points of the GDPR, let us briefly examine the different types of data that may be processed, and which need protecting.

1.3.6.6.1. Personal data

Definition

"Any piece of information relating to a natural person, identified or who may be identified, directly or indirectly, by reference to an identification number or to one or more items specific to them", such as their surname, first name, biological data, telephone number, social security number, postal address, etc. (article 2 of the French law on information technology and freedoms), constitutes personal data. In particular, this establishes the principle of the obligation to inform data subjects, and obtain their consent.

Which types of personal data are at issue?

The main types of sensitive, at-risk personal data that are already collected, or are intended to be collected, in projects relating to smart textiles and smart apparel, fall into the following categories.

8 The GDPR (around 150 pages) is free to access. We strongly recommend that readers download it from: https://gdpr-info.eu/.

Biometric personal data

Biometric data are those which pertain to the bio (i.e. living) human, and are metric in nature (that is, they are measurements). Examples include the contours or shape of the hand or fingers, fingerprints, vein patterns, temperature, face shape, iris patterns, heartbeats, etc.

> APPLICATION TO SMART APPAREL.–
>
> In applications relating to wellbeing, sport, fitness, PPE, health, etc., biometric data include "medical data collected or produced by professionals within the health system when providing healthcare to the person – i.e. during preventive health measures, diagnostics, care, or social and medico-social monitoring", measured by means of blood tests, heart rate, blood pressure, transpiration, muscle contractions, brainwaves, etc.

Behavioral data

There are a great many different types of behavioral data.

> APPLICATION TO SMART APPAREL.–
>
> Lifestyle habits, walking or running pace, stride length, position and posture of a passenger in a car seat, etc.

Geolocation and mobility data

The same is true of personal data pertaining to geolocation and mobility.

> APPLICATION TO SMART APPAREL.–
>
> Route tracking, tracking a user's movements over the course of various activities (a tracker built into a mobile application, etc.) or those of an employee for work (a PPE garment using GBP/GSM data).

Personal data collected in a business context

The problem is the same. We often need to deal with this type of scenario in examples of use of PPE in an industry.

> APPLICATION TO SMART APPAREL.–
>
> Employee absences, exposure to occupational hazards, detection of stress factors (PPE applications, etc.), journeys, etc.

1.3.6.6.2. Health-related personal data

To one degree or another, the smart apparel sector is closely connected to the monitoring of wellbeing and health, so deals with essentially personal "health" data. Health-related personal data are those data concerning the physical or mental health – be it past, present or future – of a natural person (including the provision of healthcare services), which reveal information about that person's state of health. This concept covers not only all data collected and produced while the person is receiving healthcare, but also those which, in the hands of other actors (such as app developers) provide information about the person's state of health. This definition includes information:

– *relating to a natural person* collected at the time of registering to receive healthcare services or in the provision of those services: a number, a symbol or a specific item attributed to a natural person, allowing them to be individually identified for healthcare purposes;

– *obtained by testing or examining a part of the body* or a bodily substance, including from genetic data and biological samples;

– *pertaining to a disease*, handicap, risk of disease, medical history, clinical treatment or the physiological or biomedical status of the person in question (regardless of the source of the information – for example, whether it comes from a doctor or other healthcare professional, a hospital, a medical device or an *in-vitro* diagnostic test).

This definition covers certain measured data on the basis of which it is possible to deduce some information about the person's state of health.

What are the impacts?

The category of health data is a very broad one. It needs to be assessed, on a case-by-case basis, in light of the nature of the data collected. The concept covers three types of data:

– those which are inherently health-related: medical history, diseases, courses of healthcare received, test results, treatments, disabilities, etc.;

– those which, when used in conjunction with other data, become health data, in that they allow conclusions to be drawn about the person's state of health or health risks: for example, cross-referencing a weight measurement with other readings (number of steps, calories consumed, etc.), or cross-referencing blood-pressure readings with effort measurements, etc.;

– those which become health data because of their destination – i.e. usage made of the data for medical purposes.

NOTE.–

– The law does not apply to the processing of health data for the person's own exclusive use. For example, the law does not apply to health-related mobile apps whose functions include the collection, recording or compilation of data, provided that those operations take place locally, on a computer, a tablet, etc., without connection to the outside world, and for solely personal purposes.

– The concept of health data does not include data from which no meaningful conclusions can be drawn with regard to the person's state of health (for example: an application that records the number of steps taken on a walk, without cross-referencing those readings with other data).

If a set of data is identified as health-related data, then a particular set of legal conditions applies[9], in view of the sensitivity of the data.

– The information about a degree of disability is considered health data if the degree of disability reveals that the person is handicapped, as defined in article L. 114 of the French Family and Social Action Code.

– The information about care received at a healthcare institution, subject to data processing, is qualified as health data when it reveals information about the person's state of health (e.g. admission to the specialized healthcare institution or hospital department).

– The CCAM code (this stands for *Classification commune des actes médicaux* – Shared Classification System for Medical Actions) is health data if the information stemming from that coding reveals information about the person's state of health or receipt of care in relation with a particular condition.

– The registration number in the National Registry of Natural Persons (NIR) is not considered as health data, even when it is used as a national health identifier.

---

9 The list below gives an overview of the various pieces of a legislation which may apply (note that this is not an exhaustive list, so the situation needs to be assessed on a case-by-case basis): France's Loi informatique et libertés (art. 8 and Chapter IX); secrecy regulations (art. L. 1110-4 of the French Code of Public Health – CSP); regulations on frameworks for security and interoperability of health data (art. L. 1110-4-1 of the CSP); regulations on the hosting of health data (art. L. 1111-8 and R. 1111-8-8 et seq. of the CSP); regulations on the disclosure of health data (art. L. 1460-1 et seq. of the CSP); prohibition from transferring or commercially exploiting health data (art. L. 1111-8 of the CSP; art. L 4113-7 of the CSP) etc.; multiple disability or an invalidating health condition (art. L. 114 of the social action and families code).

– A certificate of fitness to practise a sport is not, in itself, health data. However, if combined and/or cross-referenced with other information such as the circumstances in which the certificate was issued, it is considered as health data. On the other hand, a certificate of *un*fitness to practise a sport is considered health data[10].

Heath Data Hub

There are two new points which must be made on the subject of health data, the consideration of which will take us a long way in the design of patches, instituting "smartness" in apparel, and managing the devices.

France's law of 24 July 2019 on the organization and transformation of the health system, informally known as "Ma santé 2022", sets out the globalized data that can be accessed through the platform "Health Data Hub". This law gives the Hub and the *Caisse nationale de l'assurance maladie* (Cnam – National Health Insurance Fund) the power to shape and operate the *Système national des données de santé* (SNDS – National Health Data System). That system is formed of two elements:

– the main database, covering the entire population, contains all data from the *Programme de médicalisation des systèmes d'information* (PMSI – Program to Medicalize Information Systems), the *Système national d'information interrégime de l'assurance maladie* (Sniiram – National Cross-Regime Health-Insurance Information System), the *Centre d'épidémiologie sur les causes médicales de décès* (CépiDC – Epidemiology Center Investigating Medical Causes Of Death) and the information system of the *Maisons départementales des personnes handicapées* (MDPH – Regional Associations for the Disabled);

– a non-exhaustive set of databases known as the "catalog".

The Cnam is responsible for compiling these data, storing them and making the main database available. The main database is gradually supplemented by other data – pertaining to preventive healthcare, diagnostics, social and medico-social care or monitoring, data on loss of independence, data on healthcare at school or at work, data from mother-and-child protection services, and a representative sample of data on health insurance payouts made.

---

10 References: considering art. 35 and 4 of Regulation (EU) 2016/679 of the European Parliament and of the Council of 27 April 2016 on the protection of natural persons with regard to the processing of personal data and free circulation of such data, replacing Directive 95/46/EC (General Data Protection Regulation: EUCJ, 6 November 2003, C-101/01; EC, 19 July 2010, n° 317182; EC, 19 July 2010, n° 334014; EC, 28 March 2014, n° 361042) (source: CNIL).

The Health Data Hub, for its part, is responsible for enriching the main database with these other data, matching the databases in the catalog with the main database, storing data, giving access to the data of the main database and all databases within the catalog.

The databases in the catalog contain data from the above-listed sources, and the sources of the main database.

The list of state institutions which have access to the uploaded data has been expanded. Finally, "the data provided by the SNDS are attached to each individual by means of a pseudonym" and "the health professional's identification number can only be seen, if absolutely necessary, by compulsory health-insurance providers, personnel from the Department of Research, Studies, Assessment and Statistics, the Department of Social Security, the General Secretariat of Social Ministries and Regional Health Boards". Other organizations that search the database will only see the subject's pseudonym.

The text also sets out the obligations in terms of transparency, including the Hub's publication of the list of projects and their characteristics, and the list of datasets and the characteristics of the samples available.

This section has been undeniably long, but this law has very significant applicational consequences for patches and the information received and transmitted.

Evolution of teleconsultations

The Cnam has published the trend in medical interventions paid for through insurance, week by week, from January to August 2020. From the time of the outbreak of the Covid-19 pandemic, the figures show an exponential increase in the use of teleconsultation (see Figure 1.12).

Today, over 50% of doctors use this solution to liaise with patients of all ages. At the end of 2020, it accounted for 3% of all consultations.

2020 (in France)	**Teleconsultations paid for by the Cnam**
9-15 March	15,837
16-22 March	93,403
23-29 March: first lockdown imposed	527,675

Figure 1.12. *Evolution of medical interventions reimbursed by the Cnam between January and August 2020 (source: Cnam)*

Two years after teleconsultations became part of the medical interventions covered by social security in France, *Health & Tech Intelligence* provided a dedicated indicator, giving a picture of the current state of the market for teleconsultation in France, so as to more clearly track the trend in use of the technology since lockdown was imposed, and thus help to anticipate opportunities which may be seized in this thriving sector:

– evolution of the number of teleconsultations reimbursed by health insurance since January 2020 (figures from the Cnam and key figures published by Doctolib);

– evolution of the regulatory framework since Covid-19 hit (exceptional measures put in place by the Government in the context of the health emergency);

– capital increases on the market since the end of February 2020: €53 million raised by French teleconsultation companies;

– main operations (aside from capital increases) on the market over the past year: six takeovers and one merger;

– main French teleconsultation companies.

In short, teleconsultation is advancing apace, and the use of the patches described in this book would greatly help progress in this direction.

1.3.6.6.3. PIA (Privacy Impact Assessment)

Finally, in new technologies such as smart apparel, prior to any processing likely to infringe the rights and freedoms of the data subjects, a Privacy Impact Assessment (PIA) must be carried out.

A PIA is a scenario that describes an event feared by the business or people processing the data, and is based on an approach for managing the risk of impact on privacy and data protection. It is designed in the context of privacy, to help detect risks associated with an application, assess the likelihood of those risks coming to pass, and document the measures that are being taken to address those risks. In addition, the concepts of data protection listed in the regulation indicate very clearly that *privacy by design* and *privacy by default* must be implemented – the companies must take appropriate technical measures and forms of organization appropriate to the issues at hand and the rights of the people concerned.

Such impacts (if any) may vary significantly depending on whether or not the application processes personal data. The PIA framework gives operators guidance as to risk-assessment methods; in particular, it suggests appropriate measures by which to effectively, concretely and proportionately deal with any likely impact on privacy or data protection. Depending on the sensitivity level of the data and the risk of privacy impact as a result of their processing, the so-called data controller (DC – see below) must:

– carry out a full privacy impact assessment;

– produce a PIA report, setting out the life cycle of the data in relation to their nature and format, the purposes and the gains made as a result of data processing by the business of the person in question; and

– then rank these events in order of seriousness.

On the basis of the risks objectively identified, all threats likely to cause the event to come to pass must be anticipated by the data controller and the data protection officer (DPO – see below). Then, the privacy impact assessment conducted will identify whether the protective measures put in place by the company are sufficient in relation to the identified risks, and whether the remaining risks are acceptable. Then, the data controller will be able to approve or invalidate the PIA. Finally, the PIA report must be submitted to the overseeing authorities, it may even help build consumers' confidence if the company chooses to publish or otherwise disseminate it.

APPLICATION TO SMART APPAREL.–

Sooner or later, it is then necessary to carry out a detailed assessment of all risks linked to the privacy impact of smart apparel we produce, and it is mandatory to produce the PIA report – in accordance with the European norm CEN EN 16 571 (*Information technology* – RFID *privacy impact assessment process*) – which gives a detailed description of the procedure and methodology to comply with, and make sure to keep copies on hand so they can be presented in the event of questions.

NOTE.– The above paragraphs must not be taken lightly because, whether you are a creator, a designer, a start-up, an SME, etc., with shared DPOs, or huge companies with your own DPO and DC, you will inevitably have to submit to the GDPR: it is the law!

Privacy by design and privacy by default

As detailed above, the concepts of data protection are divided into two paradigms: privacy by design and privacy by default, requiring companies to take appropriate technical and organizational measures from the earliest stages of product development. The chain of security and privacy must then be examined, verified, verifiable and assured, from end to end. It is also plain to see how using the PIA process can gradually lead to improved privacy.

Example of privacy by design in smart apparel

Often, the design of a smart apparel solution (for professional PPE or for health applications among the general public) is based on a series of steps from the design stage onwards, with data being generated by the smart apparel, sent to the cloud and then, often, redistributed to various users. If security and end-to-end data protection are crucial criteria, the data will be encrypted locally in the smart apparel, before being fed back into the central system. This architecture represents the vast majority of cases in IoT solutions.

1.3.6.6.4. Personal data and smart apparel

Patches or smart apparel which, for whatever reason, are equipped with an identification system, capturing various types of data and processing that information, fall into the category of "connected things". In addition, due to the fact that the patches and/or smart apparel are worn as close as possible to the body, like a "second skin", closely in phase with the body and its biometric and behavioral properties, the data are often recorded, measured and exploited as "personal data". Therefore, companies making patches and/or smart apparel accumulate a vast quantity of data (gigabits or even terabits of data) of this type, relating to the physical wearers of these types of objects. The crucial issue, then, is to ensure consumers have trust in the market of textiles, patches and smart apparel, which are sometimes viewed as spy gadgets, tracking the wearer's movements, travel, keeping hold of their biometric data, behavioral data, private, health-related and other types of data – in short, their sensitive health or wellbeing data, and may transmit those data to practically anyone practically anywhere in the world. It is therefore important to anticipate and ensure that the smart apparel complies with the various pieces of legislation relating to personal data protection.

In addition, the majority of smart apparel is based primarily on principles of communication using RF (in HF or UHF), inspired more or less directly by technologies such as RFID, NFC, BTLE, Wi-Fi, LoRa, SigFox, IBC and others, which naturally fall within the remit of a PIA.

Industrial applications of the GDPR

The GDPR expressly mentions the right to have our data forgotten (art. 17) and the legislation (art. 33) specifies that personal data breaches must be notified to the CNIL and the data subjects. The text also sets the amount of the (enormous) financial sanctions which may be incurred in the event of failure to comply with the above regulations.

The above discussion may seem abstract and confusing, so let us now look at a concrete example of smart apparel.

EXAMPLE.– A company sells a garment, a smart jacket, which, in addition to a range of other things (such as offering comfort and wellbeing while practising sport) measures the wearer's heart rate, recorded by means of an EKG – this is biometric and behavioral data, which is duly treated as personal data. The readings are sent by an application to the user's mobile phone, which uses another application to send the data up to a cloud server (but which cloud?), located who knows where, which makes use of the data in an unknown way (for example, the data may be sold to an insurance company or, believe it or not, to a funeral provider – yes, that has happened in the past!). The records are then sent back to the user's mobile phone by some unknown route. In short, there is quite a long chain of events that must be considered.

Establishing compliance with the GDPR in businesses

In the patches and/or smart apparel industry, as in many other areas, in order to properly implement the GDPR, we must comply with the regulation. That is, we must have a Data Protection Officer (DPO) who ensures that the organization is compliant with the regulation on personal data protection. We must appoint a Data Controller (DC) within the smart apparel company (designers, manufacturers, subcontractors, etc.), keeps up with data regulations and carries out privacy impact assessments, who is co-responsible for the people in charge of respecting the regulations on data protection.

When companies process data on a large scale (for example, in PPE, in the health and medical fields), regularly and systematically monitoring people, and processing sensitive data for their main activity, they are obliged to appoint a DPO, both for the data controller and their subcontractors. Thus, each data controller must establish/put in place documentation/an internal log of processing activities (which is typically a huge document) so that, if audited, they can prove to the authorities (a posteriori) that they are compliant with the regulation.

1.3.6.6.5. In conclusion

To conclude this section, the following paragraphs are addressed to companies, designers, manufacturers, etc., both present and future, of products for this new market, so they can easily reference the main points that need to be kept in mind.

Regulations on personal protective equipment

In this book on smart apparel and the patches used in its manufacture, we cannot possibly fail to mention the branch of smart apparel designed to serve as personal protective equipment (PPE), to protect an individual from a specific danger which they encounter in the course of their work; we must also discuss the regulations pertaining to PPE. Generally speaking, in the professional sphere, the whole body (head, hands, feed, etc.) can and should be protected. Generally, PPE comes in the form of professional gear such as workwear (jackets, parkas, trousers, etc.) or professional accessories (helmets, goggles, boots, gloves, etc.).

Before proceeding further, let us present a few definitions. In European Directive 89/686/EEC, "PPE means any device or appliance designed to be worn or held by an individual for protection against one or more health and safety hazards" at work, and any accessory or ancillary device for the same purpose. This definition includes all components of smart and/or connected apparel.

NOTE.– Beware: the phrase PPE is often misinterpreted to mean *personal equipment* for protection. However, the legal definition is perfectly clear: it is the *protection* which is personal, rather than the *equipment*". The concept of personal protective equipment stands in contrast to collective protective equipment (CPE). For example, it would be possible to make only one piece of PPE available – for instance, a single pair of safety goggles, to be shared by all users at a grinding station, rather than issuing all workers with their own pair of goggles.

The PPE category of smart apparel is often subject to specific constraints pertaining to certain risky professions such as civil protection, firefighting, electrical maintenance, etc. In addition, as they are designed to be highly resistant to all forms of attack, PPE garments are very problematic to recycle.

When, to a PPE garment, we add Internet connection (IoT) and a link to a database (Big Data) in the cloud for more profound analysis of the data captured, it can be used to provide new services such as geolocation, optimization of resources, assessment of strenuousness of the workload, alerts to potential risks, etc. However, this also means complying with very stringent regulations on data management, and with the GDPR.

1.3.6.6.6. Types of PPE

According to the French Labor Code, PPE can be classified into ten or so families, depending on the area of the body which it protects, so there are potentially entire families of garments. There is also PPE for sport and leisure, which is codified in the Sport Code. To conclude the discussion on this topic, note that there are three categories of PPE, defined on the basis of the seriousness of the risks against which they protect. In addition, in order to be effective, PPE must be comfortable to wear, and not interfere with the task the user is performing (see the section on ergonomics, above).

Legally, for each type of activity, an employer is obliged to provide for the safety and protection of their employees. In France, the law requires employers to create and use a unique risk-assessment document (DUER), the purpose of which is to list all the risks inherent in each employee's activity. Once these risks are known, the employer is obliged to either eliminate them or mitigate them as far as possible. PPE is one way to achieve this. In addition, the Labor Code emphasizes that wherever possible, collective protective measures are preferable to personal protective measures, but also stresses that workers should not be unduly inconvenienced: they must only be required to wear PPE when it is absolutely necessary.

Nine general preventative principles have been established, which must be taken into account when designing smart apparel and electronic components for it: 1) preventing risks; 2) assessing the risks which cannot be prevented; 3) combatting the risks at their source; 4) adapting the work to the human operative doing it, rather than forcing the human to adapt to the work; 5) taking account of the degree of maturity and evolution of the technology; 6) replacing that which is dangerous with something that is less dangerous, or safe; 7) planning for prevention by considering, as a coherent whole, the technology, workflow organization, working conditions, social relations and influence of environmental factors – in particular, risks relating to psychological harassment; 8) taking collective protective measures and prioritizing those over personal protective measures; and 9) giving the workers the appropriate instruction.

1.3.6.6.7. Environmental and recycling regulations

When dealing with smart apparel, the questions of its environmental impact and the recycling of the "smart" parts (often electronic) at the end of their life cycle must inevitably be addressed. People often point to these issues as a barrier to the use of such technology, and it is necessary to examine this important and tricky issue, along with the regulations in force in relation to it. Here, then, are some musings to guide reflection.

Besides the fact that the presence of electronic components makes matters more complicated, in certain cases, it is possible to remove the purely electronic parts – in particular, batteries or power cells, sensors and the microprocessor to recover the copper, the silicone from the integrated circuit board, the antennas and rare metals. Nevertheless, these problems are difficult to solve because garments using composite textiles contain a range of different fibers, be they natural or artificial. However, once a wire has been integrated into the thread of a garment, it is more difficult to recover, and no industrial players are willing to unpick a textile to recover only a small weight of conductive wire, which will have little value. In order to promote recycling, the best thing to do is to focus on eco-design, which, at the time of writing, is sadly underdeveloped in the companies currently marketing smart apparel. Future technologies are not necessarily likely to simplify the problem, because the materials used will be increasingly complex, when bio-sensors and the different components are directly printed onto the textiles or integrated into the fibers, and form an integral part of the textiles, as is the case with today's patch antennas. This will mean they are difficult or even impossible to separate, and therefore to recycle, and they will become a burgeoning source of waste.

Below, we discuss certain official texts which may help light the way.

1.3.6.6.8. Waste electrical and electronic equipment

Electrical and electronic equipment (EEE) often contains substances or components which are hazardous to the environment, and there is a high potential for recycling the materials from which it is made (ferrous and nonferrous metals, rare metals, plastics, etc.). In France, the Ministry of Ecology, Sustainable Development and Energy is in charge of regulating on waste electrical and electronic equipment (WEEE). In view of the environmental issues at stake, a specific logistical channel for the processing of WEEE (i.e. its collection and recycling) has been set up, based on the principle of extended responsibility of the producers of such equipment. The directives relating to WEEE and the dangerous substances contained in such devices define the conditions whereby EEE may be sold, and set out the regulatory framework for the management of WEEE. In addition, European Norm EN 50419 relates to the marking of electrical and electronic equipment, which applies (to smart apparel and/or patches as well), as long as the equipment in question is not part of a different type of equipment. It gives an indicative list of products in each of the categories, which serves to clearly identify the producer of the equipment.

Case of garments containing electronic components

The specific case of recycling and value creating from smart apparel is addressed through the annual contribution to ECO-TLC (*Éco-organisme du textile du linge et des chaussures* – Eco-Organization on Linen and Shoe Textiles), paid by companies who market clothing and household linen. There are solutions for appropriate recycling and value creation from the electronic components (batteries, sensors, etc.), depending on whether or not they are removable, and the methods used when the components are integrated into the thread or the cloth. This issue of recycling complex products has already been identified in numerous channels, including that of medical devices containing electrical and electronic components and portable power cells. Those in charge of the relevant eco-organizations, such as the ECO-TLC, and of the various ecosystems, have produced a situation report. Even at this stage, there are two scenarios. In the first, the electrical and electronic components are integrated into the fabric, and are undetectable to the naked eye, in which case the usual procedures apply. In the second, the functional elements are removable (e.g. the casing, battery, etc.) or easily detachable, in which case, ECO-TLC sorters collect them separately and they are processed as small devices, by operators specializing in waste electrical and electronic equipment (ecosystems), or those specializing in the recycling of power cells or batteries.

EEE, as defined by the above-cited Directive 2012/19/EU, designed and installed to be integrated into a non-EEE article, is beyond the scope of ecosystems. However, the question of recycling of textiles is at the heart of the RETEX project, whose aim is to impose a global structure on the textile supply chain in the circular economy, taking action in three areas: the provision offered by economic actors in the textile sector, the management of textile products at end of life, and the demands of the market in terms of products containing recycled materials.

1.3.7. *Normative aspects*

The normative aspects are another point which must be considered. In facilitating the emergence of smart fabrics and apparel, there are two possible approaches to establishing norms – firstly, saying that innovation must be encouraged, leaving the field open for initiatives to thrive; or secondly, knowing that a lack of norms can lead to a fragmented, piecemeal ecosystem.

From the normative standpoint, this means either acting at an early stage, pre-emptively, so as to guide the market, or acting a posteriori, after the market is established, doing a little spring cleaning among the (over-numerous) proprietary systems which are present at that time. It should be noted that the market in smart apparel applications, as it stands today, is still on the borderline between these two approaches, because the true industrial market is still in the process of taking shape, so there is still time to structure the applications with compliance to certain norms.

Before delving into the subject in depth, let us briefly mention the normalization organizations which deal with textiles and smart textiles/apparel.

1.3.7.1. *The ISO, CEN and IEC, and CENELEC*

To better understand the discussion to come, know that:

– Firstly, the *Comité européen de normalisation* (CEN – European Normalization Committee) has signed agreements with the International Standardization Organization (ISO). Within the CEN, France is represented by its own normalization body AFNOR which, by delegation, is represented by the *Bureau de normalisation des industries du textile et de l'habillement* (BNITH).

– Secondly, the *Comité européen de normalisation en électronique et en électrotechnique* (CENELEC – European Committee for Electrical and Electrotechnical Nornalization) has signed agreements with the International Electrotechnical Commission (IEC).

All four of these bodies work, through introducing international norms, to promote international standardization of the trade and markets. Let us briefly examine how the situation stands at the time of writing.

1.3.7.1.1. CEN (Comité européen de normalisation) and CENELEC

In the CEN, the working group WG31 *smart textiles*, within technical committee TC 248, "Textiles and textile products", in October 2011, adopted and published a large and exhaustive technical report: CEN/TR 16298 – "Textiles and textile products - Smart textiles - Definitions, categorisation, applications and standardization needs" – setting out the main definitions and features concerning smart textiles. That norm describes the characteristics of these materials and smart textile systems.

Where the textile is a component in complex, multi-material systems, the BNITH has also established a report on the state of the efforts at normalization carried out in other technical committees of the CEN and normalization bodies such as the CENELEC and ETSI, the objective being to foster dialog on product norms and testing norms. In addition, at the request of the UIT (*Union des industries textiles* – Union of Textile Industries), the FIECC (*Fédération des industries électriques, électroniques et de communication* – Federation of Electrical, Electronic and Communications Industries) has produced an overview of these technologies, and listed the technical committees already in place at the IEC, which are likely to provide elements of normalization in the smart textiles and apparel sector. As we will see shortly, there are many.

1.3.7.1.2. IEC (International Electrotechnical Commission)

In late 2016, the Standardization Management Board at the IEC approved the formation of a specific technical committee (TC 124, Wearable Electronic Devices and Technologies), assigning it the following scope: “Standardization in the field of wearable electronic devices and technologies which include patchable materials and devices, implantable materials and devices, ingestible materials and devices, and electronic textile materials and devices”.

Discussions on the subject of Wearable Smart Devices (WSDs) have taken place in the strategic group IEC SG10, with the following three objectives: to clarify terminology and achieve an agreed understanding of WSDs; to collect use cases for health, wellbeing and the automotive sector; to affirm the principle that all wearable electronic devices must contain procedures for identification, authentication and respect for privacy.

In France, a collaboration has been established between the BNITH and those responsible for normalization at the FIEC, to put forward one or more use cases for smart textiles to IEC TC 124. The French working group recommends using the definitions proposed in 2011 in technical report TR 16298 from the CEN, to describe these materials and smart textile systems. In addition, the UIT has granted the BNITH a mandate to represent the French textile industry on the relevant normalization committees at the CEN, ISO and IEC.

1.3.7.1.3. ISO/AFNOR

In addition to the mirror committee at AFNOR monitoring the works described above, at the ISO, in the Joint Technical Committee (JTC 1), the working group WG 10 is dedicated to the Internet of Things, very close to certain items of connected smart apparel (in the medical domain, and PPE, for example). To date, this working group’s main activities have focused on general matters of architecture and generic, theoretical and academic aspects of security and privacy, which are part of the fundamentals of the systems. The earliest documents published echo the ISO 30141 – *Internet of Things reference architecture*. With regard to the concrete reality of IoT applications, there is only one Technical Report available, on use cases.

1.3.7.1.4. IEEE

In relation to ISO WG 10, the working group IEEE P2413 has produced a norm, “Architectural Framework for the Internet of Things (IoT)”, including the descriptions of numerous domains within the IoT, the definitions of abstractions to the IoT domains, and identification of the common points between the different domains of IoT. The IEEE 802.15.4 and IEEE 802.15.6 standards are currently available.

1.3.7.1.5. ETSI

For its part, in 2012, the European Telecommunications Standards Institute (ETSI) decided to develop norms in the field of IoT connectivity, which can be used in wearables and textiles/smart apparel using Long Range (LR) LTNs

(Low-Throughput Networks), generalizing two principles that are used in narrow-band (NB) transmission solutions (on SIGFOX bases) and dynamic spectrum sharing (DSS) (on the bases of LoRA-Semtech) (see Chapter 5). This is the set of documents GS LTN xxx.

There are also other normative references that must be taken into account. One such reference is GB/T 15629.15-2010: "Information technology – Telecommunications and information exchange between systems local and metropolitan area networks – Specific requirements – Part 15.4: Wireless medium access control and physical layer (PHY) specification for low rate wireless personal area networks."

1.3.7.2. *Overview of norm-setting actors*

Dating from late 2020, Figure 1.13 offers a complete overview of all normalization committees which concern, be it directly or remotely, smart apparel (the most significant are shown in blue).

1.3.8. *Applicative aspects*

In the professional world of smart apparel, applicative aspects are of prime importance, because numerous problems are linked to these aspects, stemming from difficulties with the applications and uses.

1.3.8.1. *Applicative constraints relating to "clothing"*

Let us point to the following aspects, because the fabric, the textile, and the smart garment as a whole must be esthetically pleasing, smooth, silky, lightweight, with a good drop, soft (must not chafe), strong, transparent, pliable, creasable, extendable, etc. (also see the list of finishing steps for details), and the textile/smart apparel combination must be functional, washable, pressable, reliable, of small dimensions and weight, antibacterial, hypoallergenic, etc.

To begin with, we shall give a brief and non-exhaustive list and description of other aspects and applicative constraints which must be satisfied by smart and connected apparel.

1.3.8.1.1. Prior to sale

It is crucial to undertake preventative measures with a view to explaining, training, educating, etc., future potential customers, so that only favorable sales levers apply.

Committee numbers	Committee names	Fields of activity
	IEC	
SG 10	Wearable Smart Devices (since 2015)	Wearable Smart Devices, electronic textiles, near-body electronics, on-body electronics, in-body electronics.
TC 106		Methods to assess electrical, magnetic and electromagnetic fields in relation to human exposure.
TC 124	Wearable Electronic Devices and Technologies (since 2016)	Standards for electronic devices and technologies whose materials and devices are in the form of "patches", implantables, edibles and textiles. "Links" with other technical committees (TC) at the IEC, the ISO, etc., working notably on flexibility and stretchability of the fibers and garments, and the safety of electrical and electronic devices in direct contact with the human body.
	WG1 (terminology)	Terminological definitions for wearable electronic devices and technologies.
	WG2 (e-textiles)	Measurement and evaluation methods for textile materials, devices and electrotechnical functional systems.
	WG3 (Mat)	Defining specific terms and determining the assessments, requirements and specifications for materials used in wearable electronic devices and packages, with the exception of electronic textiles. Analyzing the effectiveness of existing methods specific to the materials of wearable electronic devices and packages, with the exception of e-textiles. Developing methods for measuring and assessing materials for wearable electronic devices and packages, with the exception of e-textiles.

Committee numbers	Committee names	Fields of activity
ISO		
JTC 1/SC 41	Internet of Things and Related Technologies	
TC 38	Textile	
TC 94	Personal Safety	
TC 150	Implants for surgery	
CEN and CENELEC		
TC 162	WG 2 – PPE with electronics or ICT	M/553 Smart Textiles for protection against heat and flame
TC 206	Biological clinical evaluation of medical devices	
TC 248	Textiles and textile products Smart textiles	WG 31 – Smart textiles (2008) EN 16812 – Electrically conductive textiles – Determination of electrical resistance of conductive tracks (2016) EN 16806 – Textiles containing phase change materials (PCM) – Determination of the heat storage and release capacity (2016) EN 16806-2 – Textiles containing phase change materials (PCM) – Determination of the heat transfer using a dynamic method WG 25 – Cosmetotextiles WG 28 – Thermoregulation
	SyC AAL	
	System Committee Active Assisted Living	Committee based on Active Assisted Living (related to the environment of the elderly), working on various points including the levels and accessibility of embedded systems and their means of communication.
CISPR		
	Comité international spécial des perturbations radioélectriques – Special International Committee on Radioelectric Pollution	

Committee numbers	Committee names	Fields of activity
	ACSEC	
	Advisory Committee on Information security and data privacy	
	IPC (recognized by the ANSI)	
IPC D-70 E	Textiles committee	
IPC-2292	Combination textiles and OE (organic electronics)	
IPC-WP-024	White paper on reliability and washability of smart textile structures	
IPC-WP-025	IPC white paper on a framework for engineering and the design of e-textiles	
IPC-8921	E-textile standard for warp/weft and mesh designs	
IPC-8941	Guideline on Connections for E-Textiles	
IPC-8952	Design Standard for Printed Electronics on Coated or Treated Textiles and E-Textiles	
IPC-8981	Quality and Reliability of E-Textiles Wearables	

Figure 1.13. *Status of normalization committees dealing with smart apparel*

1.3.8.1.2. Sale

In order to prove the possibility of purchase and sale, it is often important to show the future consumers of smart apparel that the products have a "new value dimension", demonstrating that they are well founded on an applicative basis, and that the smart apparel in question is truly designed to serve a certain purpose, and is more than a mere gadget. Such would be the case, for example, with medicalized assistance. In the field, it is a matter of providing training to sales personnel and sales teams, so that rather than "selling a price" as they often do, they "sell a product, its usage, and its qualities".

Maintenance and preventive maintenance solutions

Two questions soon arise when thinking about connected textiles and smart apparel. Given that such garments have electronics on board, how can we wash and recycle a connected textile? For this purpose, we need to specifically outline these points in a set of usage instructions or a clear maintenance notice – something that is readable (rather than something in minute text) and understandable by mere mortals, without an advanced degree.

When the electronic components are directly integrated into the thread or the fabric, the question of maintenance (washing, pressing, drying) is essential, because it will have a particular impact on the item's life cycle (number of wash cycles – for example, warranty valid for 5, 20, 30 or 50 wash cycles), and consequently, on its usage, price, replacement value, etc. It must also fit in with existing practices, which are different in terms of the maintenance of clothing and household linen for private individuals (washing and pressing "at home", folding, putting away, etc.) and for professionals (businesses, dry cleaners, hotels, restaurants, etc.). The garment manufacturer must provide clear information about guarantees, durability and properties of the smart materials or textile systems to their professional service providers and their customers. They must also inform them about how to operate and store the device (folded, unfolded, flat, on hangers, etc.), how to use it, the maximum number of cleaning cycles, the quality of the amalgamation of the materials and fibers, drying (means of drying, drying time, etc.), pressing or ironing (maximum temperatures, etc.), maintenance and disinfection. In short, there is a vast range of information which must be provided.

In the context of this book, the design of soft garments for the general public, with integrated biosensors and the unintrusive electronic components associated, is no easy task, because there are numerous quality and performance criteria which must be met, but which are mutually incompatible. Put simply, we need, as usual, to square the circle of technical, industrial, financial, etc. considerations. However, over the past 50 years, in a range of fields, that circle has indeed been squared more than once. Thus, it seems perfectly realistic to expect it could be again, within a timescale of, say, five years. The main problem is the willingness to take the bull by the horns and tackle the issues head on, in today's world.

1.3.8.1.3. After-sales

The subject of "after-sales management" of smart apparel is always a tricky one. In global terms, there are two generic scenarios:

– In the first, the product is expensive, and designed to last (as is the case with PPE, for example). In this case, the smart component will need periodic software updates (possibly carried out remotely) throughout the garment's life, or it will need normal functional maintenance.

– In the second, obviously, to make life easier, certain producers opt for cheaper products (which is a rare occurrence in this particular market) – single-use, disposable items (e.g. a T-shirt for a competition), thereby avoiding the need for repair, replacement, networks of stores to offer support, etc.

Updates to software/apps

From the outset, the problem begins with how to effect software updates for smart apparel, who will perform these updates, etc., and also, how much they will cost. All solutions are, of course, feasible, depending on the underlying applications (expensive products with a short or medium life span, products with an extensive life span (PPE, military equipment, etc.)), which have varying levels of practicality, varying levels of cost, or which cannot be returned or exchanged.

Breakdown repair

Owing to the technologies employed (weaving of electrical wires directly into the material, non-removable integrated circuit boards, etc.), it is often difficult, or even impossible, to repair these products, and where it is possible, it can be highly costly, making this solution difficult and tricky to implement, both in technical and financial terms.

Replacement – service points

With after-sales service, the decision to simply replace a broken product must only be taken after lengthy static calculations for financial pre-costing and assessment of commercial risks. A different solution is to simply replace the product as a "goodwill gesture". The problem is determining whether the replacement is made "within warranty" or "out of warranty".

EXAMPLE.– You guarantee your product for 30 domestic wash cycles at 30°C. How can you know whether the product being returned to you has actually been washed 32 times, including once at 60°C. Should we question the customer's word and good faith? It is a matter of "your word against theirs", and from a commercial point of view, such a situation is not easily resolved.

1.3.8.2. *Security aspects*

To conclude this lengthy section, let us turn our attention to security aspects[11], which simply cannot be ignored, and must be dealt with when smart apparel is connected to other devices (in a BAN, for example – see Chapter 5) or to networks such as the Internet. Indeed, by the principle of applicational proximity, smart apparel (worn on the body) is very closely linked to issues of "personal data" belonging to the wearers – biometric, behavioral, geolocation data, etc., and

11 "Official" definition of security given by ETSI: "ability to prevent fraud as well as the protection of information availability, integrity and confidentiality."

therefore sensitive data, which must be handled with care (see section 1.3.6.5), and with stringent "end-to-end" security in the application. Indeed, security is one of the most important issues to be addressed, closely followed by problems of interoperability. At present, these issues are difficult to solve, and new initiatives are needed from players in the industry.

1.3.8.2.1. Weak links

Often, security is not given due consideration. How are/will the smart apparel be rendered truly secure, to prevent, or at least limit, the new opportunities for hackers that it presents? Everywhere, this topic is high on the agenda, not because industrial actors are suddenly taking more notice of the threat, but mainly from fear of the negative media coverage they will suffer in the event of problems, and the fallout from that coverage in terms of image and cost. Consider an example scenario: a piece of PPE or sports equipment whose cardio readings are hacked and sent to the wearer's insurer, and stress readings are sent to their employer (this is not a hypothetical: it has already happened in real life).

Let us briefly list some of the most usual weak links and "holes in the racket" of security, within the chain of connected smart apparel, which leave hackers a great deal of room to maneuver and carry out attacks (if it is helpful, think particularly of sports gear, PPE and medical devices).

To implement security (see Paret and Huon (2017)), it is necessary to know your enemy and define the security target. In order to do so, you must be aware of the requirements, assess the risks and consequences (for the designer and the user), know how to respond in the event of a problem, know how to and what to communicate to the customers/users, and determine the price to pay. All we need is a miracle! Thus, it is necessary to establish the desired "security target", representing two major categories of parameters: the parameters for which we wish to establish security (the concentric segments of the target), and the levels of security we wish to attain for each of them. We then need to know whether and how these goals are achievable, and if so, at what cost. The following simple and pragmatic questions should be kept in mind constantly: are the benefits worth the costs? Does the chain as a whole deliver security from end to end? Is there still a weak link in the chain, and if so, where is it?

These considerations lead us to define the use of a truly *Secure Element* – a genuinely secure solution which can guarantee encryption security at the five conventional levels: identification, authentication, message integrity, privacy and nonrepudiation.

To our knowledge, usually for reasons of cost, few smart apparel solutions implement such a high level of security, which may become a necessity depending on the quality of the data being carried – increasingly sensitive, personal, biometric, behavioral, fragile and critical – and of course, this keeps costs as low as possible so that the product can be bought by the general public.

Vulnerabilities and attacks on smart apparel

To ensure the appropriate level of security for connected smart garments and their infrastructures (networks), it is necessary to carry out risk assessments, and implement adequate protective measures. Obviously, the levels of guarantees implemented must correspond to the acceptable risk level and the likelihood of those events actually occurring.

> APPLICATION TO SMART APPAREL.–
>
> In certain applications for connected smart apparel, the data handled are often very much "personal data", in the regulatory sense of the term (biometric, heart rate, stress levels, location, etc.), relating exclusively to the individual in question. These applications include wellbeing (sleep regulation, etc.), sports (heart rate and variations, etc.), fitness (effort measurement, etc.), PPE (equipment for laborers, firefighters, military personnel, etc.), and the medical domain (detection of epilepsy, etc.). As such, the levels of security guaranteed are usually (or ought to be) defined in all software elements, all along the length of the chain, by design of the various electronic components, and the choice and applications of the integrated circuits. Finally, choices in this area will depend largely on the potential access routes available to a hacker to attack the various elements listed above.

This concludes the first chapter of this book, representing a very broad view of the extremely numerous aspects that need to be kept in mind and taken into account when making the move from a tenuous idea to a concrete reality in the form of a patch-based project. The long and winding road through the regulations, norms, applicative aspects, etc., has reached its end. This arduous journey was a necessary evil, for the reader's own good.

In Chapter 2, we shall delve deep into the numerous applicative aspects of smart, connected apparel.

2

Biosensors and Graphene Technology

In the previous chapter, all about "apparel", we discussed the functions, features, performances, etc., which the numerous types of smart apparel must exhibit, structurally. In all the examples presented, a considerable range of information and data is collected, using numerous devices known as "sensors".

In smart apparel, sensors of physical properties (stemming from mechanical, thermal, electrical, magnetic, chemical, biological, etc. phenomena and radiation), and biological biosensors are used to harvest "raw" information. This process is usually referred to as "data acquisition". There are a wide range of phenomena monitored in this process, and the resulting data are typically analog (data on temperature, pressure, magnetic fields or induction, positions, etc.). Also, in order to gather the data, sensors exploit a range of physical, chemical or biological principles (variations of resistance, inductances, capacity, induction, and effects such as the Doppler, Hall and Faraday effects, photoelectric, thermoelectric, Seebeck, piezoelectric, dilation, deformation, vibrating string principle, pH, etc.). With these techniques, it is possible to measure, without limitation, new physical properties (such as angles, stresses, forces, inertia, electrical currents, magnetic fields, flow rates, displacements, distances covered, levels, positions, pressure, acoustic waves – sound, temperature, light, etc.). It should be noted that most of these sensors can be modeled as a Z impedance (mechanical, acoustic, electrical, etc.) and a variation of the physical phenomenon being studied (measured) causes a variation of that impedance.

In this second chapter, we shall take a closer look, on both a technical and a technological level, at these principles. In particular, here, we focus on the fundamental electronic elements which are the "x" (sensors) (see Figure 2.1) that may be built into a patch or a garment to make it "smart". This aspect covers a broad segment of the field of e-textiles.

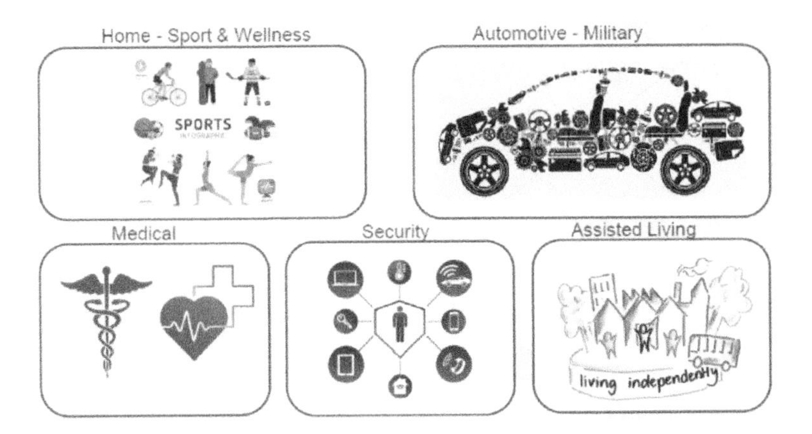

Figure 2.1. *Biosensors are everywhere. We are surrounded! For a color version of this figure, see www.iste.co.uk/paret/smartpatches.zip*

Having set out the basic plan for this chapter, it is time to sink our teeth into it. In order to do so, we shall define and envisage the principles behind numerous types of sensors, and describe the physical and technological problems relating to the design of such sensors. For the purposes of this book, we have classified sensors into:

– sensors of physical properties;

– sensors of physical properties measuring living phenomena: "bio";

– sensors of typical biological properties, "biosensors".

Thus, in this chapter, we shall introduce readers to a promising material, graphene, which, because of its advantageous mechanical and electrical properties, is in line with the materials sought after in the fields of biosensors and patches in smart apparel.

NOTE.– It is important to know that the market in sensors is growing rapidly. In the next few years, we are likely to see major changes in these components with regard to their quality, reduced dimensions (MEMS: Micro-Electro-Mechanical Systems) and new technologies), and of course, cost.

2.1. Introduction to sensors in smart apparel

Let us begin with a few observations. Over the past few years, readers cannot fail to have noticed that in related markets, such as fibers, textiles, fabrics and clothing, many of the performances are shared by numerous applications, and many of the elementary functions that are offered use similar technologies to achieve them.

Figure 2.2 summarizes the main such overlaps.

	Markets/applications							
	Fashion and arts			Sport – Health – Leisure			Medical	PPE
Parameters measured and elements involved	**High-street fashion**	**Fashion and luxury fashion**	**Furnishing and decoration**	**Wellness**	**Fitness**	**Sport**	**Medical**	**PPE**
Temperature	✓			✓	✓	✓	✓	✓
Pressure				✓	✓			
GPS position						✓	✓	✓
Motion					✓	✓	✓	✓
Acceleration							✓	✓
Inertial orientation						✓	✓	✓
Energy harvesting	✓	✓		✓	✓	✓	✓	✓
LEDs	✓	✓	✓					✓
EKGs					✓	✓	✓	✓
PPG					✓	✓	✓	✓
etc.								

Figure 2.2. *Overview of functions common to certain sensors*

Reading across the rows in the above table, we see the main points common to all of these categories. This helps limit the scope of this essentially technical and technological chapter, which sets out the principles, operation and technologies behind the main devices used in these areas (sensors of physical properties, biosensors and their technologies, etc.), which are increasingly found in smart apparel applications, and thus serve as technical and technological bases in this field.

In addition, Figure 2.3, reprinted from the textile technical report CEN/TR 16 298 offers some examples of applications by category of stimulus, on the basis of the environments, optical, mechanical, chemical, and the nature of the functional response by smart materials and smart textile systems.

Types of stimulus	Responses				
	Optical	**Mechanical**	**Chemical**	**Electrical**	**Thermal**
Optical	Photochromism (1)			Photovoltaic/ photoelectric effect	
Mechanical	Piezochromism	Dilatation (5), thixotropy, auxetics	Controlled release	Piezoelectrics	Friction
Electrical	Electrochromism, electroluminescence, electro-optics	Shape-memory, super-absorbent polymers, sol-gel/hydrogel (4)	Electrolysis		Heating due to the Joule effect or Peltier effect
Thermal	Thermochromism, thermo-opacity	Shape-memory (3)		Seebeck effect, pyroelectric effect	Phase-change (2)
Magnetic		Shape-memory, magnetostriction			

Examples of applications in smart apparel: (1) baby clothing which changes color in case of fever; (2) space suits, gloves, skiing jackets; (3) a membrane which retracts when exposed to heat; (4) filtration of artificial snow for cinema and theater; (5) a soft silicone layer in a bulletproof vest which becomes hard in the event of impact.

Figure 2.3. *Examples of applications by category of stimulus (source: CEN)*

2.1.1. *Sensors frequently used in smart apparel*

In the wake of these general remarks, let us now give a non-exhaustive list of sensors which are often found in smart apparel for health, sport, wellness, leisure, etc., and the types of measurement, technologies, applications and/or usages for these sensors (see Figures 2.4 and 2.5).

Type of final data required	Type of measurement	Technology used for measurement	Applications, uses, examples
For physical parameters			
Gas	Chemical	Gas detector	Firefighting PPE
Pressure	Mechanical	Graphene MEMs	
Temperature	Electrical	CTN semiconductive diode	Measuring external heat and body heat (in firefighting PPE, for example).
Velocity	Electrical Mechanical		Fall detector
Magnetic field	Magnetic Hall effect	Angular variation	"Dead man's switch"
Position – Altitude	Magnetic	GPS	
Acceleration	Mechanical	Displacement on the x, y and z axes. A moving plate shifting between fixed plates, moving in all spatial planes.	Recognition and monitoring of bodily posture (standing, sitting, kneeling, walking, running, etc.). Measuring virtual-reality users' activities, healthcare, sports and electronic games. Used in combination with a gyroscope.
Inertial position	Inertial position sensor	Capable of integrating the movements of a moving object or person (acceleration and angular velocity) to estimate its orientation, its linear velocity and its position. Estimation of position relative to the starting point or to the position at last calibration.	No external information input needed. Remains functional when the position cannot be based only on GPS, which is insufficiently reliable for a moving object.
Posture	Mechanical		

Type of final data required	Type of measurement	Technology used for measurement	Applications, uses, examples
For biometric parameters			
Humidity, temperature, UV levels in the external environment	Electrical, chemical		Measuring the internal and external temperature of the human body, and/or the humidity in a person's immediate vicinity.
Respiration	Electrical	Measuring volume and frequency.	
Transpiration, sweat	Chemical	Absorbent materials measuring biomarkers such as calcium, sodium and potassium content/ concentration of sweat, detection of lactic acid and pH.	A means of determining the right drinks to have in order to aid recovery. Graphene detection wires coated in a revealing solution.
Heart rate	Electrical	See EKG	Medical
Electrocardiogram (EKG)	Electrical	Electrical activity of the heart. Measuring differences in potential between electrodes attached to the skin at specific points.	May be used to diagnose heart disease or check the influence of a drug on cardiac activity.
Electroencephalogram (EEG)	Electrical	Information gathered by the electrodes, sent to an amplifier to obtain a plot graph.	Monitors electrical activity in the brain, with small electrodes attached at various points on the scalp.
Electromyogram (EMG)	Electrical	Measures electrical signals produced by the muscles during contractions and at rest.	Can be used to diagnose postural control syndrome (PCS). Nerve conduction studies are often carried out at the same time, given that the nerves control the muscles by way of electrical impulses. Muscular and nervous conditions may lead the muscles to respond abnormally.

Type of final data required	Type of measurement	Technology used for measurement	Applications, uses, examples
For blood tests			
Oximeter	Optical	Sensor clipped onto the end of a finger, an earlobe or a toe. The sensor emits a light signal which passes through the skin and measures the light absorbed by oxyhemoglobin.	Measures O_2 saturation with a non-invasive probe.
Blood sugar	Chemical, optical	Prick a finger to extract a drop of blood, placed on a test strip made of a glucose-sensitive substance. A glucometer is used to analyze the blood sample.	A non-invasive blood-sugar monitoring system is also available, using infrared technology and optical detection.
Arterial blood pressure	Electrical, mechanical	Non-invasive measurement of diastolic and systolic blood pressure, using oscillometry.	Alternatively, measure pulse transit time.
CO_2 detection	Chemical	Measure levels of gaseous carbon dioxide, to monitor variations in CO_2 level.	Also monitors oxygen concentration during human respiration.
Diabetes		Biosensor, gyroscope, insulin pump.	Sure detection of a drop in glucose levels; can send a warning to inject insulin. Thus, the problems caused by the condition are abated.

Type of final data required	Type of measurement	Technology used for measurement	Applications, uses, examples
Uric acid		Specific	Specific
Analysis of the dose of a medicine or molecule **X-emia** **of a protein** **of a drug** **morphine/heroin/ noscapine**		Specific	Specific
	Sleep tracking		
Snoring and sleep apnea		Specific	Specific
Sudden infant death syndrome (SIDS)		Specific	Specific

Figure 2.4. *Overview of commonly used sensors*

Applications	Type of sensor/actuator	Role
Cardiovascular conditions	Oximeter, heart rate monitor, electrocardiogram.	Monitoring the state of health, early warning in preparing a course of treatment.
Paraplegia	Accelerometer, gyroscope, sensors for leg position, sensors attached to the nerves, actuators capable of stimulating the muscles.	Restoring mobility.
Cancer	Nitric-acid-sensitive sensor	The sensor can be placed on suspected cancerous areas. The doctor can begin treatment as soon as a suspicious cell is detected.
Alzheimer's disease, depression, hypertension		Detection and alert to an abnormal situation in an isolated, elderly or depressed person.
Asthma	Allergen sensor	Alerting doctors or patients to the detected presence of an allergen.

Applications	Type of sensor/actuator	Role
Epilepsy	"Mobi"	The portable unit "Mobi" is designed to detect early warning signs prior to a fit (abnormal cerebral activity).
Pain relief	Stimulator	The actuator is a stimulator acting on the spinal cord, which can help reduce chronic pain.
Sight loss	Artificial retina (microsensor matrix), external camera.	Artificial retina built on a matrix of microsensors, implanted beneath the surface of the retina to translate the electrical impulses into nerve signals. The system can also take input from a camera mounted on a pair of glasses.
Hypertension	Arterial pressure sensor, medicine pump.	A medicine can be injected by a pump if a set threshold value is surpassed.
Parkinson's disease	Motion detectors, accelerometers.	Accelerometer data can be used to assess the seriousness of the tremors, bradykinesia and dyskinesia.
Post-operative monitoring	Temperature sensor, blood pressure sensor, heart rate sensor, EKG.	Avoids the need to immobilize the patient in bed.

Figure 2.5. *Overview by types of sensors commonly used*

2.2. Sensors of "non-biological" physical properties

Generally, the term "sensors of physical properties" refers to interfaces/devices made of sensitive elements which detect, transform and convert via a process, a variation in a physical property (be it mechanical, electrical, optical, light, acoustic, ambient temperature, etc.) into a variation in a signal, usually electrical, rendering the reading in the form of exploitable data. For example, a sensor of "non-biological" physical properties might express a physical measurement of pressure (in pascals) in the form of electrical voltage (in volts). It should be noted that often, instead of a sensor, we also speak of a "transducer".

2.2.1. *Types of detectors used in these sensors*

The detecting elements used by these sensors are, for example, often built with semiconductors, stress gages, MEMS, specific materials, etc. In Figure 2.6, an example is given with a simple graphene sensor (in the form of a graphene field-effect transistor – FET). The properties and qualities of graphene make it an excellent material to perform numerous sensor functions.

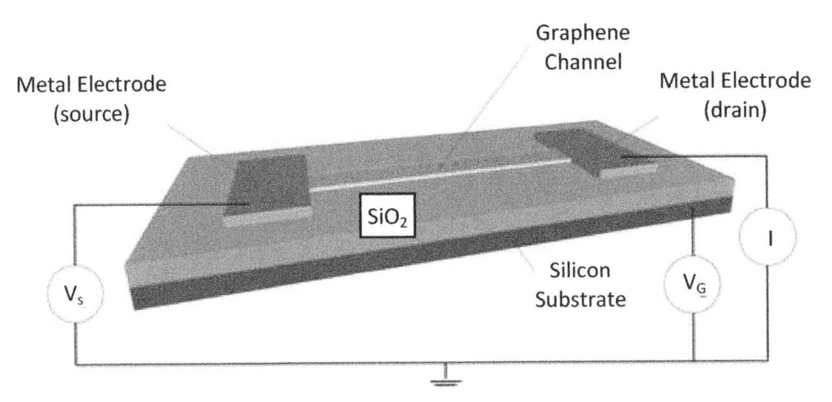

Figure 2.6. *Example of a graphene-based sensor. For a color version of this figure, see www.iste.co.uk/paret/smartpatches.zip*

Typically, the technical criteria by which a sensor is characterized are its measurement range, sensitivity, resolution, accuracy, reproducibility, fidelity, linearity, response time, bandwidth, hysteresis, and functional temperature range.

In most cases, in order to operate and produce an output signal, sensors require a (small) input of energy from the outside world. In technological terms, they are often integrated directly into the patches or smart apparel in the form of small integrated circuits, either "standalone" or directly "embedded" onto a monochip. All of these sensors have been around for a long time and are known to perform well.

2.2.2. *Examples of sensors*

Let us briefly present a few examples of conventional sensors.

2.2.2.1. *Temperature sensor*

The simplest temperature sensor, in semiconductor form, generally uses a p–n junction (a diode), polarized directly. The theory of solid mechanics, applied to the semiconductor, tells us the relation between a set value of a constant current in the junction/diode and the voltage (easily measurable) at its terminals, which depends on variations in the external temperature. Most semiconductor temperature sensors used in smart apparel work on this principle, or variants thereof.

2.2.2.2. *Pressure sensor*

The principle behind a pressure sensor is to detect variations in absolute pressure using a suspended membrane/bar, varying the impedance of a piezo-resistive component usually designed with a MEMS. A particular feature of these sensors is that they can easily be sealed with a particular gel which is resistant to the effects of chemical agents such as chlorine, bromine and salt water (for applications at sea or in swimming pools, for example), and to the soaps and detergents used for bathing or cleaning. Figure 2.7 shows a block diagram for this type of sensor.

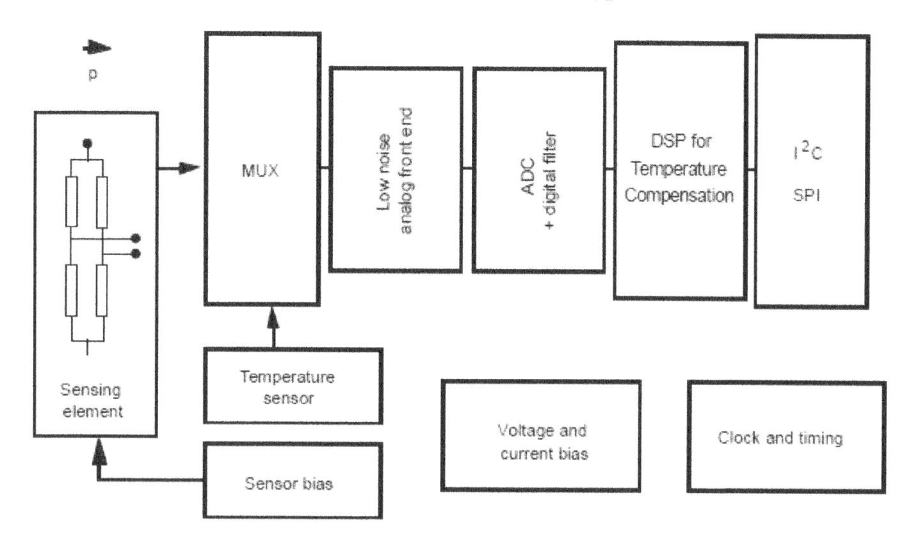

Figure 2.7. *Generic block diagram of a pressure sensor*

The specification of products as "pressure sensors" is actually rather misleading (as is often the case): in fact, they measure not only pressure, but temperature too, and are able to correct the measured pressure value as a function of the temperature, using relatively simple algorithms.

2.2.2.3. *Position sensor*

The primary purpose of position sensing, using an inertial sensor, is to measure the absolute physical orientation of an object or a person, using raw data from sensors with 9 degrees of freedom (9-DOF). This application is often highly sought after in smart apparel for PPE and military gear. To achieve it, a microprocessor is installed on a silicon monochip – for example, the ARM Cortex-M0 – with three elements/functions (see Figure 2.8): a MEMS accelerometer, a magnetometer, and a gyroscope.

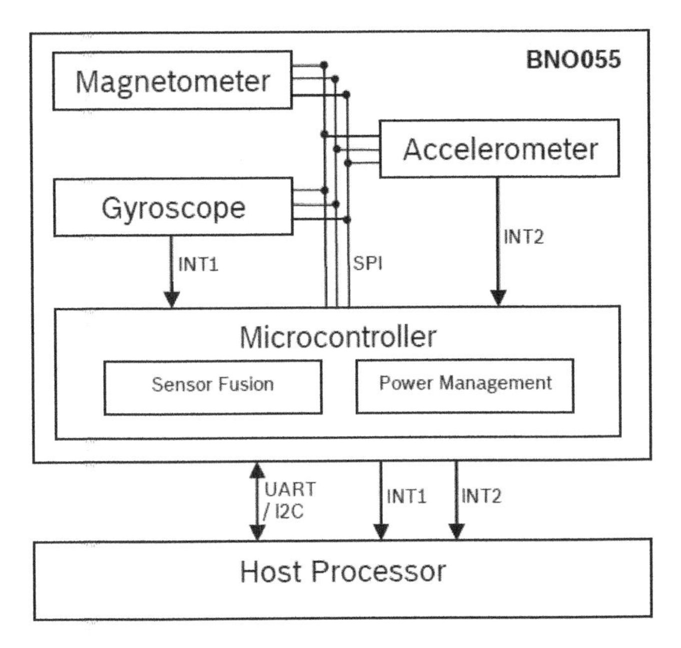

Figure 2.8. *Example of the circuit in an inertial orientation sensor (the Bosch BNO055)*

After conversion of the analog signal received, mathematical treatments and refinement of the individual raw data from the various sensors, the microcontroller integrating numerous data fusion algorithms digests the raw data in "real time", turning them into refined, usable data in the form of unit quaternions, Euler angles, linear accelerations, gravity vectors, etc. From this, the equipment is able to extract functional information for accelerometry, gyroscopy and magnetometry, and create a sensor of "3D spatial orientation", which finds the orientation and exact position without having a specific point of reference. If, to this compound device, we were to add a pressure/temperature sensor based on the piezo-resistive effect, providing data on temperature and barometric pressure/altitude, we would obtain an inertial device with 11 data axes: 3 for accelerometer data, 3 for gyroscope data, 3 for magnetic data and 2 for temperature and pressure.

NOTE.– Pressure can be used to easily determine altitude, which is highly practical when we wish to avoid the addition and/or expense of a GPS unit.

APPLICATION TO SMART APPAREL.–

At first glance, all of this may seem out of scope in this book. However, in order to design underclothes, clothes and PPE whose functions include locating people, monitoring isolated workers, ensuring someone is still alive by reading their movements, medically helping high-level athletes to optimize their performances by examining their movements and studying them in detail, or developing medical monitoring equipment and underclothes for astronauts at NASA without fixed markers for geo-location, etc., it suddenly takes on concrete meaning. Companies have grasped the fact that sensors worn on the body can capture precise biometric data within smart, comfortable and washable biometric clothing, such as heart rate, breathing rate and bodily activity. They have also developed clinically approved systems to monitor the wearer's EKG, pulmonary function and activity. This represents a financially viable and non-invasive solution for long-term monitoring of patients and subjects for healthcare, clinical research and development, sports and physical fitness, and the aerospace and defense industries.

2.2.2.4. *Smart Passive Sensors, with no power supply*

So-called "smart passive sensors" (SPS) are simple sensors measuring temperature, pressure, humidity, etc., matched on the same crystal with a batteryless RFID tag. The purpose of the tag is to provide power to the sensor element, using energy harvested from UHF signals (above 900 MHz) sent by an ad hoc RFID reader, and at the same time, to retrieve, aggregate, analyze and quickly send back the signals measured by the UHF RF reader. These wireless, batteryless sensors can be used to monitor patients over a certain distance, when it would be inconvenient to have to replace the batteries.

This economical approach is often used for applications where the constraints in terms of power consumption are particularly stringent.

NOTE.– Such sensors can be used to read data over longer distances than systems using NFC technology, and unlike with Bluetooth technology, no battery is required. Thus, designers can quickly and easily configure and deploy advanced solutions for measuring physical parameters on the fringes of a network.

Having briefly and generically discussed the "physical sensors" which can be used for smart apparel, and before moving on to a description of "biosensors", let us present a technology which lends itself well to the design of sensors and biosensors.

2.3. Graphene[1]

The time has now come for us to describe a particular technology using materials that can simultaneously serve as a support for the production of sensors and biosensors. We shall linger long over graphene which, thanks to its rare properties, offers many qualities and has aroused a great deal of interest in the fields of design and applications of health sensors and biosensors in the form of patches, usable without, with or in smart apparel for applications in health, wellness, elderly care, etc. It is not the Holy Grail, but not far off!

Graphene emerged in the technical world some eighteen years ago (2004). Since then, many research labs have made a name for themselves by focusing on how to obtain it, create it, and apply it. To present a clearer picture of the situation:

– the first part of this section is a broad overview, summarizing what is offered by the main actors in these technologies;

– the second part indicates, and more precisely delves into, the technical and/or economic interest, and which areas are promising for the concrete use of graphene in the intended applications in the field of "biosensors" and "multibiosensors". In particular, it sets these two fields apart from those of wellness, health and medicine.

To begin with, though, let us define what graphene is, and its origins.

2.3.1. *Carbon*

Carbon, a chemical element with the symbol C and the atomic number 6, is found in all forms of organic matter. The wealth of possible combinations of carbon (by means of chemical bonds) endow it with particular properties in terms of stability, chemical inertia, and a diverse range of crystallographic structures.

2.3.2. *Graphite*

Graphite (a mineral substance which is one of the natural allotropes of carbon) is a material present in nature, formed of multiple layers of graphene, superposed upon one another with 60° rotation of each plane in relation to the last. When viewed from afar, this structure makes it appear as though the layers of graphene should be fairly easy to separate from one another. However, it is very difficult to obtain graphene

1 This section of the book has been written mainly by Pauline Solère, who is in the process of completing her studies at the ESIEE (Université Gustave Eiffel), using numerous documents produced during her research on this subject at Mercury Technologies (CEO: Pierre Crégo) and dp-Consulting (CTO: Dominique Paret).

from graphite because, in order to separate these planes, we need to overcome the forces/interactions which bind them together (known as van der Waals forces – a sort of molecular glue). Consider an example: as we write on paper with a pencil, we lay down graphite in the form of a gray layer where the pencil has passed. One way to obtain graphene, then, would be to use sticky tape to remove the graphite layer by layer, so that ultimately, only one layer of carbon atoms remains on the paper: graphene. However, were we to use such a method, we would still be a long way from obtaining graphene – a single layer which is nearly transparent. Nevertheless, it was by this seemingly primitive method that graphene was first isolated, in 2004, by Andre Geim and Konstantin Novoselov. And the rest is history!

2.3.3. *Graphene*

Graphene, as we have just seen, is a flawless layer one atom thick of carbon (an elementary layer of graphite, or a monolayer), organized into a hexagonal lattice (see Figure 2.9). It therefore has the peculiarity of being a two-dimensional material (a 2D plane, occupying only the x and y dimensions), with a thickness of practically zero (the thickness of a single atom – i.e. around $^1/_{10}$ of a nanometer).

Figure 2.9. *Two-dimensional layer of graphene*

In summary:

– Graphene has a unique atomic and electronic structure.

– It is an allotrope of carbon, with a molecular bond length of 0.142 nanometers.

– In a layer of graphene, each carbon atom has molecular orbital hybridization of type sp^2, which forces the atoms into a hexagonal arrangement, and establishes three bonds with the three nearest neighboring atoms.

– The variation of energy as a function of the wave vector is linear. This linear dispersion endows graphene with remarkable properties, and means that, at the nanometric scale, it has undulations which disappear when stacked.

– Two layers of graphene may be stacked on top of one another to form "bilayer" graphene", and multiple stacked layers of graphene form graphite, with interplanar spacing of 0.335 nanometers.

– In graphite, the layers of graphene are kept together by van der Waals forces, which can be overcome by "exfoliation" of graphene from graphite (see section 2.4.1.1).

2.3.3.1. *Bilayer graphene*

Bilayer graphene is usually found in configurations where the two layers are:

– staggered and unaligned with one another;

– in so-called Bernal stacking, where half of the atoms in one layer are above half of the atoms in the next layer, as is the case in graphite;

– superposed and shifted in relation to one another by an angle of 1.1°, forming a material in which the conduction band is very narrow, and is only half fulfilled, so that electrostatic repulsion between the electrons prevents an electrical current from circulating; such a material is known as a Mott insulator. Below 1.7K, on the other hand, this bilayer becomes superconductive, meaning that it has zero resistance.

The two layers of bilayer graphene can resist a high level of mechanical stress which culminates in their exfoliation.

2.3.3.2. *Nanomaterials*

Graphene is a member of the class of nanomaterials, and has particular properties due to its structure and size, less than 100 nanometers. Its properties are different from those of graphite. Assemblies of atoms at a nanometric scale give rise to new properties – ones which are often totally different from those of the same atoms assembled into materials at macroscopic scale – for instance, in terms of their mechanical strength, chemical reactivity, electrical conductivity, thermal conductivity, light absorption, etc.

2.3.3.2.1. Nanographene

"Nanographene" refers to a finite portion of a sheet of graphene where at least one of the two dimensions is nanometric. Thus, graphene nanoribbons are greatly elongated strips of nanographene.

2.3.3.2.2. Nanotubes

Carbon nanotubes are the third crystalline form of carbon. Their structures can be compared to one or more sheets or walls of graphene, rolled up on themselves or rolled around one another. Thus, we can distinguish between single-wall (single-sheet) carbon nanotubes and multi-wall (multi-sheet) carbon nanotubes.

2.3.4. *Properties of graphene*

To sum up, the main properties of graphene are as follows.

2.3.4.1. *Electrical and electronic properties*

– The structure of its electron energy bands makes it a semiconductor with zero gap (prohibited band).

– At ambient temperature, graphene is the best known conductor of electricity, with electrical resistivity of 1×10^{-8} Ω.m, placing it among the lowest of all materials (including silver, and around 35% lower than copper).

– Graphene has an extremely high electrical current density J = I/s (a million times that of copper).

– The electronic mobility of its electrons[2] has a theoretical value of over 20,000 cm^2/V.s. Electrons move over graphene at a speed of 1,000 km/s, which is almost 150 times the speed of travel of electrons in silicon (7 km/s). This property may have applications for the production of transistors with graphene (fast transistors than those currently in production, based on silicon), and makes the material an attractive choice for the manufacture of very-high-frequency electronics.

– It is the best heat conductor at ambient temperature, with thermal conductivity of around 4,850–5,300 W/m.K, reducing the risks of overheating. It conducts heat twice as well as does diamond, and in all directions. It is thus an isotropic conductor.

---

2 The mobility of charge carriers links the mean velocity of an electric charge carrier in the medium (an electron, an electron hole, an ion, etc.) to the electrical field E.

– It can conduct an electrical current over distances a thousand times its own thickness with no resistance whatsoever.

– Graphene remains capable of conducting electricity even when it has a nominal zero concentration of charge carriers, because the electrons do not appear to slow down or become localized. The electrons orbiting the carbon atoms interact with the periodic potential of graphene's honeycomb structure, giving rise to new quasi-particles which have lost their mass. This means that graphene never stops being conductive. There are also methods to make it into a superconductor (meaning that it can transport electricity with 100 % efficiency).

– At the Fermi level, graphene has electrons whose apparent mass is zero. Thus, it constitutes the only physical system involving zero-mass fermions. One of the effects is the manifestation, under an electric field, of a quantum Hall effect at ambient temperature.

2.3.4.2. *Mechanical properties*

– Having the thickness of a single atom, graphene is the thinnest known material.

– This thinness means it has a very large exchange surface.

It is also the lightest known material. Some examples follow:

- 1 m^2 of graphene weighs around 0.77 milligrams (2,630 square meters per gram); 1 m^2 of paper is 1000 times heavier;

- a sheet of graphene (one atom thick) covering an entire soccer pitch would amount to a weight of less than 1 gram;

- graphene (d = 1.33) is six times lighter than steel (d ~ 8), and lighter than aluminum (d = 2.7).

– Graphene is so dense that it is impermeable to gases: even the smallest atom, a helium atom, cannot pass through a perfect graphene monolayer. This property can be exploited, for example, to create graphene sensors for gas detection.

– Graphene has the greatest known traction resistance (meaning the maximum stress that a material being stretched or pulled can withstand without breaking). Graphene can be stretched to around 20-25 times its original length without breaking:

- its traction resistance, measured at 130 GPa, is between 100 and 300 times greater than that of steel;

- its rupture resistance is around 4.0 ± 0.6 MPa. The Young's modulus of a flawless graphene monolayer is around 1 TPa (in comparison, A36 construction steel has an ultimate traction resistance of 400 MPa, and aramide, better known as Kevlar, 375.7 MPa), though it is six times lighter.

– Excellent flexibility, thanks to the repeating sp^2-hybridized bonds linking the carbon atoms together in a perfect hexagonal lattice. Despite its extremely flexible nature, it has sufficient rigidity and is sufficiently stable to withstand changes in its formation and the addition of other ions around it.

– In terms of elasticity, graphene has a spring constant of between 1 and 5 N/m, and experimentally, it has been found to have a second-order elastic rigidity of around 340 ± 50 N/m.

– Graphene is extremely strong and practically unbreakable, due to the strength of its carbon bonds, 0.142 nm long.

– The addition of graphene to a polymer increases the toughness and thermostability of the polymer.

2.3.4.3. *Other properties*

Other remarkable and interesting properties of graphene are:

– Its near-transparency. As it is extremely thin, and the electrons act as charge carriers with no mass and very high mobility, graphene absorbs only very little white light, because its absorption is uniform ($\pi\alpha$ – 2.3%) in the visible and near-infrared parts of the spectrum, and thus, it appears almost transparent to the human eye.

– Due to this characteristic, once the optical intensity reaches a certain threshold level (known as the saturation fluence), saturable absorption takes place (light of very high intensity leads to a reduction in absorption).

– Though it is lightweight and flexible, it is totally impermeable to gases, and thus constitutes an infallible barrier against water and gases.

– All atoms of graphene are exposed to the environment, but it is an inert material, and does not easily react with other atoms. However, graphene can absorb different atoms and molecules. This may cause changes in the electronic properties, and can be exploited to manufacture sensors or for other applications.

In summary, Figure 2.10 gives the main electrical and mechanical properties of graphene.

Surface density	0.77 mg/m²
Surface current density	10^{12}/cm² – 6 times greater than copper
Electrical conductivity	0.95×10^{6}/Ω.m – 1.5 times greater than copper
Charge mobility	2.5×10^{5} cm²/Vs – highest known value of any material
Thinnest	0.345 nm – the thickness of a single carbon atom
Strongest	200 times stronger than steel of the same thickness
Most flexible	Stretches up to 20% of its length
Best barrier	Completely impermeable (even to helium)
Best electrical conductor	1 million times more conductive than copper
Best heat conductor	5,000 W/mK in all directions (isotropic)
Most transparent	Absorbs around 2.3% of visible light

Figure 2.10. *Overview of the various properties of graphene*

2.3.5. *The usefulness of graphene in smart apparel*

In view of the properties listed above, which are directly in line with those required of smart apparel, we can envisage graphene's use in various types of smart apparel. This section gives some examples.

2.3.5.1. *Applications in "healthcare"*

The electrical performances of graphene are used for electrical detection in sensors/biosensors and adsorptions of molecules, thanks to its high electrical conductivity and the high sensitivity of its conductivity to the presence of adsorbed molecules.

2.3.5.1.1. Sensors

Today, two types of products are usually available:

– sensors using a single layer of graphene, measuring physical properties such as pressure, humidity, etc.;

– the same as the above, for the same uses, so containing a layer of graphene, but also containing a "graphene FET", which improves the sensitivity of the sensor.

2.3.5.1.2. Biosensors

The fundamental types of biosensors are the same as above (so with or without an FET), but in addition, they generally contain a substance, an analyte, which makes them "biosensors". They can then be "functionalized" in a specific biosensor for particular "healthcare" applications (sweat measurements, pulse oxygen saturation SpO_2, etc.). In addition, graphene has better biocompatibility than metal. It can therefore be used, for example, in the design of electrodes for implantation in

Parkinson's disease sufferers' brains. It also has the ability to accumulate in tumor tissues, and can therefore serve as a tool in targeted cancer treatments. In the field of smart apparel and "health", wellness, elderly care and sport, conventional sensors and biosensors (i.e. non-graphene-based) have long been used – often in sweat bands, etc.

2.3.5.2. *Mechanical applications*

The mechanical robustness and elasticity of graphene mean that it can be used for products subject to motion (the flexing of a limb, etc.) or to frequent washing. The addition of only a small quantity of graphene can improve the mechanical robustness of the materials used (all sorts of threads, tissues, etc.). It should be noted that this field of application is the first in which we see the beginnings of true industrialization of pure graphene or its various derivatives.

2.3.5.3. *Esthetical applications*

The transparency of the material is exploited for functional purposes (in wound dressings, etc.) and esthetic purposes (in clothing, etc.).

2.3.5.4. *Electrical applications*

For more detail on electric batteries (supercapacitors, ultracapacitors and energy harvesting), see Chapter 4.

2.4. Graphene and its secrets

The following sections are important; they demystify a number of obscure points surrounding graphene, which will help readers understand how to choose and use the right graphene derivative in their application for their specific sensor or biosensor. Thus, this section is divided into the following main subsections:

– the processes by which graphene and its derivatives can be obtained;

– for further information, the section offers examples of:

- graphene producers and/or suppliers,

- the price of graphene and its derivatives,

- research centers focusing on graphene,

- research centers focusing on graphene biosensors.

2.4.1. *Obtaining graphene*

To date, a great number of industrial processes by which to obtain graphene have been discovered and are in use. Each of these methods produces graphene with somewhat different properties, but nevertheless, all are known by the cover-all name "graphene". Thus, depending on the purpose of the sensors and/or biosensors, patches and smart apparel we intend to make, we must find graphene with the right properties and the right production method to obtain the desired qualities for an acceptable price. We shall now examine the numerous chemical and industrial methods by which graphene materials can be obtained.

2.4.1.1. *Exfoliated graphene*

Graphene exfoliation is a method for producing graphene consisting of:

– In the first method, removing an extremely fine layer of graphite with adhesive tape, and then repeating the operation a dozen or so times on the samples thus produced, to make them as fine as possible. Then, we place the samples on a silicon dioxide plate and, after optical identification, select those samples comprising a single layer of carbon atoms.

– In the second method, an extension of the previous method, to obtain more advanced exfoliation of the graphite. It also entails exfoliating graphite, but after "dispersing" (see below) it in a solvent using an appropriate energy source (ultrasound, microwave radiation, etc.). This second process, which is more economical, has the benefit of being very simple, but requires a great deal of time, and does not always yield exploitable graphene.

2.4.1.2. *Epitaxial graphene*

The epitaxy production method consists of synthesizing graphene from an artificial mineral composed of silicon and carbon: silicon carbide. To do so, in a vacuum, we heat silicon carbide to 1300°C so that the silicon atoms making up the external layers evaporate and, after a specific period of time, the remaining carbon atoms are reorganized into thin layers of pure graphene. This is one of the first graphene production methods which delivers high yields. However, for the moment at least, this method is extremely costly – particularly in terms of energy.

2.4.1.3. *Graphene production by CVD*

Another commonly used method to produce graphene is Chemical Vapor Deposition (CVD). This method uses high-temperature catalytic decompositions of a carbon-based gas (methane, ethylene, etc.) and deposition of the atoms on a metal – usually copper, nickel or iridium. The optimal temperature for the reaction depends on the type of gas and metal. We can distinguish two main families of reaction.

2.4.1.3.1. Graphene deposition on copper

On metals such as copper, the decomposition of the carbonated gas produces carbon atoms which remain on the surface, as a result of their very low solubility in metal, and interact to form a layer of graphene on the surface.

> EXAMPLE.– Graphene is produced by the catalytic decomposition of methane, CH4, at 1000°C, in an atmosphere of diluted hydrogen on copper leaf 25 μm thick (at a purity of 99.8%).

2.4.1.3.2. Graphene deposition on nickel

On nickel metals, once the carbon produced has been diffused into the metal at a high temperature, the significant variation of the solubility of carbon in the metal as a function of the temperature means it is expelled onto the surface when the temperature falls. This technique generally produces several layers of graphene. This method has already been developed to produce carbon nanotubes industrially.

2.4.1.4. *Graphene from PECVD*

Another method to produce monolayer graphene is plasma-enhanced chemical vapor deposition – PECVD. Using plasma, this method deposits a layer of graphene on a nickel or copper substrate.

> EXAMPLE.– A mixture of gases, at least one of which is a compound of carbon, is heated to produce plasma. This heating may take place in a vacuum chamber. To ensure success of the chemical deposition in PECVD, it is essential that the plasma is made up in the right proportions, with great accuracy, using extremely precise flow measurement instruments. The slightest error in the plasma production may cause defects in the graphene sheet produced.

2.4.1.5. *Flash graphene*

This production method uses any material containing carbon, which is transformed into graphene in under a second, by an electrical pulse. The device tends to be fed with carbon black, but can also handle food waste, plastic or even wood chippings. Ground down into a powder, the input chosen to feed the device is placed inside a ceramic or quartz tube, and lightly compressed. The tube is then placed between two electrodes, which heat the material to 3000K in less than 100 ms, by means of a high-energy electrical pulse (for example, 7.2 kilojoules of energy are needed to produce a gram of graphene, with a yield of up to 90% when a pure source, such as coal, is used). The material produced, known as "flash graphene", has better properties than its counterparts. As the compression of the sample, the voltage and the duration of the electrical pulse can be controlled, flash graphene contains very few defects. This method is low in cost, quick, environmentally friendly, and can be used to create good-quality graphene from a limitless range of carbon sources.

2.4.1.6. *Graphene oxide (GO)*

As graphene is costly and relatively difficult to produce, research has been conducted to find efficient but less costly ways to manufacture it, and to use graphene derivatives or similar materials. Graphene oxide (GO) is one such material. It is a monatomic-layered material, made by extreme oxidation of graphite, which is abundant and inexpensive. Graphene oxide is considered to be easy to work with, as it is "dispersible" in water and other solvents (see below), and can even be used in the production of pure graphene. However, graphene oxide is not a good conductor, but there are processes by which its conductive properties can be improved. It is generally sold as powder, in a dispersion or as a coating on various substrates. Graphene oxide films can be deposited on practically any substrate, and then turned into a conductor. For this reason, graphene oxide is particularly well suited for the production of transparent conductive film, such as those used in flexible electronics, solar cells, chemical sensors, etc. Graphene oxide is less costly and easier to produce compared to graphene. It can also be easily mixed with different polymers or other materials, and improve the properties of composite materials, such as traction resistance, elasticity, conductivity and more. In a solid form, flakes of graphene oxide attach to one another to form flat, thin and stable structures which can be folded, crumpled and stretched.

2.4.1.7. *Reduced graphene oxide (rGO)*

Graphene oxide is a form of graphene which includes oxygen functional groups, and has advantageous properties, which may differ from those of graphene. By reducing graphene oxide, these oxide functional groups are eliminated, giving us reduced graphene oxide (rGO). There are various methods of producing rGO, which are generally economical and simple, but reduced graphene oxide generally contains more defects and is often of poorer quality than graphene produced directly from graphite. Nevertheless, it may be of sufficient quality for a wide range of applications, with more attractive prices and easier production processes.

2.4.1.8. *Other methods of obtaining graphene*

2.4.1.8.1. Double process

The double process is a means of killing two birds with one stone: on the one hand, it reduces carbon dioxide emissions linked to industry, and on the other, it produces graphene. The process involves heating the mixture of carbon dioxide and hydrogen to 1000°C, using a metal plate as a catalyst (for example, an alloy of copper and palladium). With this technique, we obtain graphene and water.

2.4.1.8.2. Optical pencil

The method uses epitaxial graphene, and is an ideal technique for creating large sheets of graphene. However, it has its limitations when it comes to printing a shape, a circuit or a precise design after evaporation. "Optical pencil" technology, in relation to the use of graphene sheets, aims to overcome one of the major obstacles: how to produce graphene perfectly reliably and on a large scale. An ionic pencil, emitting a focused beam of gold (Au) ions, is used to create a very precise marking, some 20 nm thick. This technique lends itself well to the exact tracing of complex designs involved in the production of graphene integrated circuits. The method can also be used to produce a graphene nanothread. It uses a mineral with a monocrystalline form which is almost exclusively artificially produced: silicon carbide, which can be considered a semiconductor, although its polycrystalline form is classed as a ceramic.

2.4.1.8.3. Optical forging

The technique known as "optical forging" involves using a pulsed laser beam to sculpt a sheet of graphene so that it is no longer merely flat. The samples exhibit different properties compared to a normal sheet of graphene.

2.4.1.9. *Chemical dispersion: reminder of the definition*

When we speak of graphene technology, as we have just seen, we often use the term "dispersion". At this point, we need to take a look at some basic definitions in chemistry.

2.4.1.9.1. Suspension

In chemistry, a *suspension* is a colloidal dispersion (a mixture) in which a finely divided product is combined with another product, the former being so finely divided and thoroughly mixed in that it will not quickly be redeposited. Most suspensions encountered on a day-to-day basis are suspensions of solids in liquid water.

2.4.1.9.2. Solution

A *solution* is a homogeneous mixture obtained by dissolving one or more chemical species, known as *solutes* (these chemical species may be liquids, gases or solids), in a liquid called a *solvent*. During *dissolution*, the solute gradually disperses throughout the solvent, until together, they form a homogeneous mixture called a solution. In concrete terms, the two phases have become blended. When the solute is no longer visible to the naked eye, it is completely dissolved in the solvent: at that point, only one phase remains.

The choice of solvent depends greatly on the solute with which it is to be mixed. The more important the dissolution, the better will the solvent need to be. It should be noted that dissolution may take place naturally, but sometimes, it needs to be assisted or speeded up by heating the mixture or stirring. For example, sugar (= solid solute) and water (= solvent) form a good solute/solid pair, because when mixed, they form a single phase. On the other hand, oil (= liquid solute) and water (= solvent) or sand (= solid solute) and water (= solvent) do not form good solute/solvent pairs, because in each case, the two phases do not mix. We then speak of non-homogeneous mixtures.

NOTE.– In the case of two liquids in solution, the solvent is always the liquid present in the greater quantity, while the solute is the liquid in the minority.

Aqueous solution

In view of the elements defined above, an aqueous solution is simply a solution in which one or more solutes (in solid, liquid or gaseous form) are dissolved in water (= the solvent) to form a homogeneous mixture.

2.4.1.10. *Graphene dispersion*

Aqueous dispersions of graphene are of interest, because they allow us to manipulate the graphene without danger to the environment for applications in coatings, composites and other materials. The dispersion of graphene in water and certain other solvents with tensioactives, polymers and other dispersants, shows that almost completely exfoliated graphene can be obtained at concentrations of between 0.001 and 5% per weight in water. The practice consists of energy activating graphene in various solvents with various stabilizers, followed by centrifugation to isolate the "right" components in the dispersion. Such approaches do not have practical applications, and often involve a 90–99% waste of graphene. However, alternative approaches omitting centrifugation yield dispersions of 0.5 to 5% per weight of graphene, with higher yields.

2.4.1.10.1. Graphene water

The single layer of carbon atoms which makes up graphene is often still difficult to isolate. It is difficult to produce and too costly to consider real applications at the industrial scale. To get around these problems, a strategy to produce graphene in liquid form has been devised, as has been done for carbon nanotubes. The fundamental idea is to intercalate potassium ions between the levels of a graphite structure to spontaneously dissolve under the influence of entropy, on the basis of neutral solvents. The production method is to plunge the graphite into a highly oxidizing mixture of sulfuric acid and potassium permanganate to obtain graphite oxide which is exfoliatable in water; the trade-off is that the material's conductivity is lower. Using degassed water (we simply remove the naturally dissolved gas in it), we can obtain dispersions of graphene, exclusively the monolayer, with concentrations of 400 m^2/liter, starting with a graphite salt to ensure it is soluble in certain solvents, yielding a solution of "graphenide" (the presence of negative electronic charges such as in sodium chloride). However, these solutions can only be handled in an inert atmosphere – that is, in an environment with neither oxygen nor water. This solution is simply exposed to air and injected into the degassed water before the graphene reaggregates into graphite. To do this, we use a neutral solvent, degassed water known as "graphene water", in which the OH– ions naturally present in water become fixed and adsorbed to the surface of the graphene, instead of the gas molecules which were removed from the water. Thus, the sheets of graphene are electrically charged by the adsorption of the OH– ions, with the same electrical charge and repel one another to ensure they do not reaggregate into graphite, but this time, the dispersion is aqueous and stable in air. Graphene water is the first single-layer graphene formulation, in the form of an aqueous dispersion stable in air and easy to handle. The graphene thus obtained can be exploited in the same way as conventional graphene. It also has the advantage of being cheaper and simpler to produce.

EXAMPLE.– The French company Carbon Waters has developed aqueous dispersions of graphene which are polyfunctional, biodegradable and biocompatible. This helps to alleviate the issues with powdered graphene, with stable dispersions, with a high level of performance because of the quality of the material produced (1 to 8 sheets, with fewer than 0.05% defects). Graphene water can withstand temperatures of 400°C. Their electrical conductivity lends these materials favorable electrostatic or electromagnetic properties. Salt fog ageing tests reveal that corrosion begins after 500 hours.

2.4.1.10.2. Graphene nanoplates (GNP)

Graphene nanoplates are generally synthesized by micro-mechanical cleavage of bulk graphite and can only produce graphene flakes in limited quantities, mixed with graphite stacks. One method to produce bulk graphene nanoplates is plasma exfoliation.

2.4.1.10.3. Graphene-coated textile threads

The aim is to produce conductive textile threads which are washable, flexible, inexpensive and biodegradable, using graphene, in large quantities, using existing textile machinery and therefore incurring no increases in production costs. To do this, graphene flakes and dispersions are produced, to select the best graphene formulation for applications in textiles. We then use a high-speed thread dyeing technique to coat the thread with graphene-based inks. These threads are then included in a woven structure to produce textiles.

2.4.1.10.4. Graphene foams

Graphene foams are generally produced by growing graphene using CVD on a 3D metal structure. The metal is then removed, leaving only the 3D graphene foam. A graphene foam is solid, conductive and useful in numerous applications, including in sensors and purification or absorption materials. Such graphene foams are somewhat similar to graphene aerogels, in which the liquid part of the gel is replaced with a gas (usually air).

2.4.1.10.5. Graphene aerogels

Graphene aerogel is a material composed of lyophilized carbon atoms and graphene oxide. It is also known as aerographene, and is believed to be the least dense solid in existence (see Figure 2.11).

Figure 2.11. *Graphene aerogel. For a color version of this figure, see www.iste.co.uk/paret/smartpatches.zip*

Graphene aerogels are reasonably elastic, and can easily return to their original shape after some degree of compression. In addition, the low density of graphene aerogels renders them highly absorbent (they can even absorb over 850 times their own weight). Aerogels are a particular class of open-celled foams which exhibit numerous unique and advantageous properties, such as low mass density, continuous porosity and extensive surfaces.

2.4.1.10.6. Graphene inks

The production process of graphene ink can be used to make a range of ink formulations to fit the requirements of the various printing processes and substrates used. At the time of writing, the majority of such inks are aqueous, eco-friendly and non-toxic. They are used as printable coatings of graphene and graphene-related materials (GRMs), which offer multifunctionality (high electrical and thermal conductivity, fire retardant properties, UV protection, etc.), produced by means of a high-pressure homogenization process which offer high yield and uniform size. There are various types of standard graphene inks, differentiated by the deposition method, from inkjet printing to silk-screening.

2.4.1.10.7. Graphene paste

A malleable, non-flammable graphene paste has been developed by transforming graphene oxide into a plastic material which has deformability.

2.4.1.11. *In summary*

As shown by the numerous sections above, depending on their own sensitivities, histories, cultures, etc., numerous methods and variants of methods have been developed to produce graphene and its derivatives. Each of these, aiming for the Holy Grail of ideally balanced techniques, performance and cost, have varying performances and acquisition/production costs. Consequently, they are used in a variety of fields of application.

NOTE.– Before embarking on the development of a sensor/biosensor or a textile/clothing application with graphene, it is essential to carefully define your targets in terms of performance and cost.

Figure 2.12 summarizes the main performances of the industrial graphene production methods.

Methods	Processes	Advantages	Disadvantages
Exfoliation	Exfoliation involves removing a very fine layer of graphite with adhesive tape, and then repeating the process multiple times on the samples produced, to make them as thin as possible.	Mass production. Economical.	Production time. The graphene produced is not always usable.
Epitaxial growth	At high temperature and in an ultra-strong vacuum, silicon carbide breaks down, and the evaporation of the silicon atoms from the surface leaves a surface rich in carbon atom, which form a layer of graphene.	Good-quality graphene. High production yield.	Surface area of the graphene depends on the size of the graphene wafer. Energy-hungry process.
CVD	A carbonated gas in placed in a chamber at a high temperature, where it breaks down owing to a metal catalyst – usually copper or nickel.	Large expanses of metal foil can be used to obtain large surface areas. Easily reproducible.	Possible structural defects.
PECVD	A mixture of gases, at least one of which is a carbon compound, is heated to produce plasma.	This method can use carbon dioxide as a base material from which to create graphene, which is beneficial in combating pollution.	An error in the plasma may cause defects to appear in the sheet of graphene.
Flashing	Ground into a powder, the chosen source is placed into a ceramic or quartz tube, and compressed slightly. The tube is placed between two electrodes, which send a high-energy electrical pulse, heating the material to 3,000 kelvin in under 100 milliseconds.	Capable of creating good-quality graphene from any carbon source. Inexpensive, quick and eco-friendly.	The method is not widely developed. It is not used on an industrial scale at the time of writing.

Methods	Processes	Advantages	Disadvantages
GO	One of the methods used to produce graphene oxide is the Hummers and Offeman method. The oxidizing agent is a mixture of concentrated sulfuric acid, sodium nitrate and potassium permanganate. The whole reaction process requires 1–2 hours.	Less costly than pure graphene. Abundant, and easy to produce.	Not a good conductor.
rGO	In the procedure of chemical reduction, GO is dispersed in water, and then treated with hydrazine hydrate. The reduction takes place at 100°C by an overnight one-step reflux process. After reduction, the GO is transformed into rGO. It is then rinsed repeatedly with water and ethanol, filtered, and then air dried to obtain the solid product.	Easy and economical production method.	Contains more defects. Lesser quality than pure graphene.

Figure 2.12. *Performances of industrial processes to produce graphene*

We shall now look at the main producers and suppliers of graphene or its derivatives.

2.4.2. *Graphene producers and suppliers*

On our planet, there are a very large number of graphene producers and suppliers.

To offer readers a concrete idea of the number and diversity of these producers and suppliers (and this very long list is in fact only the start of it), we have selected only a few producers of graphene for you to note. They are grouped according to geographic region (see Figure 2.13). It should be noted that in terms of industrial production, China, South Korea, the United States and the United Kingdom are some way ahead in the production of this material.

In Asia		
China	Sixth Element Inc.	R&D on powdered graphene, produces flakes of graphene, graphene oxide and graphene-related materials.
	Ningbo Morsh Technology	Production, sale and application technology.
	HuaGao MoXi	12 grades of graphene materials, including powdered graphene, graphene suspension, and graphene oxide.
	Shenzhen Xiwang Technology	Graphene heating, covering thermotherapy, smart apparel, and smart household linen.
	Tunghsu Group	Display materials, optoelectronics, heat-conductive materials.
	Jinan Shengquan	Graphene fibers and composite materials.
	Leadernano Tech L.L.C	Application of graphene and nanomaterials.
	Graphene Technology	Mass production and development of applications of graphene nanoplates.
South Korea	Carbon Nano-Material Technology	Refined graphite and graphene materials, carbon.
	Samsung	Flexible screens, connected apparel and other items, or replacement of silicon.
In Europe		
United Kingdom	BGT Materials Limited	Fully personalizable high-quality graphene on various substrates, inks and membranes.
	Cambridge Nanosystems	Graphene materials (and single-walled carbon nanotubes). At present, the company produces high-quality graphene flakes (without metal).
	Textile Two Dimensional	Graphene-based inks and other 2D materials for textile applications. Offers graphene-based inks, printing and integration services.
France	Carbon Waters	Graphene water.
Spain	Graphenea	Single-layer sheets of graphene, bilayer graphene, multilayer graphene, graphene oxide and other materials on a wide variety of substrates.
	Graphenano	Graphene-based paint. Graphene-enhanced polymer batteries. Graphene composites, graphene-based sensors and photovoltaic solutions.
	Graphene Nanotech	Epitaxial graphene grown on SiC substrates.

In Europe		
Italy	Nanesa	Inks, pastes and graphene-based materials, silver-based conductors. Offers graphene flakes, graphene oxide and graphene-based thermal blankets.
	Directa Plus	Superexpanded powdered graphite using fine nanographite.
Sweden	Aninkco	Graphene-based conductive inks and other 2D materials.
Netherlands	Applied Nanolayers ANL	High-quality graphene and other 2D materials.
In the Americas		
United States	First Graphene	Graphene oxide, flakes of graphene oxide and flakes of graphene.
	Graphene One LLC	Contiguous filaments, discontiguous fibers, graphene-based fibers for the textile industry. The products are bacteriostatic, deodorizing and UV absorbent.
	Grolltex	Large surface-area monolayer CVD graphene.
	TCI America	Nanoplates, aggregates of nanoplates or graphene oxide.
	XG Sciences	Graphene flakes, nanoplates produced from graphene sheets, conductive inks, coatings.
	ACS Material	Monolayer graphene, graphene oxide,, graphene nanoplates, carboxyl graphene and graphite oxide.
	Garmor	Graphene oxide flakes.
	Carbon Gates Technologies	Graphene flakes (PNB), large nanoplates.
	Graphene Frontiers	Atmospheric pressure chemical vapor deposition (APCVD). Graphene synthesis.
	Graphene CA	Graphene paste, graphene flakes, graphene compounds and enhanced composites.
	Graphene Technologies	Synthesis of graphene (and other carbon-based materials) from carbon dioxide.
	Cheap Tubes Inc.	Carbon nanotubes and graphene-based products, CVD graphene sheets, PNB and graphene ink.
	AzTrong	(GO and rGO) in the form of inks, powders, suspensions and films.
	IBM Research	Silicon replacement.
Canada	Graphene Leaders	Producers of GO and rGO.
	Group NanoXplore Inc.	Range of graphene-based solutions.
	Elcora Advanced Materials Corp.	Exploits, treats and refines graphite. Research into production of applications of graphite and graphene.

Figure 2.13. *Some examples of graphene producers and suppliers*

2.4.3. *Market price*

As indicated in the previous sections, it is, as yet, still tricky and costly to produce graphene in large quantities. In 2008, it was estimated that to produce one square meter of graphene would cost €600 billion. Since then, increased productivity has led to a sharp drop in price. While we often speak of the technical and geopolitical problems linked to the use of rare earth metals, graphene exhibits the enormous benefit of being carbon-based – carbon being present in an inexhaustible supply, and cheaply.

Materials	Derivatives	Sale price (early 2021)
Graphene powder	Depending on thickness and quantity	\$50–\$700
Graphene	Sheets	\$90–\$130
	In solution	\$5–\$250 depending on quantity
Graphene oxide		\$75–\$1000
	Highly concentrated	\$450 (400 ml with 10 g of GO)
	Film	Between \$53 and \$80
	Paper	
	Single-layer (SLGO)	~\$150.00/g
	Powdered	Per 100 mg (\$19) or 1 g (\$99)
	Dispersions	\$90.94–\$148.58
Reduced graphene oxide (rGO)		\$200/g
		\$143.87
	Powdered	\$150–\$300
	Paper	\$109.81
CVD graphene	Depending on the size of the film and the substrate used	\$75–\$950
		\$42 on copper \$129 on PET
	Single layer on a sheet substrate	\$149.74
Graphene aerogel		\$125
Graphene nanoplates (GNP)		Between \$0.45 and \$15.00/g

Figure 2.14. *Examples of the sale price of graphene and its main derivatives/compounds*

In terms of rough estimates of the sale price of graphene and its main derivatives/compounds, Figure 2.14 (from early 2021) gives the results of our research (the values reflect average prices across a range of suppliers). Obviously, the price depends on the size of flakes, the degree of purity, any intrinsic structural defects and the quality of the graphene, the functionalization, the amounts ordered, and many more factors besides.

2.4.4. *Research around graphene*

Every day, research centers and universities make significant discoveries. However, very few of these discoveries have been followed through to their natural conclusion. In particular, the discoveries are often not exploitable by industry, and therefore accessible to the public in large quantities and at an affordable price.

NOTE.– Figures 2.15 and 2.16, compiled as a snapshot in late 2021, give an overview of the state of research, firstly on graphene as a material, and secondly on biosensors. These should merely be regarded as starting points for readers to conduct their own research for further information.

2.4.4.1. *Research on graphene*

Figure 2.15 shows the numerous research centers which have conducted studies on graphene.

Europe		
Europe	Graphene Flagship	Research efforts cover the whole of the value chain, from production of the materials to the components and integration into systems, and targets a number of specific objectives exploiting graphene's unique properties.
United Kingdom	Payper	Special paper with integrated NFC antennas.
	University of Manchester	Smart textile which keeps the wearer cool during hot weather, and warm during cold weather.
France	Grapheal	Dressing capable of monitoring the changing state of a wound (for example, due to diabetes).
	IEMN	Active transdermal patch to deliver insulin when needed.
	CEA	A substrate for the deposition of a film of graphene.
Germany	Institut Fraunhofer	OLED electrodes.
Norway	NTNU	Flexible systems, capable of incorporating numerous items, from the television to smart apparel.
Denmark		Production of electronic components.
Finland	Nokia	Production of flexible screens for mobile phones.

The Americas		
United States	Michigan Technological University, Houghton	Transforming solar energy without losses in yield.
	MIT and Harvard, Cambridge	Replacing silicon with graphene in computer chips.
	Rice University, Houston	Cathodes of batteries which can be recharged in 20 seconds and retain 90% of their capacity.
	Northwestern University, Chicago	Graphene anodes.
	Graphene Enterprise CA, New York	Graphene-enhanced ceramic-based coating, with an integrated antimicrobial additive for the protection of various types of glass.
Asia		
China	Jiangmen Dacheng Medical Equipment	High-performance graphene fabric.
	Jiangsu College Engineering & Technology	Woolen-type fabric made from a mixture of wool, camel hair and graphene-based nylon, and a manufacturing process to obtain a fabric containing chain threads composed of wool, graphene nylon and fine camel hair. The fabric is lightweight, smooth, shiny and comfortable.
	Jiangnan University	Conductive fabric. The deposited film has a compact surface and adjustable thickness – this is advantageous in the construction of an intentionally rough surface. It does not impact the feel of the fabric, and can widely be applied to functional textiles.
	Hangzou Gaoxi Technology	High-strength graphene/nylon fibers.

Figure 2.15. *World centers of graphene research*

2.4.4.2. *Research on graphene-based biosensors*

Figure 2.16 shows the numerous research centers which have conducted specific studies around graphene-based biosensors.

There are many other graphene-based sensors: in particular, MEMS, pressure sensors, pH sensors, environmental contamination sensors, gas sensors, DNA sensors and more.

Asia		
China	University Tsinghua, Beijing	Carbon dioxide laser to write on Kevlar textiles with graphene, synthetic polyamide fiber used to produce personal protective equipment (PPE) and military equipment capable of taking an electrocardiogram (EKG) or detecting toxic gas. The laser burns and depolymerizes the Kevlar fibers and the carbon atoms recombine to form graphene.
	Shanghai Institute of Microsystem and Information Technology	Reduction of the contact surface between graphene and polymers in order to increase sensitivity. Fiber-based fabric sensors, capable of recognizing movements in real time. These systems are capable of reading a pulse from the wrist.
Singapore	Nanyang Technological University (NTU)	Graphene-based sensor to detect light across a very broad spectrum.
Japan	University of Tokyo	Film based on silicone oxynitrite and parylene, attached directly to the body. Highly accurate, extremely thin and ultra-flexible, it forms a sort of "e-skin", taking real-time measurements of a number of biological properties, including blood oxygen level and heart rate. It is also impermeable to oxygen and water vapor from the air, and therefore has a lifespan of several days.
Europe		
United Kingdom	University of Manchester	Graphene-based conductive thread: washable, extremely flexible, inexpensive and biodegradable, using existing textile machines and therefore not increasing production costs. Production of wearable textile sensors using graphene. High-speed thread dyeing to coat a textile thread with graphene-based inks. These threads are then woven into a structure to function as a flexible sensor.
France	University of Strasbourg	Pressure sensor, flexible, cheaper and more sensitive, measuring pulse, etc.
	University of Grenoble, CNRS and CEA	Biosensor based entirely on membrane proteins.

<table>
<tr><td colspan="3">Europe</td></tr>
<tr><td></td><td>Institut d'électronique de Montpellier, TatiTag</td><td>Biosensors for real-time detection of pathogenic bacteria.</td></tr>
<tr><td>Ireland</td><td>AMBER Centre</td><td>The combination of graphene with G-Putty has produced a material which is highly sensitive to deformation, whose electrical resistivity increases rapidly when pressure or deformation is applied to it. This resistivity returns to normal levels over time. Sensor is hundreds of times more sensitive than a normal sensor.</td></tr>
<tr><td>Spain</td><td>Graphenea</td><td>Graphene field-effect transistors (GFET) and graphene-enhanced surface plasmon polaritons (SPPs) for applications in biosensors.</td></tr>
<tr><td>Sweden</td><td>Chalmers University of Technology</td><td>Graphene-based photodetector sensitive to high frequencies, capable of detecting waves with frequencies near to a terahertz.</td></tr>
<tr><td colspan="3">The Americas</td></tr>
<tr><td rowspan="6">United States</td><td>California Institute of Technology (Caltech)</td><td>Sensing levels of cortisol, the hormone associated with stress and anxiety, in a reliable, quick and non-invasive manner.</td></tr>
<tr><td>Caltech</td><td>Electronic skin which can connect multisensors to monitor a range of physiological parameters.
Biocell uses an organic molecule, lactate, which is a byproduct of muscle activity and is found in sweat.
Measures concentrations of urea, glucose and ammonium ions contained in sweat, and its pH.</td></tr>
<tr><td>University of Illinois</td><td>Detecting those specific segments which are biomarkers of cancer in the DNA and RNA secreted by tumors into the person's blood.</td></tr>
<tr><td>University of North Carolina</td><td>Harvesting thermal energy from the human body and using it to supply a tracker.</td></tr>
<tr><td>University of Houston</td><td>Material, sufficiently soft to be woven into a fabric, but having detection capabilities that can serve as an early warning system for injury or illness.</td></tr>
<tr><td>MIT – Massachusetts Institute of Technology</td><td>The way in which plants respond to various stresses such as damage or infections, monitoring plant "stress" levels from a smartphone. Indeed, a plant sends out a "distress signal".</td></tr>
</table>

The Americas		
United States	University of San Diego	A miniaturized device stuck to the skin which has the purpose of measuring the peripheral arterial pressure in the wrist or the foot, but also the central pressure on the aorta. It is based on ultrasound technology.
	University of California (UCLA)	An adhesive "smart" patch capable of administering the correct dose of insulin by first calculating the blood glucose level of a diabetic user.
	University of New York	Ultrasensitive and ultra-small electrochemical sensors, with homogeneous and predictable properties.
	Rutgers University, New Jersey	Graphene-based sensor capable of detecting inflammation in the lungs and which could improve asthma assessment, diagnosis and treatment. A smart tracker made of a flexible biosensor capable of raising an alarm as to environmental exposure and monitoring the main health markers.
	Stanford University	Wireless sensor stuck to the skin with a dressing, using Wi-Fi to send data to a receiver affixed to a garment. Can be used to monitor patients suffering from sleep disorders or heart problems. The BodyNet patch has an antenna which harvests incoming RFID energy from a receiver placed on the garment to supply its sensors; the antenna is as flexible and stretchy as the skin itself. It is made by silk-screening using metallic ink, on a rubber-based patch. The drawback of this new system is the need to keep the sensor and receiver close together. In their experiments, the researchers attached a receiver to a garment just above each sensor. The objective is to integrate the antennas into the garment so that signals can be sent and received no matter where a sensor is placed.

Figure 2.16. *World centers of biosensor research involving graphene*

2.4.5. *Patents and intellectual property*

Every day, across the world, numerous innovative patents including those concerning graphene are filed. A survey/report from the UK Intellectual Property Office (IPO) – unfortunately published in 2015 – on the number of patents filed in relation to a particular innovation stated that in 2014, 29% had to do with graphene, and were filed by Chinese innovators, with universities strongly represented among the applicants. In terms of the number of patent families filed, the ten most prominent applicants in 2015 were Asian, dominated by the South Korean firm Samsung, with nearly 500 patent families. China accounts for 47% of graphene-related patents filed in 2012, followed by the United States, South Korea, France, Russia, Spain, Germany, Austria, Sweden and Canada. Judging by these patent applications, we can deduce, for example, that the Chinese are greatly interested in fields relating to patches for eye-related problems. However, very few patents have been filed for a health-related patch system.

2.4.5.1. *Relevance of existing patents*

The figures shown in Figure 2.17 (which are, admittedly, some way out of date at this point) provide an idea of the types of patents published (some being applicative patents) and the numbers thereof between 2005 and 2014.

Number of patent families	13,355		
Number of patent publications	25,855		
Publication year range	2005-2014		
Peak publication year	2014		
Top applicant	Samsung (Korea)		
Field choices	**Field name**	**Number of entries**	**Coverage**
People	Inventors	23,284	99%
Applicants	Patent assignees	5184	99%
Countries	Priority countries	43	100%
Technology	IPC sub-group	7095	99%

Figure 2.17. *Patents filed and published between 2005 and 2014*

It should be noted that, in 2021, China was a long way ahead of any other country, followed by the United States and Korea. The most active companies in this field at the time were Samsung, Ocean's King Lighting Science and Technology, Korea Advanced Institute of Science and Technology (KAIST) and IBM (see Figure 2.18).

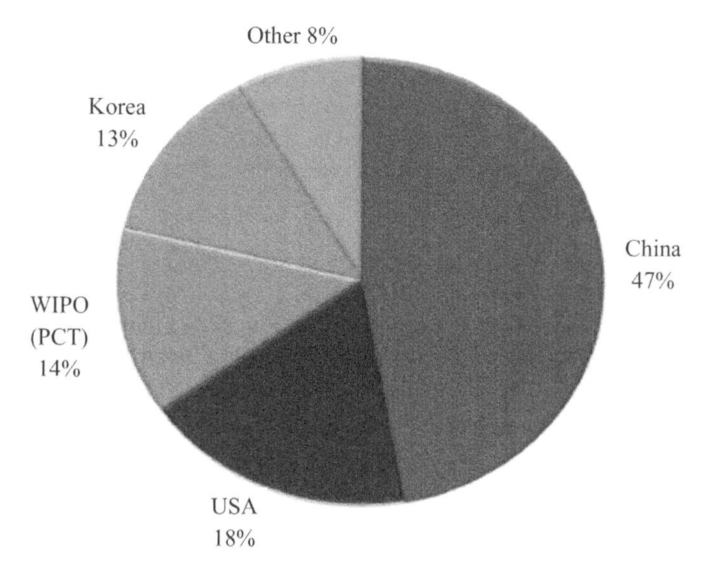

Figure 2.18. *Geographic distribution of patents filed in relation to graphene technology. For a color version of this figure, see www.iste.co.uk/paret/smartpatches.zip*

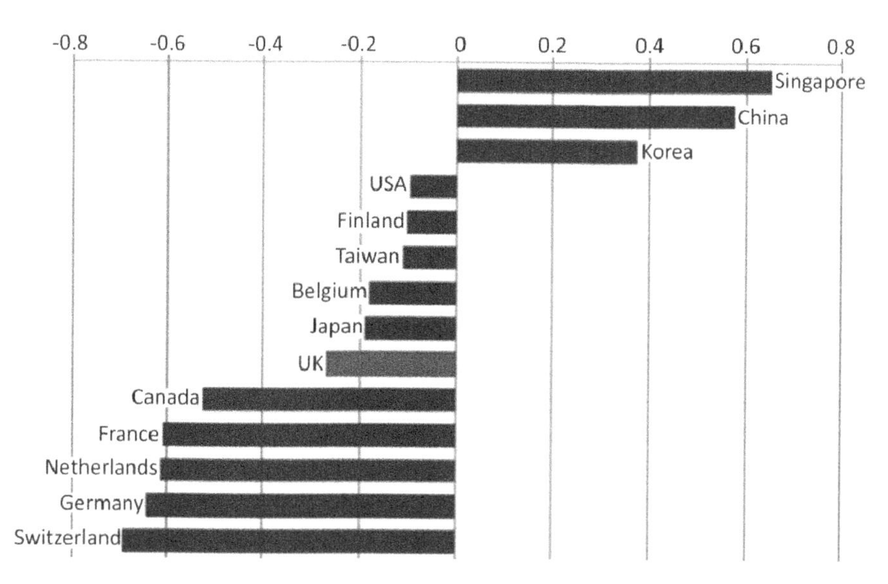

Figure 2.19. *Relative index of patent specialization by country. For a color version of this figure, see www.iste.co.uk/paret/smartpatches.zip*

2.4.5.2. *Patent specialization*

Note that Asia greatly dominates the field in specialized patents, while Europe appears to be focusing primarily on fundamental research. To make up some ground in this domain, in 2012, Europe launched an ambitious program known as the "Graphene Flagship", and since then, a number of European companies have become "figureheads" in this sector.

Numerous European universities and research labs have instituted development programs relating to the material itself and to sensors – biosensors in particular. While the rate of patent filing is dropping sharply, we are beginning to see numerous specialists emerging in the fields of batteries, paints, biosensors for health and sensors in textiles. However, it is likely to be around a decade before some of these technologies mature sufficiently to be taken up by industry.

2.5. "Bio" sensors

In the wake of this lengthy (and absolutely necessary) detour to look at the technological side of graphene, this next section looks at physical sensors connected to "bio" materials – that is, to living things – and at true biosensors. We look at how they are designed, their qualities and their performances, based around this material as a medium. Broadly speaking, they can be divided into two categories: "physical" sensors without analyte, connected to bio tissue, and actual "bio-based" sensors – biosensors – which do have analyte.

2.5.1. *"Physical" sensors connected to bio materials, without analyte*

Things become more complicated with the types of sensors that we are about to discuss, as they are part of the same family of "physical" technologies as those discussed above, but without any form of intermediary process, they are directly linked to measure a physical value from a living thing: for example, body temperature. The term or the associated adjective "corporal/bodily" changes the game, and refines the type of measurements carried out specifically with "bio" measurements, and in particular, opens the door to the interpretation of a vast range of subtle aspects. In addition, the market trend in diagnostic systems with formats and qualities which would be referred to as "wellness, health-related, clinical, etc." is moving towards very small mechanical dimensions, meaning that sensors can be integrated into all sorts of smart garments and connected wearables for day-to-day use. These technologies are defined by:

– Firstly, the desire on the part of designers of wearable wellness systems (with applications in fitness, sport, etc.) to go further and further towards acquisition of data which closely adhere to medical standards.

– Secondly, doctors' and healthcare institutions' desire for the miniaturization of systems capable of rigorously analyzing physiological signals, in order to make them wearable, as close as possible to the body, offering continuous measurement of medical data, 24/7.

To illustrate this situation, let us consider two examples.

2.5.1.1. *Example 1: heart rate and pulse oximetry*

In the wellness and health market, numerous semiconductor manufacturers produce integrated circuits designed to measure photoplethysmography[3] (PPG) signals from the wrists, the fingers and the earlobes. The objective is to precisely measure heart rate and rhythm, any variations in those readings, and the pulse oximetry[4] (SpO_2) by means of optical sensors. These measurements are generally taken using an assembly of various elements:

– red and/or infrared LEDs used to modulate the pulses necessary to measure heart rate and rhythm and the level of blood oxygen saturation (SpO_2);

– another set of LEDs, photo-electrical detectors and optical elements associated with a low-noise electronic system capable of filtering out ambient light;

– a high-yield, low-consumption voltage converter and a system to precisely interpret the readings.

This electronic platform generally comes with a free-to-use software algorithm, to calculate heart rate and SpO_2, mated with a proprietary embedded piece of micro-software.

2.5.1.2. *Example 2: preventative healthcare through smart apparel*

To detect heart rate, respiration and arrhythmia, for acquisition on the chest or wrist of biological signals for an EKG and bio-impedance signals[5] (BioZ), it is possible to combine these systems with a so-called Analog Front-End system (AFE – see Chapter 3). By recording cardiac data, beat after beat, these solutions can be used to produce biodetector devices capable of precisely monitoring a patient's vital signs, in order to identify certain symptoms in advance, prevent health problems, and design smart underclothes to track and monitor physiological parameters

3 Photoplethysmography: a non-invasive functional vascular exploration technique.

4 Estimating the amount of oxygen in the blood.

5 Measuring the resistance of biological tissue.

24 hours a day, 7 days a week (see the example of Chronolife discussed in the conclusion of this book). As a final point, though, it should be noted that the solution usually does not comply with the ECG-IEC 60601-2-47 clinical standards.

2.5.2. *"Bio" sensors (biosensors), with analyte*

Biosensors are analytical tools, systems or devices which interpret biological information and render it as a quantifiable signal. They differ from other analytical systems in that they are independent, integrated devices which act as receivers, and can select and collect quantitative or semi-quantitative information. Overall, they are made up of two parts (see Figure 2.20):

– Firstly, an immobilized biological element, the analyte.

– Secondly, a transducer/sensor as described previously (for example, with a MEMS, on graphene, etc.), which converts the physical signal from the biochemical element into quantifiable and understandable information.

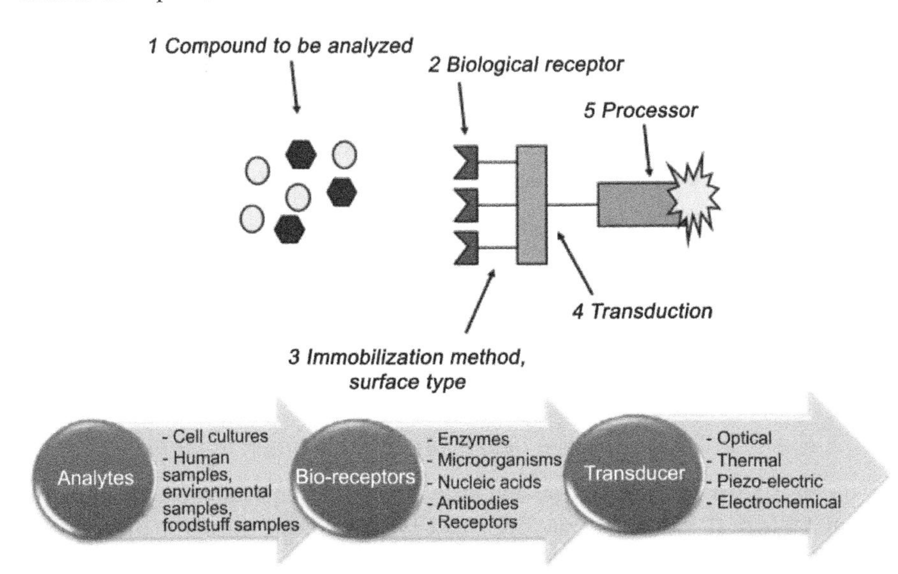

Figure 2.20. *Generic examples of biosensors. For a color version of this figure, see www.iste.co.uk/paret/smartpatches.zip*

2.5.2.1. *Nanosensors*

We speak of a nanosensor when the sensor is composed of a nanometric transducer, which generally uses nanomaterials or nanostructured materials to operate. There is growing interest in nanosensors because generally, for an equivalent level of cost, the nanomaterials, from which they are built, are more sensitive than traditional materials. Carbon nanotubes, for example, are materials which have very large surface areas exposed to the environment, making them highly sensitive to the outside world, and therefore advantageous for the production of minuscule nanosensors which consume very little energy.

2.5.2.2. *Analyte*

What is behind the term "analyte"? There are several frequently used meanings of this term which are accepted.

In analytical chemistry, the term "analyte" is defined as a chemical species or substance, or the amount to be determined of a chemical species or a chemical component present in the sample chemically analyzed during a test. Thus, it is a component or chemical species, a chemical substance whose presence we wish to determine, along with its concentration, identifiable and quantifiable properties, by means of a process of chemical measurement. The analyte is a component (an element, a compound or an ion) of interest in a sample separate to chromatography.

In semantic terms, analytical procedure invariably seeks to measure the properties of one or more analytes, as an analyte cannot be "measured" as such. For example:

– We cannot measure a table (the analyte), but its height, its width and various other properties can be measured.

– We cannot measure hydrogen, DNA, etc., but we can measure parameters such as their molecular weight, radioactivity, concentration etc.

– Similarly, we cannot measure blood glucose, but we can measure glucose concentrations. In this example, "glucose" is the analyte, the component, and the "concentration" is the titrant (i.e. the measurable property).

– In an acid–base titration, the analyte is the acid being measured, and the base, of known concentration, is the titrant.

EXAMPLE.– Professor Joe Bloggs has titrated the analyte to find the proper dosage. This gives rise to a signal which, after processing, can be directly correlated with the concentration in solution of the target analyte.

For the sake of simplicity, in plain language, the "property" is often disregarded, particularly when the omission does not cause ambiguity in terms of the "measured" property. However, as precision is so often necessary in the scientific method, the word "analyte" can be used to emphasize certain aspects of the analysis. Identification of the analyte is a crucial element in defining an analytical problem, typically in the fields of metrology or chemometrics. It can be determined by absolute or relative methods. The information obtained from a sample of analyte may be qualitative (if there is a known quantity of analyte in the sample), quantitative (determining the proportion of the substance) or structural.

2.5.2.2.1. Composition of a biosensor with analyte

Let us now examine the structure of a biosensor. Each type of biosensor has a specific biological component, which acts as a semi-biological detector, and an electronic component which transmits the signal via the transducer. It is a detection device, made-up of multiple elements (see Figure 2.21).

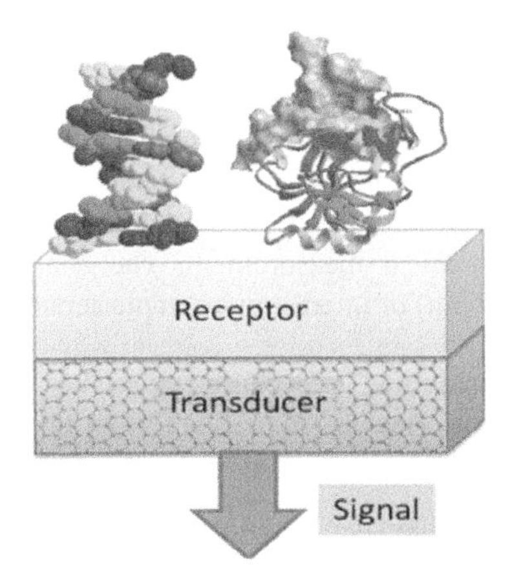

Figure 2.21. *Biosensor with analyte. For a color version of this figure, see www.iste.co.uk/paret/smartpatches.zip*

– The *sample to be studied*: water, air, soil, etc., biological material (tissue, micro-organisms, organites, cellular receptors, enzymes, antibodies, nucleic acids, genetically modified organisms, or GMO-based material, etc.).

– The immobilized *biological element* (the analyte – for example: an antibody, an enzyme, DNA, micro-organisms or a nucleic acid), which recognizes a target substance (e.g. the antigen, a substrate, the complementary DNA sequence) within a complex environment.

– The *sensor element* (the "detector" element) (typically an electronic chip), which works on the basis of principles as widely varied as physico-chemical, electrochemical, optical or piezoelectrical principles, or more rarely magnetic or thermometric principles. Its purpose is to detect physico-chemical changes in the form of signals (presence/absence) – biochemical, physical (analog) or chemical – in a medium (inside or outside of the human body), and emitting a biological signal.

– The *transducer*, which translates the physico-chemical changes (for example, photon emission, pH variation, mass variation) picked up by recognition between the bio receptor and the target molecule, into a measurable electrical signal which can be correlated to the concentration of the target substance in the medium. The signal resulting from the interaction of the analyte with the biological element is turned into another signal, which can be more easily measured and quantified.

Finally, there is a fourth electronic element – the associated signal bioprocessor, examined in Chapter 3, which allows the readings to be sent in the form of digital data, files or databases. The bioprocessor usually has a memory, whose purpose is to store series of readings and successive findings over time.

Real-world examples of graphene-based transistors and biosensors

Figures 2.22 to 2.24 show a few examples of the principles of the production of graphene-based biosensors and transistors.

2.5.2.2.2. The different types of biosensors with analyte

Biosensors can be classified into different types, and grouped according to the types of analyte used, or according to the way in which they work.

Classification on the basis of the type of detection element (analyte)

For example:

– *Enzymes*, our proteins with a high level of selectiveness for a particular substrate, upon which they effect a catalytic change. Enzymes may be affixed, in various ways, to the surface of a transducer by adsorption, covalent assembly, or occlusion in a gel or an electrochemical polymer.

– *Antibodies*, also known as *immuno-sensors*, are produced by B lymphocytes in response to antigenic stimuli such as foreign bodies or microbes. Antibodies are particularly sensitive to changes in pH, in ion concentration, chemical inhibitors and temperature. These antibodies may be immobilized on the surface of a transducer by a covalent assembly made by conjugation of an amino, carboxylic, aldehyde or hydrosulfuric group.

– *Micro-organisms* are also employed in biosensors to determine oxygen consumption or carbon dioxide consumption in an environment, using electrochemical techniques. Micro-organism biosensors have the benefit of being less costly than the ones based on enzymes or antibodies, and are also more stable. However, they may be less selective.

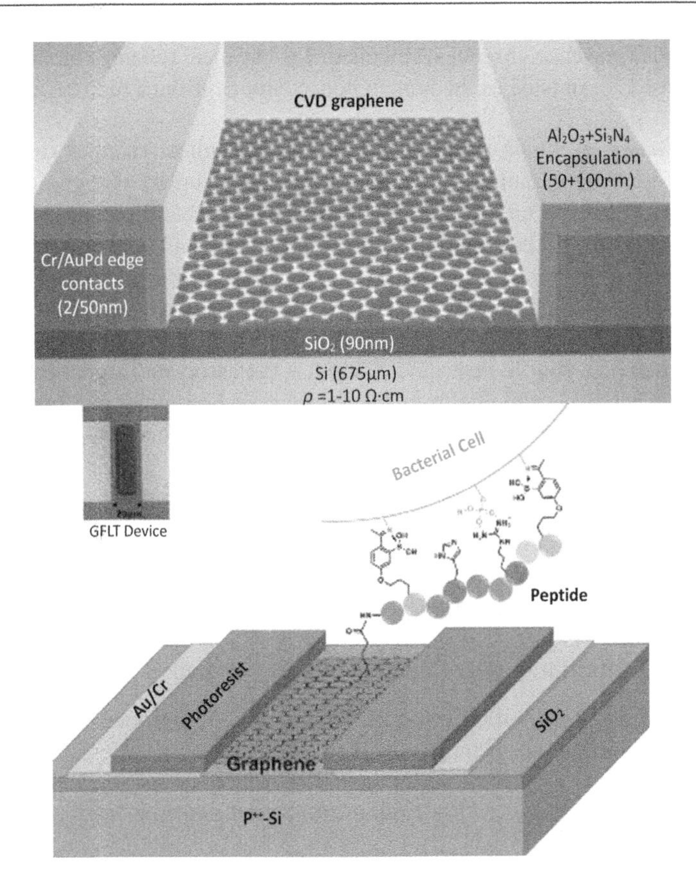

Figure 2.22. *The principle employed in producing a graphene-based transistor. For a color version of this figure, see www.iste.co.uk/paret/smartpatches.zip*

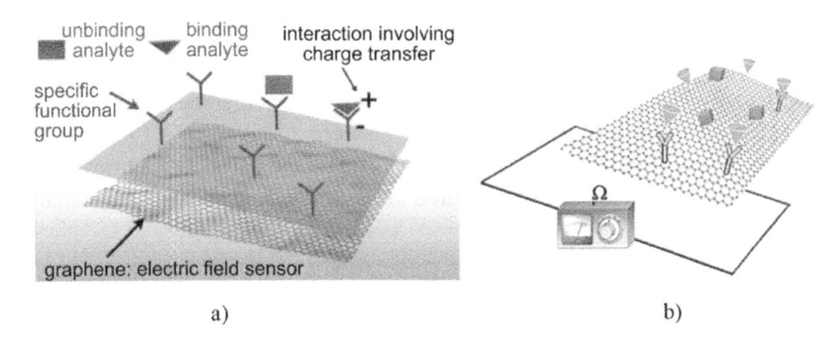

Figure 2.23. *a) Fundamental principle behind graphene-based sensors; b) measurement of current variation by varying the resistance of graphene. For a color version of this figure, see www.iste.co.uk/paret/smartpatches.zip*

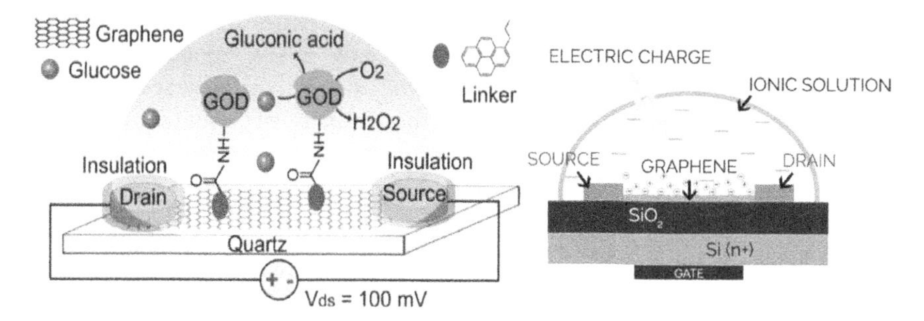

Figure 2.24. *Principe behind biosensors based on graphene FETs. For a color version of this figure, see www.iste.co.uk/paret/smartpatches.zip*

Other bio-elements may be used for biosensors, including organelles, nucleic acids or biological tissues.

Classification on the basis of the way in which biosensors work

Biosensors can also be classified into different types depending on the way in which they work. For example (see Figure 2.25):

– If the bioelement latches onto the analyte, the detector is classified as an *affinity sensor*.

– If the bioelement and the analyte come together to bring about a chemical change, which can be employed to measure the concentration of a substrate, then the detector is called a *metabolic sensor*.

– If the biological element combines with the analyte and does not chemically alter it, but converts it into an auxiliary substrate, then the biosensor is referred to as a *catalytic sensor*.

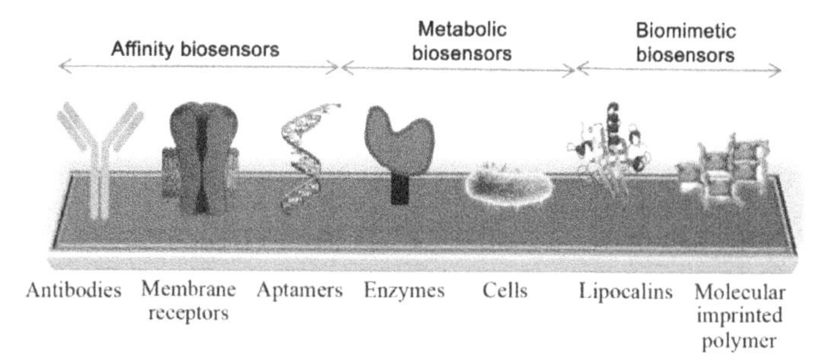

Figure 2.25. *Representation of the different types of biosensors (source: A. Montrose, PhD Thesis, University of Toulouse). For a color version of this figure, see www.iste.co.uk/paret/smartpatches.zip*

As we shall see in Chapter 3, bioprocessors used in patches are, in fact, miniature electronic biological detection labs, measuring only a few square millimeters, capable of identifying, in a short space of time, for example, the presence of DNA or particular proteins, but also variations in the gaseous or liquid environment, measuring glucose, glutamate, cholesterol, hemoglobin and many more factors.

Examples of analytes

Figure 2.26a shows a few examples of analytes and their applications. Figure 2.26b offers examples of pertinent quantitative parameters for the selected analyte of the sensor and biosensors.

Analyte	Enzyme	Application
Glucose	Glucose oxidase	Diabetes monitoring
β-hydroxybutyrate	β-hydroxybutyrate dehydrogenase	Ketoacidosis diagnosis
Lactic acid	Lactic acid oxidase	Athlete tracking
Cholesterol	Cholesterol oxidase	Cardiovascular disease

a)

Analyte	Electrode material	Limit of detection	Linear range
Pb^{2+}	Glucose	0.02 μg/L	0.5–50 μg/L
cd^{2+}	Glucose	0.02 μg/L	1.5–30 μg/L
H_2O_2	Graphene/AuNPs/chitosan	180 μM	0.2–4.2 mM
H_2O_2	Reduced graphene oxide	0.05 μM	0.01–10 mM
Dopamine	Glucose	NA	5–200 μM
Dopamine	Glucose	2.64 μM	4–100 μM
NADH	Ionic liquid-graphene	5 μM (ethanol)	0.25–2 mM
Glucose	Graphene/Au/Nafion	5 μM	0.015–5.8 mM
Glucose	Reduced graphene oxide	2 μM	0.01–10 mM
Glucose	Graphite nanosheet/Nafion	NA	0.2–1.4 mM
Glucose	N-doped graphene	0.01 mM	0.1–1.1 mM

b)

Figure 2.26. *a) Examples of analytes and applications; b) examples of analytes. For a color version of this figure, see www.iste.co.uk/paret/smartpatches.zip*

Let us now come back to the concrete way in which a biosensor is built.

2.5.2.2.3. *Structures of biosensors*

The active sites of a biosensor for a selective analyte and the adjacent surface are the points at which the chosen biological events take place, generating a signal which is then measured by the element of the transducer. In the case of the biosensors which we are going to examine, the element of the transducer is typically graphene, which transforms the transducer signal into an electronic signal (and sometimes amplifies it using a graphene FET). Then, the software included in the biocontroller (see Chapter 3) converts the electronic signal into a material parameter which can unambiguously be interpreted. It is then presented to the user by an interface.

> EXAMPLE.– One of the most widely used systems is the blood-glucose analyzer. It uses the enzyme glucose oxidase, which releases electrons when it interacts with glucose. An electrode captures those electrons and translates the resulting signal into a readable level (blood sugar).

Electrochemical biosensors

Electrochemical biosensors are a class of biosensors which work by using an electrochemical transducer. They can detect biological materials such as enzymes, entire cells, specific ligands and tissues, but also modifications not involving enzymes.

There are a range of types of electrochemical biosensors, distinguished by their means of biological selectiveness, their means of signal transduction, or a combination of the two. The receptor may be selective for a biocatalytic event such as a reaction which is catalyzed by macromolecules, including enzymes, or for a specific bio-affinity, such as the interaction between an analyte and a macromolecule, independently of their biological environment. Screening methods for electrochemical transducers include:

– amperometry, based on measurement of the electrochemical current resulting from the oxidation or reduction of electroactive substances;

– conductometry, which measures the conductivity in enzymatic reactions and biological membranes;

– potentiometry, which uses the difference in potential between an indicator electrode and a reference electrode in the absence of any current flowing between them. Most often, pH electrodes are used, but ion-selective electrodes may also serve.

Finally, electrochemical biosensors can measure analytes using field-effect transistors. Ion concentrations are determined by placing an ion-selective membrane in a transistor, which becomes a biosensor when added to a biocatalytic or bioaffinitive layer.

Graphene biosensors with analyte

In the early sections of this chapter, we discussed the general nature of the revolutionary material graphene, at great length. Let us now look, specifically, at its applications to biosensors.

– Graphene's large specific surface provides numerous possible bind sites or anchor sites for the target molecules and manufactured metal nanoparticles, sometimes used to improve the detection efficiency of biosensors. The different domains in a sheet of graphene (and the ionic regions if metal nanoparticles are present) allow the graphene sensor to interact in different ways, which can increase its sensitivity in comparison to non-graphene-based biosensors.

– Graphene's excellent charge mobility and electron transfer rate facilitate the movement of electrons between the target biomolecule and the sensor's electrode, making it a useful material in biodetection applications and the design of biosensors.

– Graphene's excellent mechanical, electrical and optical properties are often used for biodetection applications. Its high mechanical strength, Young's modulus, flexibility and traction resistance are key properties in the manufacturing process, as the surface is sufficiently robust to be easily used to modify other surfaces.

The fundamental principle on which a biosensor works is as follows: on a sheet of graphene, when biomolecules are adsorbed, a change in charge carrier density brings about a change in the electrical conductivity. This, in turn, generates a measurable electrical response which can be used to determine whether or not a molecule has been detected. Graphene is used as an electron transfer material, because its large specific surface area means that all carbon atoms are exposed to the target biomolecules, enhancing the system's sensitivity. In conclusion, graphene can also be used as an electrical transfer medium in sensors, because its charge carrier mobility is high and its electrical conductivity facilitates the movement of charges and electrons across the sensor. In addition, it is capable of acting as an electron transport material, being physically connected to the biomolecule and moving the electrons by means of bond interactions.

Figure 2.27 shows a few examples of the stages in the manufacture of a biosensor using graphene. As the figures indicate, graphene is used as an impedance material to functionalize the surface of the electrode in a biosensor. The π network delocalized from the graphene generates π-π stacking interactions, which can anchor other π-conjugated molecules to the electrode. This provides a method to modify the surface of the graphene in biosensors without impacting its desirable electrical conductive properties.

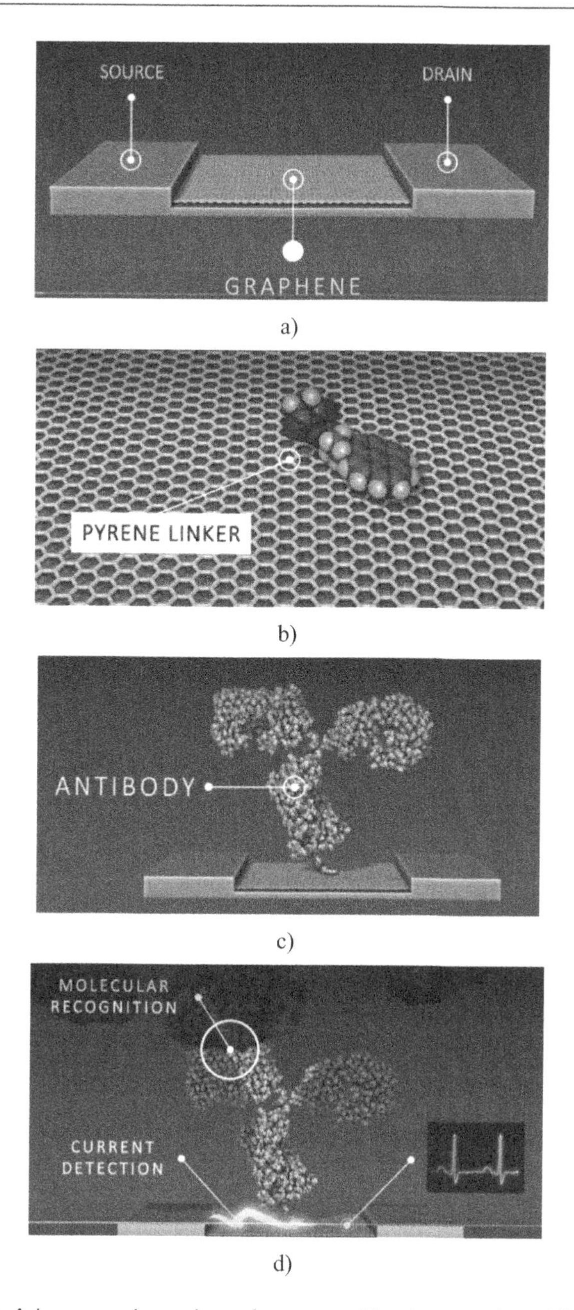

Figure 2.27. *a) A lone graphene-based sensor without an analyte; b) contribution of a pyrene linker; c) contribution of the antibody (analyte) on the graphene sensor; d) creation of a signal after molecular recognition by the analyte (source: Grapheal). For a color version of this figure, see www.iste.co.uk/paret/smartpatches.zip*

Graphene FET

Graphene can be used directly to produce field-effect transistors in biosensors. Thus, certain models of graphene-based sensors include an FET with a graphene channel to improve the device's sensitivity by interfacial charge traps, which act as external distribution centers and harm the transport properties. When the target analyte is detected, the value of the current across the transistor changes, sending a signal which can be analyzed to determine a range of variables. To do this, two types of "graphene FET" – "back gate" FETs and "floating gate" FETs – can be used in biosensors.

– In back-gate FETs (see Figure 2.28), the change of voltage between the gate electrode and the source/drain electrodes alters the conductivity of the graphene, which enhances the sensitivity.

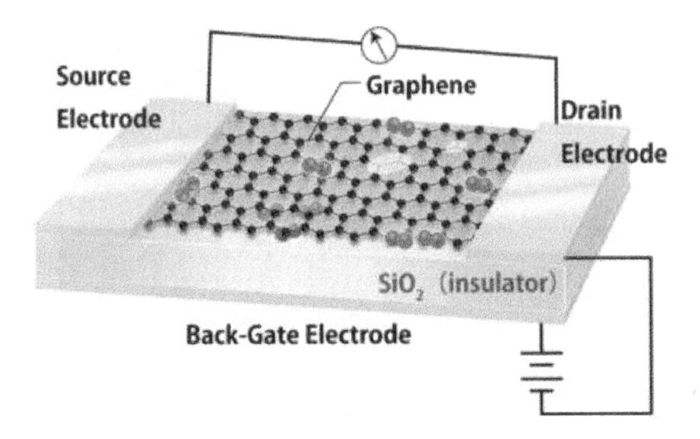

Figure 2.28. *Back-gate graphene transistor. For a color version of this figure, see www.iste.co.uk/paret/smartpatches.zip*

– In floating-gate FETs (see Figure 2.29), the graphene is highly sensitive to a change in the surrounding liquid, either in terms of ion density or a surface charge, which yields a higher detection sensitivity in relation to other floating-gate FET biosensors without graphene.

In these components, the operation is based primarily on the way in which the conductive and insulative layers of the FET are stacked. In a biosensor, the insulative part is left empty, so that the substance being analyzed can infiltrate this space. Liquid fills a space of around 0.5 μm thickness on the surface of the component, without touching the graphene (see the example given in Figure 2.30).

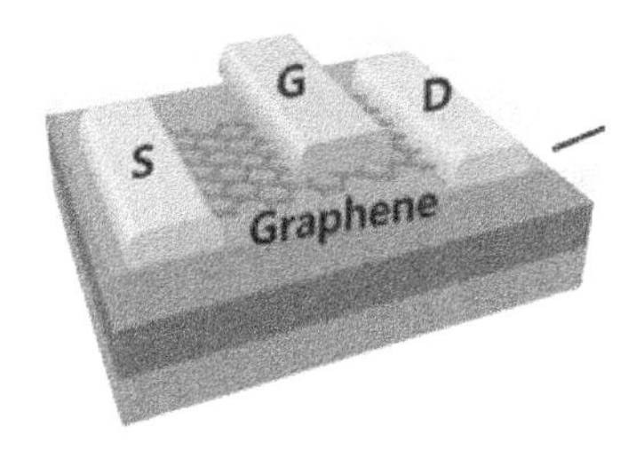

Figure 2.29. *Floating-gate transistor using graphene. For a color version of this figure, see www.iste.co.uk/paret/smartpatches.zip*

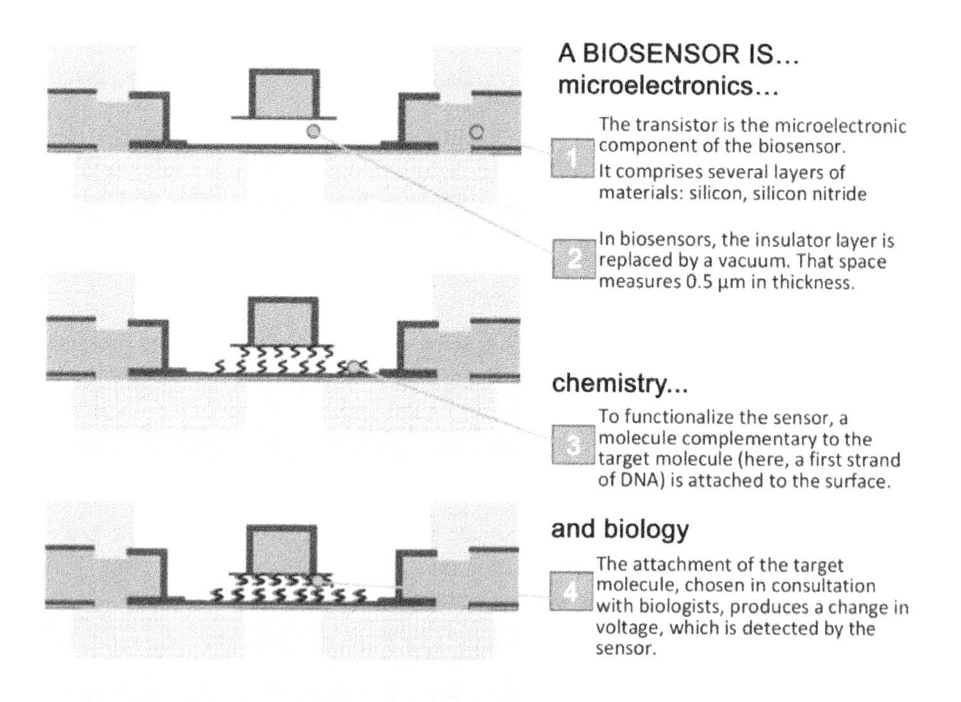

Figure 2.30. *Technological example of a biosensor with a graphene FET. For a color version of this figure, see www.iste.co.uk/paret/smartpatches.zip*

The surface of the graphene sheet can also be designed to be selective for a single biomolecule, and can detect electronic changes in the surrounding liquid medium by two different mechanisms: electrostatic triggering and surface transfer doping.

– Electrostatic triggering mechanisms use the density of electron holes, which leads to a change in the overall conductivity of the graphene, and this can be detected.

– Surface transfer doping occurs when a dopant or biomolecule causes a shift in the voltage of the graphene network, leading the Dirac point[6] to shift to the left by means of a charge transfer mechanism, which represents the transfer of electrons from the biomolecule to the graphene matrix.

Such mechanisms induce a response which can be detected and measured. In other words, the electron transfer gives rise to the detection of the molecule.

Graphene also has significant potential in allowing the development of electrochemical biosensors based on direct electron transfer between the enzyme and the electrode surface.

Graphene can be used to make smaller, lighter, more sensitive sensors, capable of detecting smaller changes in the material, which work more quickly and might even be less expensive than traditional sensors.

Example 1: sensors with DNA or proteins

For many years, experiments have been being carried out with DNA and proteins by "functionalizing" the sensor – that is, by attaching to it a strand of DNA which is complementary to the target base sequence, or the specific antibody for the target protein. Of course, in order to do so, it is crucial to first understand the specific nature and role that each molecule plays.

The two parts are attached to one another by immersion. To detect fixation, the biosensors exploit the fact that the biological molecules carry a charge, and detection is based on a shift in the voltage being applied to the transistor:

– In the case of DNA, the biosensor's response is relatively simple:

- if the complementary strand is present, a difference in voltage will be observed;

- if it is not, the voltage level remains the same. This is the all-or-nothing principle. The difference in voltage (by a few hundredths of a mV) can be measured with an ordinary voltmeter.

---

6 A Dirac point is characterized by contact between two electronic bands with a relation of linear dispersion, which illustrates multiple features of the Dirac equation for particles with zero mass.

– In the case of a protein biosensor, things are a little different. Not only can such a biosensor detect the presence of the target protein; it can dose it as well. The functionalization of the sensor is crucially important, because the reaction needs to be reproducible and sufficiently sensitive to detect minuscule quantities. Dosing on a blood protein (transferrin) is capable of detecting quantities ten times smaller than those detected by conventional analyses, and selectiveness tests to see how the biosensor behaves in a complex environment have shown positive results. The ideal is to be able to work directly on a drop of blood, and even to combine different sensors, each specific to a different protein, to perform a complete biological analysis.

This type of biosensor has proven effective in measuring pH: it is 4–5 times more sensitive than conventional devices.

Example 2: fluorescence sensors

There are also two new main mechanisms which determine the transfer of photons and phonons through graphene: *electrochemiluminescence* and *fluorescence*.

– *Electrochemiluminescence* (ECL) (without graphene) has two mechanisms: annihilation and coreaction. These two mechanisms involve intermediary products which undergo reactions, transferring electrons to the surface of the electrode. They elevate the electrons to higher states of energy, so light is given off as they fall back into their initial state. In graphene, two opposing mechanisms take place: enhancing ECL and extinguishing ECL. Graphene's intrinsic electrical conductivity and mobility favor the quantum yield of luminophore. The electrical properties of graphene encourage an electron transfer which, per unit time, means a higher concentration of intermediary species are raised to higher states of energy. These properties, coupled with graphene's large specific surface area, lead to a high concentration of sites for the luminophores and the target biomolecules, which increases the sensor's sensitivity. Graphene (and in particular, graphene oxide) may also suppress the mechanism of ECL due to the transfer of resonance energy. There is a critical point for the intensity of ECL. Below this point, incorporation of graphene into the sensor increases the intensity of ECL. Above this point, the intensity of ECL decreases, even if graphene continues to be added into the matrix.

– Graphene also has the capacity to either create *fluorescence* or to extinguish it. Graphene can be used in sensors to detect the fluorescence imposed on a sheet of graphene (or GO) by photo-induced charge transfer and fluorescence resonance energy transfer. Graphene (and GO) can also be used as an energy extinguisher for organic fluorophores and nanomaterials.

EXAMPLE.– A fluorescent DNA nanosensor is based on the hybridization of two single-strand segments of DNA (ssDNA). One ssDNA segment is marked with a fluorescent dye and the other is the complementary DNA strand corresponding to the target DNA. This method requires optical detection; consequently, it draws upon the property of optical extinction of graphene-based materials to improve the visualization and detection of the target DNA. The fluorescence-marked DNA can be immobilized by direct adsorption of the DNA probe onto the graphene-based surface by means of π-π interaction between the cyclical structure of the DNA bases and the graphene surface. An example of graphene oxide-based fluorescence biosensors has been produced with multicolored DNA probes to detect different specific DNA sequences. This GO-based multiplex DNA sensor has low background fluorescence and an excellent emission signal from specific targets when hybridization occurs. Another widespread use of the fluorescence detection approach, which may also use graphene-based materials, is fluorescence resonance energy transfer (FRET, or Förster resonance energy transfer). In this detection method, initially, the fluorescence-marked DNA probe is extinguished in the surface of the graphene-based nanomaterials by FRET, which deactivates the fluorescent signal. When the probe bonds with the target DNA, the fluorescent molecule is released with the ssDNA from the surface of the graphene, and the fluorescent signal is activated for optical detection. With this detection method, the sensor is reusable.

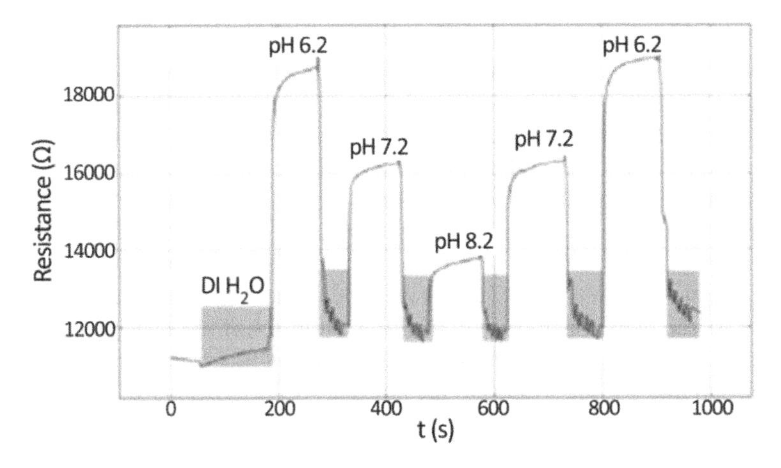

Figure 2.31. *Response of a biosensor for pH detection, over time. For a color version of this figure, see www.iste.co.uk/paret/smartpatches.zip*

Example 3: sensors with enzymes

NOTE.– Let us return, for a moment, to the immobilization of the bioreceptor on the surface of the transducer, which plays an important role in the production of a biosensor. It impacts directly on the sensor's performances: its response time, its specificity, its selectivity or its lifespan. In addition, while the biological material is immobilized, the bioreceptor must retain its recognition properties, notably in the case of proteins or antibodies, but also its catalytic activity in the case of enzymes. Finally, it is important to ensure that the chosen immobilization method allows for reproducible, homogeneous deposition. Immobilization of the bioreceptors also represents a financial interest in industry. Notably, in the case of biosensors made with enzymes (enzymatic sensors), immobilization enables the biological material to be retained between different measurements, and ensures that the biosensors can be continually reused (see Figure 2.31): they are not altered at the end of the reaction process, and are therefore reusable catalysts, which significantly decreases production costs. Furthermore, these molecules are involved in the metabolism of all living things..

This raises the following questions.

2.5.2.2.4. Critical points surrounding analytes for biosensors

Before ending this section, let us take a brief look at some critical points that are also worth examining before embarking on an industrial project involving biosensors:

– the right choice of analyte;

– the availability and cost of the analyte on the market;

– the analyte deposition method (see earlier on in the chapter);

– will the biosensor be single-use?

– can the biosensor be used multiple successive times, and if so, how many times?

– what is the recovery time needed for the analytes between two successive measurements?

– what is the reliability of the analytes between two successive measurements?

– what is the reliability of the analytes over time?

– what are the resistance and stress parameters of the analyte exploited by the biosensors? For example, if the garment is washed (mechanical forces, repeated vibrations, pressure, temperature, number of cycles, detergents, solvents which may include neutralizers, anti-caking agents, surfactants, saponifiers, dispersants, ultrasound cleaning, etc.).

2.6. Applications of graphene in smart apparel

2.6.1. *Biosensors and patches*

Conventional health monitors are able to track biological markers such as heart rate, temperature, blood sugar, distance walked and other raw measurements, but a more fundamental understanding of an individual's state of health, stress and performances is needed for the collection of medical data, or for applications in healthcare, sport, the military domain or high-performance equipment. In particular, metabolic markers such as electrolytes and other biological molecules offer a more direct indicator of human health, which allows for a more accurate assessment of sporting performances, health and safety at work, clinical diagnosis and management of chronic health conditions.

At present, there are already a range of graphene-based sensors capable of measuring numerous parameters, which can provide information about a patient's state of health. For example, biosensors monitoring glucose levels, pH, proteins, drugs, codeine, paracetamol, UV, Covid-19, etc. Some such sensors are discussed in the following sections.

2.6.1.1. *Sweat biosensor*

Like other bodily fluids, sweat is a goldmine of information as to what is going on inside the body, which is useful for health monitoring. In addition, it is easily accessible and can be collected in a non-invasive manner. Furthermore, the markers which can be read from sweat correlate closely to blood plasma levels, making it an excellent substitute diagnostic fluid. Here follow two examples of solutions.

2.6.1.1.1. Example 1: a sensor without graphene

A biosensor *not using graphene,* resembling a dressing, which collects sweat.

The purpose of this wearable biosensor is to sample the sweat and use a simple color-change system to indicate the quantities of various components present in it, helping to diagnose health conditions.

First of all, a super-hydrophobic silica suspension is covered with a flexible polyester film. Then, micro-wells are drilled into the silica layer, to capture transpiration in the bottom, and are filled with dyes whose color changes with pH or with the concentration of chloride, glucose or calcium. Finally, with an adhesive support, a bandage is stuck to the skin of the person performing exercises, to hold the biosensor in place. The subject's sweat accumulates in the micro-wells and the stains change color. By analyzing the colors – for example, with a smartphone – we can determine the pH of the sweat, between 6.5 and 7.0, with a chloride concentration of around 100 mM and traces of calcium and glucose. It may be a primitive system, but it works.

2.6.1.1.2. Example 2: transpiration sensor

A flexible electronic patch can be sewn/integrated into clothing to detect and analyze various markers present in the person's sweat.

In these patches, detection uses special flexible threads, coated with conductive graphene inks; different coatings lend the threads different functions. For example, lactate can be detected by coating a thread with an enzymatic detection material including the enzyme lactate oxidase; a pH detection thread is covered with polyaniline, which reacts to acidity; and so on.

Thus, the biomarkers present in the sweat are measured in real time: in particular, ions of sodium and ammonium (electrolytes), lactate (a metabolite) and acidity (pH). The device platform is also sufficiently versatile to include a wide range of sensors capable of monitoring nearly all markers present in sweat. The measurements taken may have practical diagnostic applications. For example:

– sodium levels in sweat can give an indication of the person's state of hydration and electrolyte imbalance in a body;

– lactate concentration can be an indicator of muscle fatigue;

– the levels of chloride ions can be used to diagnose and monitor cystic fibrosis;

– cortisol, a stress hormone, can be used to assess emotional stress, metabolic function and immunity.

This type of sensors/patches, which act by direct contact with the skin, can be used to monitor an individual's health and performances in a ways which are not only non-invasive, but entirely discreet, as the wearer may not even feel it or be aware of it. Thus, users can monitor a broad range of markers during physical exertion, helping to predict peaks and slumps in performances during competitive physical exercise. For example, by monitoring their electrolytic and metabolic response. The sensors can detect variations in levels of analytes at intervals of 5 to 30 seconds, which is sufficient for most requirements for real-time monitoring during a range of physical conditioning measures, from physical exercise with a diet specially suited to the performance, to people who were not physically active and did not have specific dietary restrictions. Similarly, these patches can be used to diagnose and monitor acute and chronic health problems or to monitor health at work, or during sporting performance. On a technological level, the sensor's network of wires is integrated/sewn into a garment or into a patch, equipped with a microprocessor and a circuit to communicate wirelessly with a smartphone.

2.6.1.2. *Glucose biosensor*

Graphene-based sensors capable of detecting glucose are a well-established field of activity. Numerous types of graphene electrodes can be used in glucose sensors, which, today, contain numerous graphene derivatives, composed of both biological and non-biological materials. Depending on the type of sensor, the sensitivity may range from 0.64 to 1100 $\mu A \ mM^{-1} \ cm^{-2}$, and the linear range can vary between 0.05 µm and 32 mm.

The two main types of graphene electrodes used in these sensors are enzymatic and non-enzymatic electrodes. Typical glucose sensors are developed with enzymatic components. Numerous variants of graphene can be used; however, pure graphene and reduced graphene oxide (rGO) are the most commonly used graphene derivatives in glucose detection. They form a wide range of compounds with metal nanoparticles, polymers and conductive polymers, which are used to modify the surfaces of the electrodes.

Two main factors are used in glucose detection: the effective transfer of electrons between glucose and the graphene, and the presence of a catalytic material. The interaction of glucose with graphene produces clear redox[7] peaks, which provides an effective electron transfer system. In view of its large surface area, graphene has a high capacity for glucose to adhere to its surface. Thus, graphene-based glucose sensors practically always use metal nanoparticles on their surface, given their capacity to improve sensitivity, electron transfer and response times.

2.6.1.2.1. Example 1: sensor using lacrimal fluids

To measure glucose concentration from human tears, a disposable, inexpensive, non-enzymatic detection strip has been developed, using a copper–graphene nanocomposite. The electrodes of these sensors contain surface modifications of the copper nanoparticles, because the size and distribution of those nanoparticles play an important role in optimizing the sensor. A higher concentration of copper increases the signal output, owing to the increased number of reactions between the copper ions and the glucose molecules. The incorporation of graphene into the matrix creates a uniform distribution of the copper nanoparticles by controlling electrodeposition under the influence of applied voltage. Using graphene oxide (GO, rGO, etc.), the functional groups on the surface help keep the copper nanoparticles in place, which facilitate a uniform distribution of exposure.

---

7 An oxidation–reduction (or "redox") reaction is a chemical reaction involving the transfer of electrons.

2.6.1.2.2. Example 2: FET-based sensor

Field-effect transistors depending on a graphene solution, modified with an enzyme, can be used as high-performance glucose sensors. These glucose sensors are a type of "solution-gated graphene transistors" (SGGT), which can provide sensible detection in real time with a high throughput. They can also function at low levels of voltage and in aqueous environments, which are two highly prized properties in the detection of biomolecules. These sensors' detection mechanism hinges on interactions between the biomolecules and the sensors' channels, or gates. These glucose sensors' gates are composed of CVD graphene, and the electrodes of the gates are modified with the enzyme glucose oxidase (GOx), biocompatible polymers and platinum nanoparticles. The sensors work by oxidizing glucose (a reaction catalyzed by GOx), which generates hydrogen peroxide at the gates. The hydrogen peroxide is then oxidized, which regulates the effective gate voltage applied to the transistor. The sensors detect variations in voltage, and record a detection signal. The high level of sensitivity of these sensors makes them ideal for non-invasive glucose sensors, detecting glucose in human bodily fluids such as saliva.

2.6.1.2.3. Example 3: diabetes monitoring

Let us begin with a few important figures. It is estimated that in 2007, there were around 250 million diabetes sufferers in the world, 6% of whom were between the ages of 20 and 79, and 80% lived in developing countries. By 2025, the number of diabetics is expected to rise to 380 million, or 7% of the adult population. In children and young people, type-2 diabetes is on the rise, and is becoming a public health problem.

Across the world, somebody loses a lower limb to amputation every 30 seconds; up to 70% of these amputees are diabetics, and up to 85% of amputations take place as the result of an ulcer. In developed countries, up to 5% of diabetics have a foot ulcer. They consume 12–15% of health resources devoted to diabetes. In developing countries, this figure could even be as high as 40%. Better care can be administered to the feet when personal care by an informed patient is supported by a specialized, multi-disciplinary team. It has been shown that a multi-disciplinary approach can decrease the amputation rate by 50–85%.

NOTE.– In numerous countries, there is an urgent need for podiatry training programs. The people running a podiatry service must be required to have a certain minimum level of experience and equipment, so that the patients are not exposed to unnecessary risks when undergoing treatment with unregulated, unqualified and ill-equipped practitioners. Investment in a diabetic foot care program may be one of the most effective approaches in terms of the cost/effect ratio, if such a program is outcome-focused and applied correctly. Economic studies have shown that strategies which lead to a 25–40% reduction in ulceration and amputation are financially viable, and can even lead to savings being made.

To make progress towards the monitoring and treatment of diabetes, with a view to preventing amputation and danger to the patient's life, it is possible to use patches directly applied to the sores resulting from diabetes, which measure how the condition is evolving by continuously measuring the pH (see the example of the company Grapheal presented blow). Another potential option is a connected diabetic shoe.

2.6.1.3. *Biosensor measuring pH*

The acidity parameter pH can be monitored and measured using a graphene-based biosensor on the surface of the body, which is a non-invasive technique. Such a biosensor would use a floating-gate field-effect transistor. The supple and flexible nature of the biosensor and its associated patch mean that they can be placed anywhere on the body (for example, on joints, etc.), for real-time pH monitoring. A sensible balance between the performance (sensitivity), production cost and ease of use of a sensor is the main criterion determining the use of such a device.

2.6.1.3.1. Example: graphene-based pH sensor

The sheets of graphene for this sensor are produced by mechanical exfoliation of bulk graphite using adhesive tape, and deposited at random on a layer of SiO_2 285 nm thick on a silicon plate. The thickness of the SiO_2 layer has been chosen because it provides the best visibility of the graphene sheets. Then, a highly focused beam of Pt^+ ions is used to sweep the desired areas of the sample surface. The organic precursor of the metal platinum Pt (trimethyl(methylcyclopentadienyl)platinum) is introduced by a gas injection system during the ion beam sweeping, ensuring the ionized metal is deposited in a carefully controlled manner. Thus, rows of Pt can be manufactured directly on the desired area of the graphene to a precision of around 20 nm. Then, two other rectangular electrodes (200 µm × 10 µm × 100 nm) are put in place. To extend the platinum contact electrodes, two millimetric test buffers are produced, by laying down a layer of silver on the slice using high-purity silver conductive paint (SPI 05001-AB), working with a stereomicroscopic channel.

To detect pH, a process of wetting and drying is employed. Once the device's resistance in air has stabilized, the first drop of pH buffer is carefully placed on top of the graphene sheet. After a few minutes, when the resistance has become stable, the drop of pH buffer is carefully aspirated away. Then, further drops of buffer with different pH values are applied and removed in the same way, multiple times. In these experiments, buffers with pH values between 4 and 10 have been tested. Over the course of the experiment, the resistances of the device, with a constant current of

10 μA, were monitored and recorded in real time by a semiconductor analyzer. Examples of the results are shown in Figure 2.32.

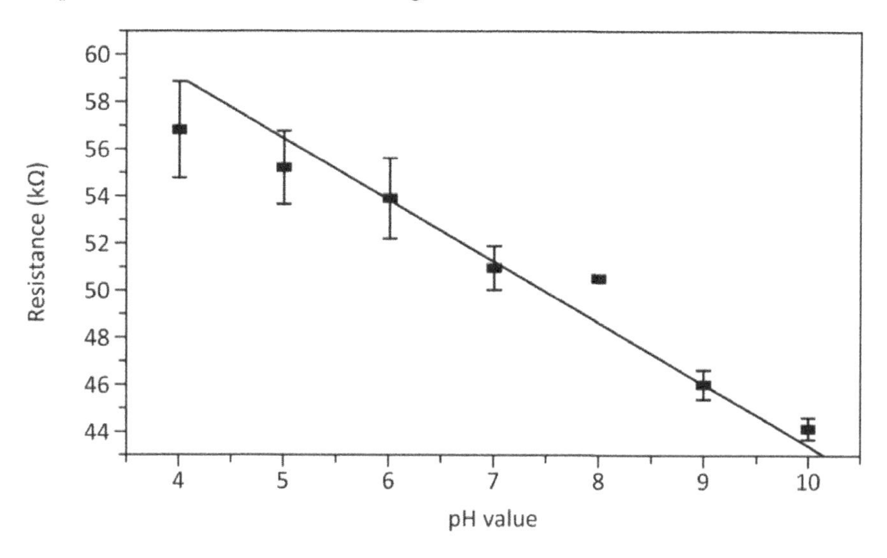

Figure 2.32. *Variations of the resistance as a function of pH*

2.6.1.4. *SpO_2 biosensor*

The same holds true when measuring the degree of blood oxygenation by pulse oximetry: a non-invasive method used to estimate the percentage of oxygen bound to hemoglobin in the blood. The approximation of SaO_2 is referred to as SpO_2 (peripheral arterial oxygen saturation).

A pulse oximeter is a small device which clips onto the body (usually the end of a finger, an earlobe or a child's toe) and transfers its readings to a central unit, either via a cable or wirelessly. Used in conjunction with a graphene-based light-sensitive sensor, the devise uses light-emitting diodes to measure the absorption of red and near-wave infrared light. The difference in light absorption between oxygenated and deoxygenated hemoglobin is exploited to calculate the percentage of oxygenated blood.

2.6.1.5. *Protein biosensor*

Numerous protein sensors without graphene suffer from a lack of flexibility, making graphene an excellent material for composite protein sensors. Graphene-based protein sensors are used to detect complex proteins such as antibodies and biomarkers, to be used as diagnostic testing tools.

2.6.1.5.1. Example 1: nano-FET graphene-based protein biosensor using CVD-cultivated graphene to detect biomarkers of thrombin

Unlike other FETs which are generally made with exfoliated graphene, serving to detect proteins, the use of CVD-cultivated graphene allows for greater evolutivity, an easier manufacturing process, a larger detection zone, and a reusable product. Such products also have advantages which they share with other, similar sensors, such as low noise and high transconductance.

The sensor is used in real time to detect the bonding (and debonding) of biomarkers of the protein thrombin, using variations in the electrical current produced by the bonding and debonding mechanisms. These sensors are also capable of measuring the bond kinetics during the bonding and debonding processes. This sensor can be regenerated simply by rinsing with a buffer solution, which eliminates any proteins bonded to the surface. The sensor surface also contains a DNA aptamer which is specific for bonding to thrombin. These aptamers have a half-life of 10 hours (meaning that 50% of the aptamer will be eliminated after 10 hours), but the device itself has a lifespan of over a week.

2.6.1.5.2. Example 2: protein sensor using thermally reduced graphene oxide (TRGO) and gold nanoparticles conjugated with antibodies

The sensor's response, like that of numerous biosensors, occurs when a protein binds to the conjugate nanoparticles or antibodies, which produces a change in the conductivity of the TRGO sheet. The signal is recorded by an FET and direct current measurements. There are numerous techniques by which these sensors are made: in particular, electron-beam lithography, the methods of dispersion and suspension, and multiply repeated coatings. These sensors have far greater sensitivity than numerous other carbon-based protein sensors.

2.6.1.5.3. Example 3: bioelectronic sensor to detect fluorescent proteins using a graphene FET with biological and inorganic functional groups

The sensor works by detecting tagged proteins (for example, polyhistidine). The device gives an electrical reading for each given protein, by measuring the optimal excitation wavelength for the proteins.

When the excitation wavelength is known and the surface is rendered multifunctional, these sensors have the potential to be used as diagnostic tools in the future, to detect a wide range of protein species.

2.6.1.6. *Drug biosensor*

Morphine, noscapine and heroin are the three main alkaloids in heroin samples. The most abundant components of opium are morphine and noscapine, which are directly extracted from the poppy.

A Glassy Carbon Electrode (GCE), modified with graphene nanosheets (GNSs) is used for simultaneous identification of morphine, noscapine and heroin. It has excellent electrocatalytic activity with reduced overpotentials across a wide range of pH. In addition, a GCE modified with an electrochemically reduced composite film, doped with GO-MWNTs (Er-GO-MWNTs) is used as a morphine sensor. It combines the excellent conductivity of MWNTs and Er-GO with the filmogenic properties of GO. In addition, hybrid rGO-Pd exhibits a stronger current response to the oxidation of morphine in comparison to non-modified rGO, with a low detection limit of 12.95 nM.

2.6.1.7. *Codeine biosensor*

Graphene and the GCE-modified film Nafion has been produced to detect codeine. It has a superb analytical performance and improved application for detecting codeine in samples of urine and cough syrup. The sensor's high electrocatalytic activity with respect to codeine has been attributed to graphene's exceptional electrical conductivity and the high charging capacity of codeine on the surface of the electrode. In addition, a nanocomposite of nanoparticles of Gr and $CoFe_2O_4$, modified with CPE, proves to be an ultra-sensitive electrochemical sensor for codeine and acetaminophen, with low detection limits of 0.011 and 0.025 μM. The proposed method is immune to the effects of interference from glucose, ascorbic acid, caffeine, naproxen, alanine, phenylalanine, glycine and others. In addition, a quick, simultaneous identification of tramadol and ACOP (see below) has been performed with CPE modified with nanoparticles of $NiFe_2O_4$/graphene, with a low detection limit of 0.003 6 and 0.003 0 μM.

2.6.1.8. *Paracetamol biosensor*

Paracetamol, or acetaminophen (ACOP), is a long-established drug and one of the most widely used over-the-counter medications in the world. It is non-carcinogenic, and is an effective aspirin substitute for patients who are allergic to aspirin. It is used to reduce fever, relieve a cough, cold, and mild to moderate pain. It is also useful in treating osteoarthritis, protects against hardening of the arteries, provides relief to asthmatic patients and protects against ovarian cancer. It is therefore highly important to develop a simple, quick and accurate method for detecting ACOP. An electrochemical ACOP sensor based on the electrocatalytic activity of functionalized graphene was presented earlier in this chapter. A graphene-modified GCE increases sensitivity to the presence of paracetamol, with a low detection limit of 32 nM and satisfactory recovery of between 96.4% and 103.3%. In addition, graphene nanocomposites – SWCNTs or MWCNTs – modified by a GCE nanocomposite (SWCNTs-GNS/GCE or MWCNTs-GNS/GCE) have been used as new simple electrochemical sensors, highly selective for tyrosine and paracetamol. The sensor has ideal characteristics, such as a large effective surface area, high porosity, more reactive sites, excellent electrochemical catalytic activity and applicability in human blood serum and pharmaceutical samples. Very low detection limits of 0.19 μM and 0.10 μM have been recorded for tyrosine and paracetamol.

2.6.1.9. *UV biosensor*

Flexible, transparent and disposable patches can be used to monitor the skin's level of sun exposure, by means of an ultraviolet (UV) sensor.

NOTE.–

– UV radiation is classified into UVA (wavelengths of 315–380 nm), UVB (280–315 nm) and UVC (200–280 nm). The patch described below has a plastic hood including a bandpass filter to block the undesired spectrum of light (> 400 nm). A graphene-based photodetector, sensitized with graphene quantum dots (GQDs). Changes in the intensity of UV radiation in the environment are recorded by the integrated GQD assembly, bringing about a variation in the resistance of the graphene, which is then converted into a digital signal by the integrated microcontroller. The patch, which can be attached to clothing or stuck directly to the skin, provides an effective UV detection system to monitor the harmful effects of sun exposure. Direct cutaneous absorption of UVB and UVC rays creates genotoxic substances because of the high levels of energy these forms of radiation carry (4–6 eV). Measurements at 285 nm have been performed to imitate the irradiation which causes sunstroke, and correlated on the irradiance scale with UV levels of 25 mW/m^2 in terms of the Diffey-weighted mean. The UV patch is capable of recording changes in irradiance with resolution of 0.1 mW/m^2, and the UV patch's output signal is linearly dependent on the intensity of the incident UV light.

– The connection technology used by the patch (see Figures 2.33A and B). This patch uses a high-frequency NFC link (Near-Field Communication) (see Chapter 5) and works without a battery. Wireless data transfer and power supply use an NFC connection from a smartphone, when it is brought into proximity with the UV patch. On the smartphone screen, an application displays the real-time environmental UV readings, remaining recommended exposure time, and alerts the user once they have reached a set threshold of sun exposure. Thus, these patches actively monitor changes in the UV index in the surrounding environment.

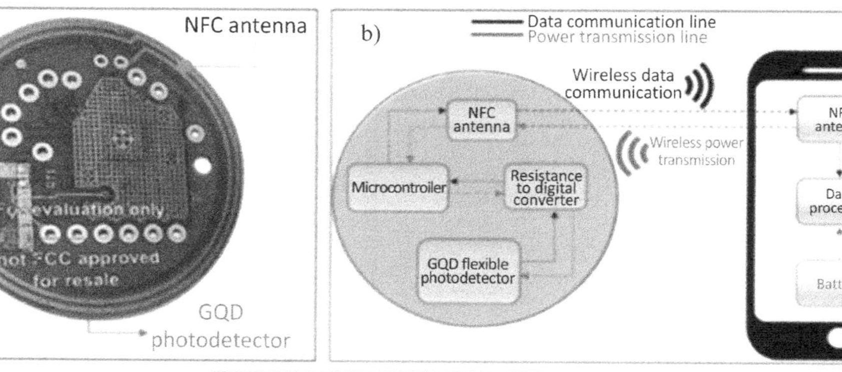

c)

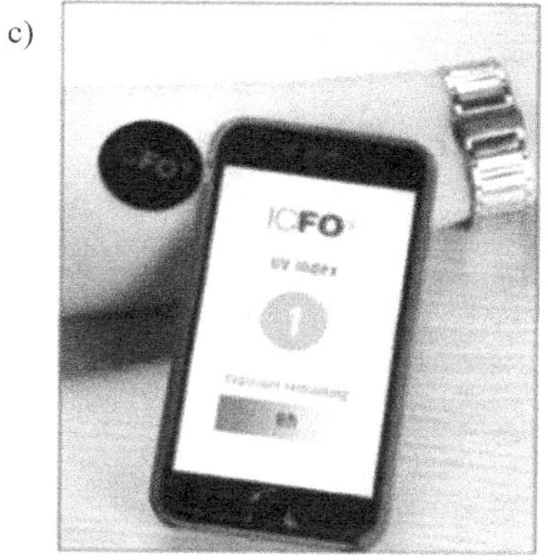

Figure 2.33A. *The patch and its connection to a smartphone (source: ICFO). For a color version of this figure, see www.iste.co.uk/paret/smartpatches.zip*

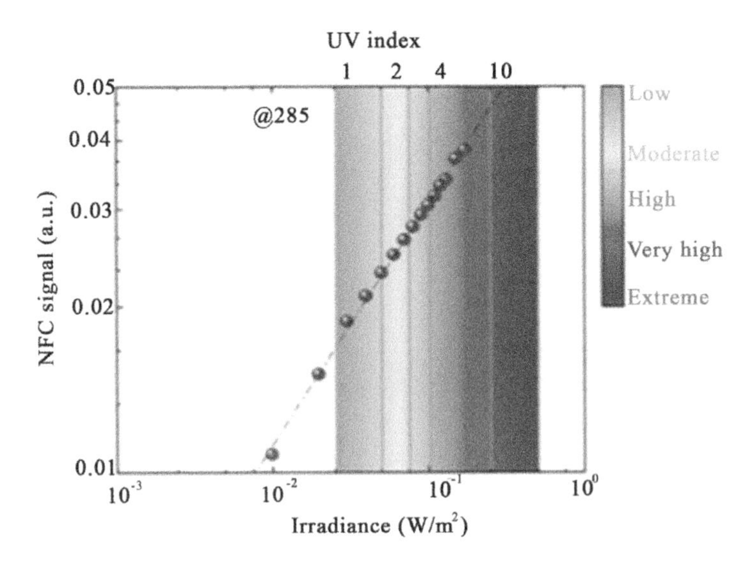

Figure 2.33B. *UV response and display (source: ICFO) (cont.). For a color version of this figure, see www.iste.co.uk/paret/smartpatches.zip*

NOTE.– The same basic NFC technology used by this UV patch can be developed to produce numerous devices yielding continuous, precise measurements of various vital signs (e.g. heart rate, respiration rate and pulse oximetry).

2.6.1.10. *Covid-19 biosensor*

In 2020, due to the worldwide coronavirus pandemic, there was a surge in interest in molecular sample analysis. In this field, the metamaterials of periodic and artificial structures which can be used as biosensors, exploiting microfluidic technology.

A biosensor based on a field-effect transistor has been developed, using a sheet of graphene with high electronic conductivity. This biosensor can detect whether a person is carrying Covid-19 from nasopharyngeal swabs in under a minute, with no prior preparation of the bio-samples. To achieve this result, on graphene, we attach antibodies for the spike protein SARS-CoV-2. When a purified spike protein of the SARS-CoV-2 virus is in the vicinity of the sensor, the binding to the antibody (the light blue Y shapes in Figure 2.34) alters the level of electrical current. That electrical current, then, can be interpreted as an indication that the coronavirus is present.

For example, the company Grapheal, based in France, developed a diagnostic test based on a saliva sample to quickly detect Covid-19 infection. This electronic test is an antigen immuno-test: it is the presence of the virus itself, not the immune response, which indicates whether a patient is carrying the virus. The result can be recorded electronically on the device, which speeds up diagnosis and allows for secure digital transfer of the data relating to the result.

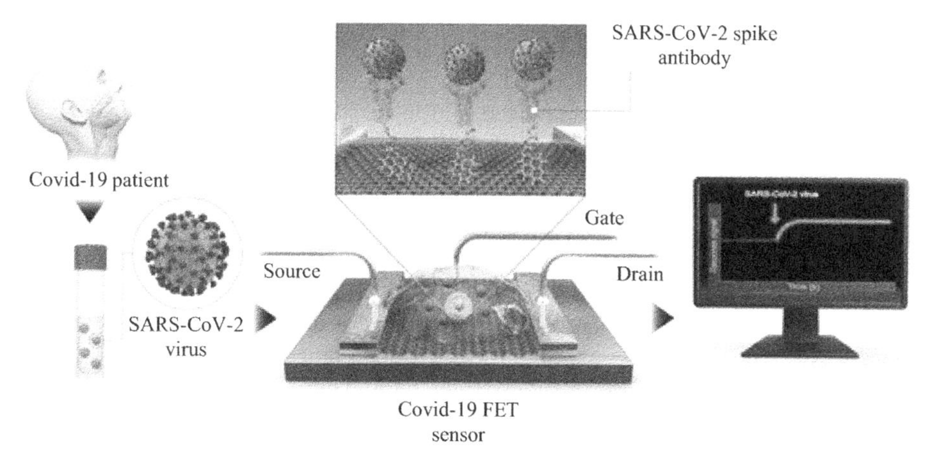

Figure 2.34. *Example of Covid-19 measurements using a graphene-based biosensor (source: Grapheal). For a color version of this figure, see www.iste.co.uk/paret/smartpatches.zip*

Another very small biosensor, based on graphene and testing for flu and Covid-19 antibodies, is capable of distinguishing between Covid-19 and flu, and/or testing for both simultaneously. Part of the device is sensitive and reacts to flu antigens, and another area is sensitive only to Covid-19 antigens. The test is performed using saliva samples, using a graphene-based biosensor to detect iron deficiency in children by combining the sensor with anti-ferritin antibodies. By changing the antibody attached to the sensor, it is possible to turn the platform into a dual Covid-19/ flu test.

Around the globe, numerous enterprises are currently engaged in developing medical masks to combat Covid-19, which also contain graphene. For example:

– The AECC Beijing Institute of Aeronautical Materials (AECC BIAM) has created a facemask that is a powerful tool in the fight against Covid-19, and is more sustainable than other solutions. The fabric used is composed of graphene and polypropylene, along with other materials used to make ordinary facemasks. It remains effective for 48 hours; beyond that point, its filtration efficiency drops by 4%.

– In Hong Kong, laser graphene-coated disposable surgical masks have been developed; they are self-sterilizing and waterproof. Droplets detach automatically from the mask matrix and fall to the floor. The same is true of the companies

PlanarTech and IDEATI, Bonbouton, and LIGC Applications (with the Guardian G-Volt). Some of these have patents pending, but others have already been launched on the market. Certain businesses like Grolltex, in the United States, are adapting their existing graphene biosensors, which have previously been used to detect strains of flu and other viral strains, to detect Covid-19.

2.6.2. *Multisensors: multibiosensor patches*

In this book, the concept of a "multisensor monopatch" – that is, multiple sensors built on a single graphene support – is envisaged. It is fairly narrow, but could easily be extended.

EXAMPLE 1.– A measurement of temperature combined with a pH biosensor, themselves coupled with management software, aimed at demonstrating the characteristics and capabilities of graphene-on-polymer flexible technology for health/comfort monitoring.

EXAMPLE 2.– The solution discussed below combines various hardware and software components, producing an electrochemical multisensor capable of simultaneously measuring four markers: immunoglobulin G (IgG) proteins, immunoglobulin M (IgM) proteins, C-reactive proteins, which play a role in assembling new viral particles, and nucleocapsid proteins, whose levels indicate the seriousness of an illness.

2.6.2.1. *Application: detection of SARS-CoV-2 virus*

The multisensor is composed of four separate electrodes, and a further shared electrode which surrounds the other four (see Figure 2.35).

Figure 2.35. *Example of a "multisensory patch" (source: California Institute of Technology). For a color version of this figure, see www.iste.co.uk/paret/smartpatches.zip*

The functional structure of the sensor is situated on graphene electrodes, and comprises a linker, an antibody and tagged antibodies. A linker is needed to stabilize the graphene surface and link to an antibody or to another protein.

To record C-reactive proteins and nucleocapsids, an antibody is adhered to the linker, which binds specifically with the desired protein; then, another tagged antibody is added to improve the signal. This structure, known as a "sandwich", is highly sensitive because of the presence of two antibodies. The structure for detecting IgG and IgM is somewhat simpler, comprising a section of actual SARS-CoV-2 virus, to which the antibody binds, and a tagged antibody.

Because of the presence of the second antibody, this structure is referred to as "indirect". In addition to measuring markers diluted in a buffered solution, the sensor tests real biological fluids: blood plasma and saliva. In both cases, the presence of the virus and the seriousness of the illness are assessed and reliably determined (see examples of the results in Figure 2.36).

Other biomolecules such as cells, electroactive analytes, dopamine and uric acid (amongst others) can now be detected by various graphene-based sensors. Graphene can be used, for example:

– As a biocompatible substrate to improve cell adhesion and growth so as to detect cell populations. Graphene oxide, with a negatively charged surface, can be used to interact with positively charged poly-L-lysine, resulting in a biocompatible interface which promotes cellular adhesion.

– A composite film of rGO and carboxymethyl chitosan with molecules of folic acid anchored to the surface can be used to detect tumor cells which have a folate receptor.

– To interact with an individual's biological system, or "talk to their cells". Graphene itself does not communicate as such, but a layer of graphene placed between the layers of synthetic phospholipids does the job. This demonstrates the versatility of this material, and its ability to meld with our own biological systems.

2.6.3. *Beyond sensors and beyond "bio"*

2.6.3.1. *Usages in textiles: finishing of textile fibers, tissues and threads*

Let us cite two examples of fabric finishing processes.

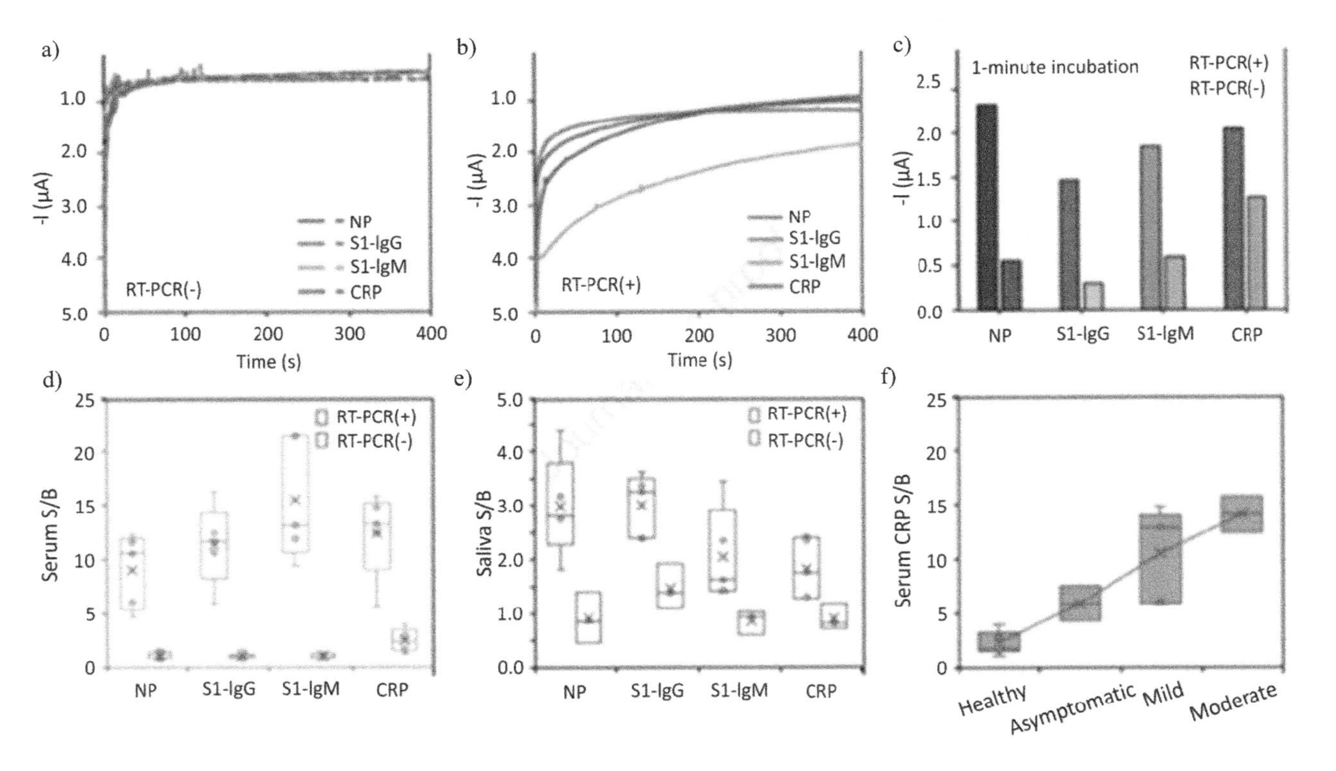

Figure 2.36. *Experimental results obtained using real biological fluids (source: Rebeca M. Torrente-Rodriguez et al. 2020). For a color version of this figure, see www.iste.co.uk/paret/smartpatches.zip*

NOTE ON FIGURE 2.36.– *Change in current across the cathodes for healthy subjects (a) and for ill patients (b); histograms comparing the healthy and sick individuals for each type of marker (c); comparison of the results for blood plasma and saliva (d, e); relation between the level of C-reactive protein and the seriousness of the illness (f).*

– Thermal conductivity of the thread: to improve the thermal conductivity of a textile (thread) and increase its capacity to dissipate body heat and make the garment more comfortable to wear, graphene is grafted to the textile fibers (see Figure 2.37).

Figure 2.37. *Example of denim garments. For a color version of this figure, see www.iste.co.uk/paret/smartpatches.zip*

– Fluidity of a garment: similarly, graphene's high capillarity greatly reduces friction with air and water, helping to improve the wearer's sporting performances.

2.6.3.2. *Industrial usages*

Because its sale price is still rather high, graphene is very often combined in very small quantities (of the order of a few percent) with one or more other materials: for example, plastic, carbon fiber or rubber. Let us here give a few examples, depending on the desired applicative properties.

2.6.3.2.1. Improving the mechanical strength of material

Examples of applications: strengthening the carbon fibers of a bike frame, reducing the thickness of the resin holding the fibers together, improving the scratch resistance and shock absorption of vehicle bodies and/or hoods, the capacity to recover the original shape after a crash, etc.

2.6.3.2.2. Mechanical wear resistance

Examples of applications: wear resistance of tire grips for mountain bikes and control of the behavior and deformation of the grips when braking and supporting, resistance to deformation and sudden changes in temperature of a road surface ("paved"). In addition, the material is 100% recyclable and has fatigue resistance that is 250% higher than that of conventional tar, drain covers made from recycled tires, enriched with graphene, which are more resistant (capable of withstanding stresses of up to 40 tons), invulnerable to corrosion, etc.

2.6.3.2.3. Thermal resistance (increased by around 30°C)

Examples of applications: better handling of high temperatures by carbon fiber in disc brakes, whilst also reducing weight.

2.6.3.2.4. Light weight

Examples of applications: producing more lightweight (and stronger) composite materials for airplane fuselages, motorbike engine covers, etc.; the quest for lightweightness is not the same as fragility, because breakage resistance is doubled; improving the lateral rigidity of a wheel and increased solidity; progress in making wheels from graphene; better vibration damping; improving overall driving comfort; very high stress resistance at the spoke anchor point; improving the "slip" of boat hulls and garments because of graphene's excellent hydrophobic properties.

2.6.3.3. *Uses in energy*

2.6.3.3.1. Batteries

EXAMPLE 1.– Graphene, formed of a single layer of carbon atoms, undulates and deforms with changing temperature. The thermal motion of the graphene induces an alternating current, which it is possible to harvest with an electronic chip, to provide unlimited low-voltage direct-current power supply to small devices and sensors. The direct current can be stored in a supercapacitor for subsequent use. For this objective, we need to miniaturize the energy harvesting circuit and replace a low-power battery (see Chapter 4).

EXAMPLE 2.– The integration of graphene, replacement of the graphite electrode with an electrode covered in a fine layer of graphene, and modification of the composition of the electrolyte of the batteries rapidly improves the yield. Now, batteries using graphene polymer cells are able to achieve an energy density three to five times higher than conventional lithium-ion cells. Samsung has managed to create a lithium-ion battery which stores energy in graphene pellets, which, for equal volume, are able to store 45% more energy, and recharge five times faster, than current technologies. This battery is lighter, safer (with better heat and pressure resistance), and more flexible in terms of integration.

2.6.3.3.2. Supercapacitors

The goal, firstly, is to store a larger quantity of energy/coulombs ($Q = i.t = CV$) than traditional batteries, in graphene-based supercapacitors; secondly, it is to store energy more quickly, in the space of a few seconds. The most common applications are for new models of urban surface transport and also batteries for mobile phones or small IoT elements such as health patches, as mentioned multiple times throughout this book (also see Chapter 4: Energy Harvesting).

2.6.3.4. *Uses in electronics*

Graphene has excellent electrical connectivity and could, in time, replace silicon to create an entirely new generation of transistors with unparalleled performances among today's technologies. In addition, it plays a role in the manufacture of transparent and flexible screens.

2.7. Conclusions on graphene in smart apparel

2.7.1. *Benefits to the applicative constraints of smart apparel*

The design of flexible, smart apparel for the "general public", equipped with biosensors and the associated non-intrusive electronic components (using capacitive-mode IBC, as we shall see later on) is no easy task, because there are multiple requirements in terms of quality and applicative performance which must be satisfied, but which are often mutually incompatible!

In Chapter 1, we briefly listed some general constraints. We shall now return to that list and add to it, taking account of some exacting and interesting specific points about the used of graphene technology, as discussed above.

2.7.1.1. *Lightweight*

In the field of smart apparel, for producing a next-generation "healthcare" patch, we have seen that graphene is extremely lightweight, and this lightness is an advantage, as the patch must be discreet and comfortable for users to wear.

2.7.1.2. *Strong*

The patch also needs to be strong, and practically tear-proof. Indeed, if the smart garment is to be work all day long, it must not degrade during use, or tear as a result of friction or folding, for example. We have seen that graphene is very strong indeed, and practically impossible to tear.

2.7.1.3. *Transparent*

Graphene's transparency may be useful, or even required, for esthetic reasons. For example, to create a discreet patch which blends in with the color of the skin, but also for other applications such as solar panels or transparent conductors.

2.7.1.4. *Foldable*

Foldability is an important point when it comes to putting clothes away on shelves or in cupboards. As explained above, because of the strength of the covalent bonds between the carbon atoms, graphene has extremely high traction resistance, and can be stretched up to 20 or 25 times its original length without breaking. The rupture resistance is one of graphene's most important mechanical properties.

2.7.1.5. *Friction-resistant*

Friction resistance needs to be taken into consideration in relation to the wearing of the garment, or washing in a domestic machine, etc. Thanks to the repeated sp^2-hybridized bonds between the carbon atoms in a perfect hexagonal network, graphene has excellent flexibility.

2.7.1.6. *Comfort when wearing*

All the properties of graphene, a two-dimensional material only the thickness of a single carbon atom, mean it can be used to create a patch which is small, flexible, supple, lightweight and thin which, if stuck to the skin, is capable of tracking the patient's movements. This means the patch is comfortable to wear. The only remaining source of discomfort may be the electronic part (due to its size, rigidity, etc.).

2.7.1.7. *Batteryless power supply*

Graphene-based systems are recognized to be promising platforms for the storage of energy from patches, resulting from energy harvesting techniques, in view of their excellent electrical and thermal conductivity, and their exceptional mechanical properties.

2.7.1.8. *Power supply duration suitable for the application*

The characteristics of graphene make it an ideal material from which to produce electrodes for battery sources of energy and supercapacitors for patches: its extremely high mechanical strength and chemical resistance offer a favorable life span and little loss of capacity after multiple charging/discharging cycles. The thinness of the graphene sheets ensures there is a large exchange surface. It is this ability to exchange ions which impacts on the performances of batteries, both in terms of capacity and of charging and discharging rates. Its excellent conductivity reduces the risk of overheating, meaning graphene-based batteries can be charged more quickly.

Perforated, silicon-doped graphene sheets can be used to replace traditional graphite anodes, increasing the capacity of the batteries somewhere between three- and tenfold. In order to do this, it is possible to combine graphene with silicon particles, which increases the energy storage capacity by a factor of 10: 3,200 mA h/g as opposed to 300 mA h/g with conventional lithium-ion batteries.

2.7.1.9. *Washability and °C: number of wash cycles*

When we speak of a smart garment for professional use or for ordinary consumers, in which the patch is not disposable (or the electronic components are incorporated into the garments), one question which regularly comes up is its maintenance. "Will it be able to stand being washed? If so, how many wash cycles can it survive? Do parts need to be removed in order to wash the garment?" These are among the questions which – quite legitimately, obviously – tend to be asked. In general, smart apparel and fabrics are – or should be – washable, with the exception of disposables. It is therefore generally accepted that they need to be able to stand being washed at home so they can be reused thirty or so times.

Smart apparel manufacturers have developed solutions to insulate the electronic parts, allowing the garments to be put through the washing machine (protecting the sensors, etc., with resins, varnishes that are as supple as possible and waterproof). In the case of textiles using conductive fibers, these fibers are insulated from the outside world by a sheath, meaning the central conductive wire is insulated. However, one of the most significant problems lies in the "basic" mechanical connection of all the electronic parts, whatever their shape, soldering them together, etc. Here, once again, the solidity and watertightness are generally assured by a resin top coat (and therefore, greater thickness), or else another type of insulator may be used. This means that the government can undergo a certain amount of washing without any problems… provided the users remember to take the battery out first.

With respect to industrial washing as governed by European Norm EN 15797, linen undergoes a cycle of different treatments: disinfection, prewashing, rinsing, softening, washing, etc. Larger items or ones requiring specific treatment (notably, jackets) are kept in separate washing machines.

Beyond all these fundamental considerations, in the knowledge that watching programs are chosen depending on the degree of soiling and the type of garment to be treated, in the area of maintenance, the following main variables need to be taken into account when dealing with a patch.

2.7.1.9.1. Water temperature in domestic and industrial machines

What are referred to as "domestic" water temperatures are somewhere between 30 and 60°C for linen and underclothes. With respect to the "industrial" cleaning of certain garments (shirts, PPE-type workwear, industrial dish towels and hand towels, bedsheets, etc.), temperatures of the order of 85°C are often reached, and the components used need to be able to withstand these temperatures without giving up the ghost!

2.7.1.9.2. Steam temperatures

Maintenance may be carried out, either with domestic "steam" irons, or in professional presses and dry cleaners.

2.7.1.9.3. Operation times

The duration of washing has chemical impacts on the textiles, fabrics and electronic components, which depend on the products used, and mechanical forces, primarily due to agitation of linen over certain periods of time.

2.7.1.9.4. Chemical products used

The same remark as above applies: the type of chemical product used is also important (bleach, detergent, corrosive agent, or descaler).

2.7.1.9.5. Mechanical effects

Stresses due to the mechanical effects of the washing process, described below, are often of greater consequence than whether the smart apparel is washed in water or some other form of liquid… even soapy liquid.

2.7.1.9.6. Linen agitation

In laundering linen, it is usual to subject it to some form of agitation. This may be done manually with a brush or similar implement, or mechanically, for example, with alternating-motion rotating drums, etc. Obviously, this generates considerable mechanical stresses (friction, creasing, tear-away, etc.); we must also consider the weight of the garment on the conveyor belt, which the fabric and its "electronic" components must be able to withstand. We must take care to prevent mechanical breakages of certain electronic elements, which are mainly of two types:

– fractures due to wrenching forces, at the (micro)welds, resulting from thermo-compression of the connections of the pads of the patch to the connections of the integrated circuits (for information, these are often gold wires 10 μm in diameter, and the pad measures around 30 × 30 μm²);

– breakages of the connections themselves, which are very thin and supple, during agitation, due to the mechanical forces applied by the moving masses of wet linen. Obviously, it is possible to use more robust connections (for example, using stainless steel), which are far less flexible, but it is forbidden in the medical domain, where accidents can occur due to the presence of magnetic fields of 2-3 teslas in an MRI.

2.7.1.9.7. Spinning and drying

Usually, industrial spin drying takes place under pressure, at around 17 bars (except for items with zips and plastic buttons). Industrial clothing is dried and smoothed in a tunnel finisher by high-temperature steam jets (at around 150°C), which gives the clothing its final appearance.

2.7.1.9.8. Ironing/pressing

From the very outset, it is important to distinguish between conventional ironing, as might take place in a home, and professional pressing, using rollers (calenders) – for example, for industrial sheets. When the textiles in question contain electronic components, the following aspects must be taken into account:

– The risk of breakage of connections as a result of the rigidity and thinness of the components (problems may be encountered if presses or calenders are used on such garments, for example). These problems are well known, and solutions have been found and put into practice, with the use of chips for traceability in the professional sphere.

– The risks of chemical alteration and oxidation of the components when they contain metals, and risk of corrosion as a result of the high temperatures (to recap, the temperature of an iron's sole plate may vary between 130 and 240°C).

Obviously, for maintenance solutions, it is possible to be prudent, and choose to:

– protect the sensors and other electronic components with resins, varnishes, polymers, etc.;

– for private individuals, avoid temperatures higher than 40°C for "domestic" washing;

– use less corrosive chemical products (detergents, solvents, etc.);

– opt for suitable durations of washing, drying and pressing.

Let us now return to the question of the use of graphene. The capillary effect of the interatomic dimensions of graphene makes it transparent when wet, impermeable to water, and pliable. Thus, the same is true of a surface which is covered with graphene: the main interactions between the water and the surface (intermolecular forces known as van der Waals forces) are not greatly changed. Therefore, it is not incompatible with a washing machine cycle. This property could potentially hold interest for industrial applications: for example, a coating with a monolayer of graphene could protect a surface from oxidation without altering the wetting properties. An example of such an application is heat transferred by condensation, using copper. This material is an excellent heat conductor, but becomes oxidized when exposed to water. A graphene coating would provide protection from oxidation, and increase the efficiency of heat transfer by 30 to 40%.

2.7.1.10. *Small dimensions*

The ideal format for a patch would be a square or circular mechanical format, around the size of a €1 or €2 coin, capable of accommodating an NFC antenna made of single or bilayer graphene.

2.7.1.11. *Mechanically extendable dimensions*

The surfaces of the moving parts of the body (joints, etc.) and the apparel in relation to the patch are not necessarily flat. Therefore, the patch needs to be extendable to a reasonable degree without impacting the operation of the NFC antenna. The use of an extendable material such as graphene, which can be stretched up to 20–25 times its original length without breaking and withstand the crumpling of the mesh, of the elastic fabric, etc., is a plus.

2.7.1.12. *Inexpensive nature*

It almost goes without saying that the price of such technology needs to be as low as it possibly can be. Whether it is a disposable or a reusable patch, it must be available at an affordable price, so it can be bought on a day-to-day basis. Today, the price of graphene remains high, as it is difficult to synthesize and to produce in large quantities. However, day after day, progress is being made which will drive down the price.

2.7.1.13. *Easy to integrate with fibers and fabrics*

Capacitive IBC technology (see Chapter 5) can be used either to produce an external graphene component or patch, or to integrate the patch into the fibers and fabrics.

2.7.1.14. *Integratable with industrial machines/tools*

Today, we have the potential to produce and use tons of graphene-based conductive threads by using existing industrial textile production machines and processes, with no increase in production costs.

2.7.1.15. *Suitability for conventional finishing processes*

Graphene-based patches can easily withstand the three types of material production processes needed for the textile applications intended here: thread preparation, dyeing and finishing (mechanical and chemical processes, see Chapter 1).

2.7.1.16. *Antibacterial*

Graphene has other interesting properties. Amongst them is its antibacterial effect: bacteria cannot grow on it. It is for this reason that the material is used in antibacterial masks, for example.

2.7.2. *Graphene-based biosensors in smart apparel*

As detailed throughout the previous sections, at the time of writing, numerous biosensor solutions and designs use graphene, which:

– already exist, with sensitivity levels similar to or better than other non-graphene biosensors, and may have beneficial properties such as mechanical flexibility and robustness, high conductivity and enhanced selectiveness for certain molecules and certain applications;

– were developed by research labs at universities or start-ups, each having their own avenues of research and (extremely) lengthy specific production processes;

– in addition, and above all, to date, have not undergone concrete testing, being tested mainly in laboratory environments and/or on a small scale. Thus, some doubts remain as to industrial solutions which will ultimately be adopted.

Today, one of the major problems in the use of graphene in industry is how difficult it still is to produce. There is a very great difference between the samples available and the real-world need for graphene in the manufacturing materials of the future. A great deal of research is being done, and discoveries are being made, with a view to addressing these issues. However, as it currently stands, it is difficult to release a graphene-based product on an industrial scale, because it is only produced in small quantities, of middling quality, and highly costly. Today, inferior-quality graphene, containing defects, is found in the commercial sphere at an affordable price, but it is only suitable for use in certain applications. In addition, its cost needs to be viewed in relative terms, because two major graphene-production techniques are becoming more widely accessible. Thus, the next stage will be when there are numerous graphene composites available, and we need to increase and optimize their levels of sensitivity and selectivity, so they are able to select and distinguish different biomolecules, to move production forward to commercial levels.

2.7.3. *Critical points that must be examined at the very outset*

To conclude this chapter, despite the wonderful physical and technical properties that graphene-based devices currently offer, in the near future, we still need to more closely examine a number of critical points in relation to these biosensors. In particular:

– Is a graphene-based biosensor, with its analyte, intended to be single-use, or can it be used multiple times (and if so, how many)?

– How long do the analytes take to recover between two successive measurements?

– How reliable are the analytes between two successive measurements?

– How reliable are the analytes over an extended period of time?

– What are the parameters of resistance and stress surrounding the analyte and the points at which it is fixed to the biosensors: for example, during washing (when the garment will be subjected to considerable mechanical forces, repeated vibrations, high pressure, temperature, multiple cycles, detergent, cleansing solvents which may contain neutralizers, surfactants, saponifiers, dispersants, anti-foam agents, ultrasound cleaning, etc.)?

The next major step will be the arrival of new and numerous graphene composites, so their sensitivity and selectivity can be fine-tuned to drive production to commercial levels, in order to bring these biosensors and these solutions to higher levels in terms of production and improved sensitivity. The capacity to select and distinguish between different biomolecules is a very real possibility in the near future.

The foregoing has been an undeniably lengthy discussion of the mechanical and chemical workings of the raw information yielded by sensors. The next stage is a much more material process of amplifying, cleansing, digitizing, refining and transmitting the data, taking care to supply the whole system with enough energy to make it work. If we do all that, we will have made a "patch", which we can then seek to integrate into a garment (into the fiber, the fabric, stuck on, sewn in, etc.) to make it a "smart" garment.

That is what we shall look at in the second part of this book, which focuses on the technical and technological aspects of smart apparel.

PART 2

Biocontroller

3

Bioprocessors

In the previous chapter, we completed our presentation of the possible techniques and technologies for graphene-based "sensors and biosensors". Today, there is an extremely broad range of these types of products, and to put it simply, practically anything can be recorded and measured. Some of these sensors/biosensors operate in an "all-or-nothing" mode (a binary state: 0 or 1, open or closed). They majority, though, capture analog physical or biological signals, with very low values, corresponding to a person's physiological activity, or the actions/activities of their body. Generally, such signals need to be amplified, purified and quantified, with a certain degree of precision, in analog and/or digital format.

The second part of this book, discusses the analog and digital handling of these data. As shown by Figure 3.1a, it is made up of three main blocks, quite distinct from one another. We will examine each in detail in turn:

– a "bioprocessor", which includes two specific subsets (see Chapter 3) – the *Analog Front-End* (AFE) and the *Central Processing Unit* (CPU);

– "power supply" to the patch, and its management (Chapter 4);

– "management of communications" that the patch must conduct (Chapter 5), either with the world outside the body (Out-of-Body Communication, OBC) or within the body (Intra-Body Communication, IBC).

Let us begin by looking at the "bioprocessor" that is included in a "biocontroller".

3.1. Overall structure: "AFE (Analog Front-End) + CPU (Central Processing Unit)"

After the biosensor, which we discussed previously, the "bioprocessor" is the second vital node in a patch (see Figure 3.1b).

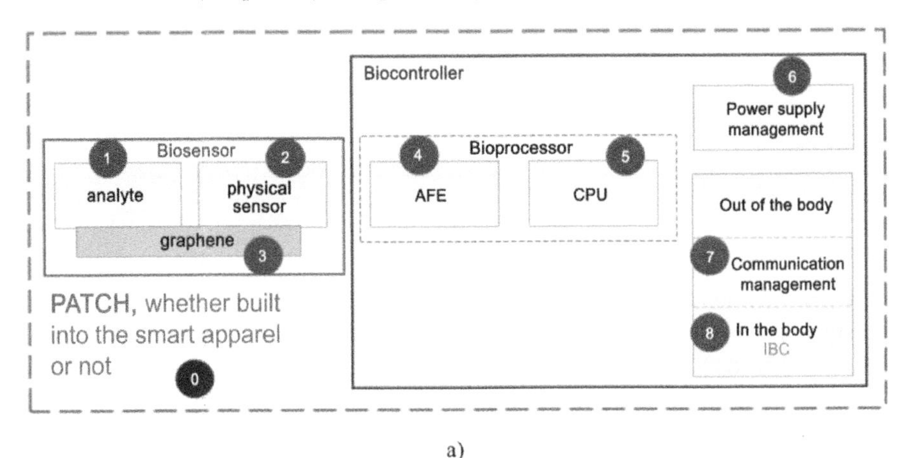

a)

b)

Figure 3.1. *a) Reminder of the block diagram of the patch; b) the bioprocessor (AFE + CPU). For a color version of this figure, see www.iste.co.uk/paret/smartpatches.zip*

The bioprocessor is formed of two entities:

– an Analog Front-End (AFE) system – in charge of processing the analog signals generated by the analog sensors and/or biosensors;

– a Central Processing Unit (CPU).

It is here that the data are formatted, constructed, digitized, structured and given meaning, before they can be delivered to the outside world.

The bioprocessor may have two structures: the older "all-in-two" structure, or the more modern "all-in-one" structure.

3.1.1. *"All-in-two"*

As stated above, the AFE and CPU are fundamentally distinct structures, and thus are normally formed of two integrated circuits. In principle, this is true.

3.1.1.1. *AFE*

Analog Front-End circuits include multiple analog interfaces, designed to recover signals from multiple biosensors – for example, bio-electrical impedance analysis (BIA), photoplethysmography (PPG), electrocardiography (EKG), skin temperature, galvanic skin response (GSR), fall detection, etc. (there are many applications!). The signals from the bionsensors then need to be amplified, cleaned up, filtered, etc., and then digitized using A/D converters.

3.1.1.2. *CPU*

For its part, the CPU includes A/D converters, microcontroller units (MCU), digital signal processors (DSP), Random-Access Memory (RAM), Flash, E2PROM, communication circuits (BTLE, NFC, etc.), power-management integrated circuits (PMIC), etc. In recent years, as technology has evolved, the AFE + CPU system has become an integrated monocircuit, and in five years, this arrangement will be practically universal in the market, which is good news for the intended applications in smart apparel, in view of the constraints in terms of volume, dimension, desired flexibility, energy consumption, etc.

3.1.2. *"All-in-one"*

On a technological level, after going through the various solutions, more or less discrete, mentioned earlier, the solution of an "integrated bio-processor, AFE + CPU" is becoming/is the classic solution for applications in health, wellness, etc.:

the "all-in-one" for the industry. In today's world, the silicon surface of a bioprocessor's integrated circuit represents only around a quarter of the size of that of a system made of two or more discrete components, which is ideal for small modules, offering a multitude of options in designing new devices. In fact, to deal with current demands (measuring fat mass, skeletal muscular mass, heart rate, skin temperature, stress levels, various combinations of these factors for a range of new use cases, etc.), a standard monochip bio-processor is capable of processing numerous analog biological signals without the need for other circuits such as external AFE systems.

Numerous integrated circuit manufacturers, such as Analog Devices, Maxim, NXP, STm, Samsung Electronics and many others offer their own specific bioprocessors.

3.2. The AFE

3.2.1. *Functions of an AFE*

The purpose of an Analog Front-End system (AFE) is to recover the weak analog signals output by biosensors (generally ranging from a few mV to a few hundred), amplify those signals using low-noise amplifiers (LNAs), applying various thresholds for detection levels, filtering them, smoothing them (using low-pass, high-pass or bandpass filters), cleaning them up (getting rid of parasites, peaks, noise, artifacts, etc.), and either presenting them directly to the CPU in analog form, or digitizing them locally before sending them to the CPU for mainly digital processing.

3.2.2. *The numerous possible types of AFE*

In the previous chapter, we learned about numerous types of sensors/biosensors devoted to measuring a range of parameters, making use of a wide variety of specific measuring methods.

On the basis of the parameters and functions listed up until the previous section, and depending on the intended applications of the biosensors and the target markets, big-name integrated-circuit manufacturers (Texas, Maxim, Analog Devices, NXP, Micro Chip, etc.) and smaller businesses (Melexis, etc.) offer ranges of products which are often highly specialized/specific.

3.2.2.1. *For the medical market*

In the context of the professional-level medical market, there is practically a reference integrated circuit with AFE for each type of parameter that can be measured, with all the required precision, in accordance with all the norms in force, etc.: in short, circuits for professionals, taking measurements in a hospital environment, with all the correct equipment. Consequently, there is a single type of patch for each type of application. The idea of a "multisensor patch" soon becomes unrealistic for a reasonable price, except in very specific professional fields.

3.2.2.2. *For the health, leisure and sport (etc.) markets*

The same manufacturers also supply components for very good products – a little less professional, very versatile, able to simultaneously measure numerous parameters (though of course, not all). These are the types of integrated circuits found notably in smart watches and smartphones. With these solutions, it becomes possible to produce "multibiosensor patches" at a reasonable price, using a single medium for all the sensors: graphene. It is these types of patches that we will be focusing on in this book.

3.2.2.3. *Concrete examples of AFE systems*

To illustrate this section, below are some typical examples of AFE circuits on the market for applications ranging from consumer electronics to professional equipment, in the fields of healthcare, sport and fitness (e.g. heart rate, EKG, respiration), monitoring of vital signs, responsive telemedicine, etc. Indeed, in structural terms, all these circuits belong to the same family. The details on the differences between them and their peculiarities can be found in the data sheets, available on the manufacturers' websites; readers are recommended to consult these for further information. These circuits can also be used in high-performance multichannel data-acquisition systems. In short, these systems are of interest to us in this book, looking at the applications of smart apparel in sport, PPE, healthcare, wellness and leisure.

A generic Analog Front-End integrated circuit generally presents the following characteristics and specific features (see the block diagram in Figure 3.2):

– multichannel input (1 to 6) with simultaneous sampling;

– a range of differential input voltages: ±400 mV;

– highly sensitive detection (3 mVpp, typically);

– programmable-gain amplifiers (PGAs);

– delta-sigma analog-digital converters (ΔΣ) on 24 bits (ADCs);

– data rate up to 25.6 ksps;

– a single analog input pin for the detection signal;

– low power consumption, with Flexible Power-Down and Standby modes;

– for an EKG, Wilson and Goldberger Terminals;

– High Modulation Depth Sensitivity (lower than 8 %);

– Demodulated Data;

– Received Signal Strength Indicator (RSSI), Input Carrier Frequency: 125 kHz, typical Input Data Rate: 10 Kbps, maximum, Internal Configuration Registers;

– operating temperature range from –20°C to 85°C.

3.2.2.3.1. Examples of block diagrams of AFE circuits

Examples of block diagrams are presented in Figures 3.2, 3.3 and 3.4.

3.2.2.3.2. Example of the AFE in an electrocardiography application

During athletic training, tests and medical exams, it is often necessary to record various physiological signals, such as temperature, blood pressure, respiration rate, cardiac activity (by means of an EKG), brain activity (EEG), etc. Let us here consider a concrete example of an AFE that is often in demand: that of an EKG.

NOTE.– Take careful note of the term "EKG sensor[1]" and of its derivative, "heart-rate monitor", often encountered in advertising for consumer electronics for wellness, fitness, sport, etc. To be perfectly clear, to begin with, there is no such thing as an "EKG sensor". There are multiple sensors of analog signals which generate electrical potentials, from which the resulting voltages, once they have been electronically filtered, fused, then processed with complex logical and mathematical algorithms, will give rise to a new signal (or signals), representing an electrocardiograph. From that EKG representation, we can summarily derive the value of a regular ("sinusoidal") heart rate, and then determine a heart rate expressed in number of beats per minute. We can then claim to have invented the heart-rate monitor which is so dearly beloved of manufacturers/sellers of watches, bracelets and sporting equipment that are far from being "medical devices".

1 Dictionary definition: an electrocardiogram is a graphic representation, on paper, of the heart's electrical activity. The electrocardiograph is a device used to create an electrocardiogram. An electrocardioscope is a device which displays the plot on a screen. Electrocardiography, EKG, is the recording technique associated with a graphic representation.

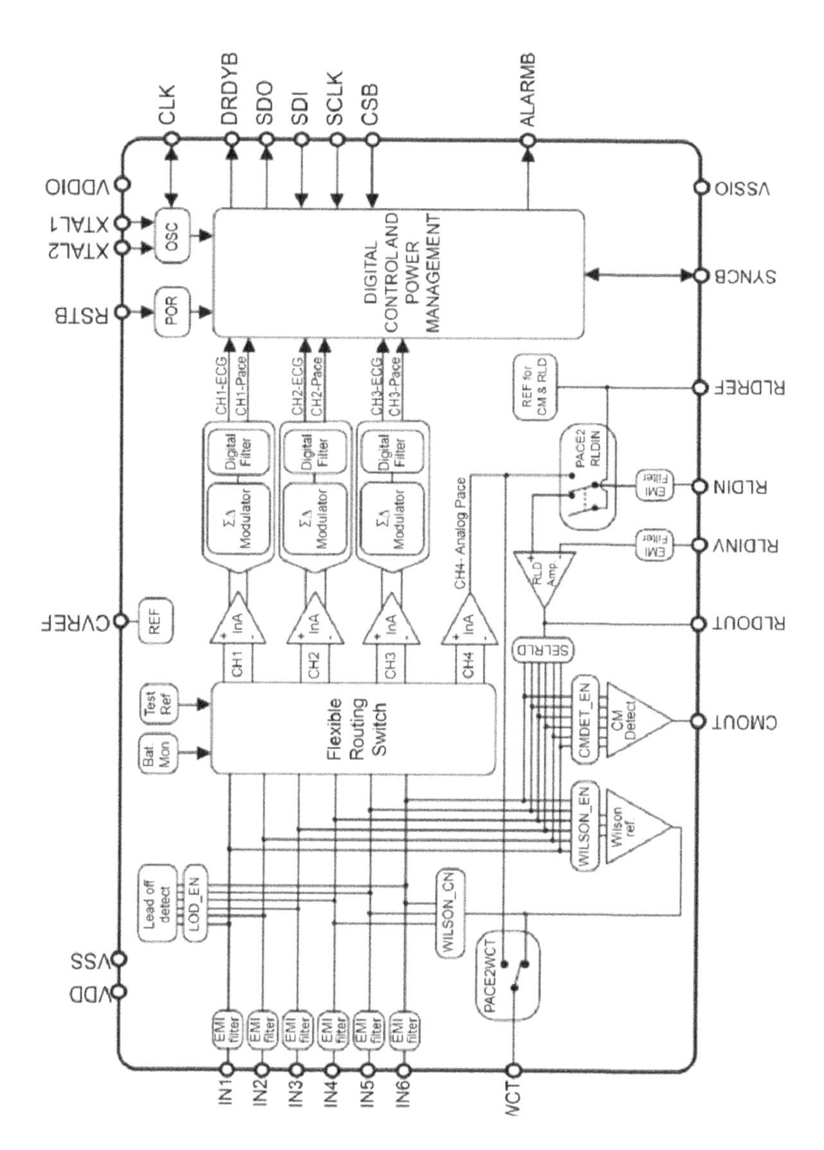

Figure 3.2A. *Block diagram of the integrated circuit ADS1293 Texas Instruments*

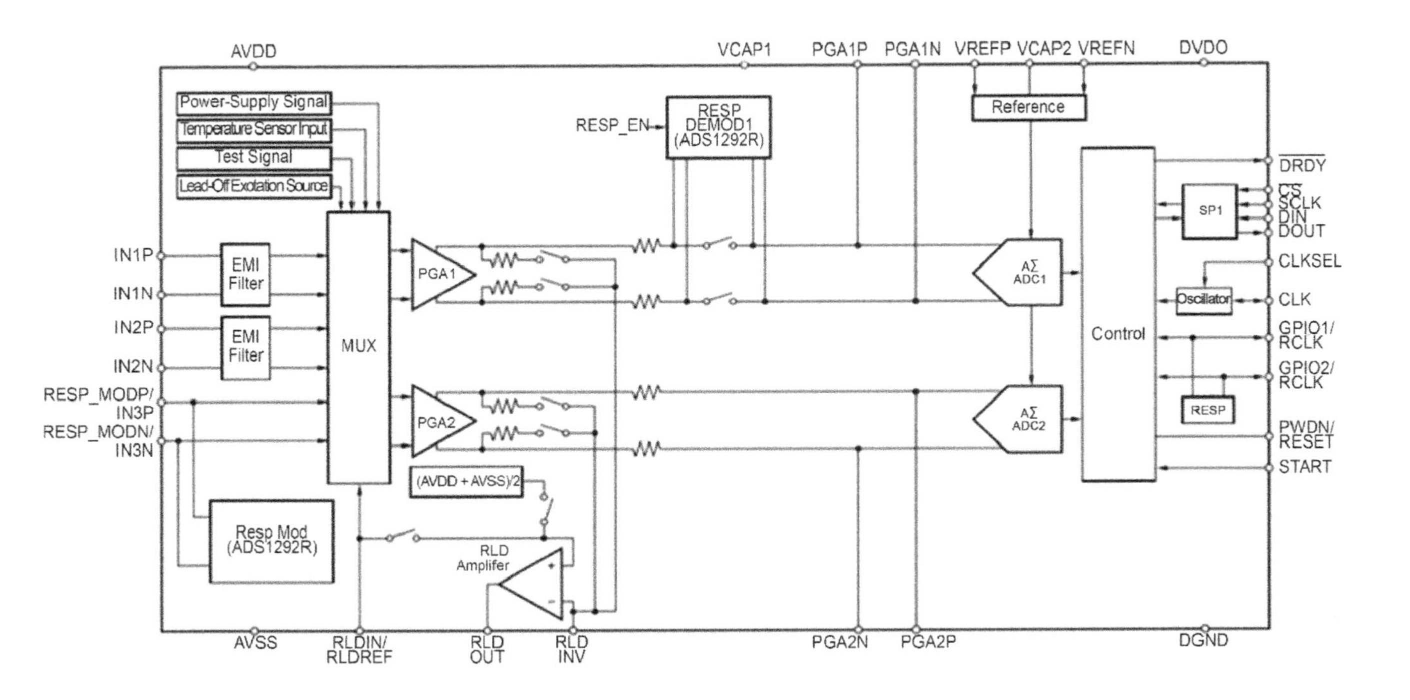

Figure 3.2B. *Block diagram of the integrated circuit ADS1293 Texas Instruments (cont.)*

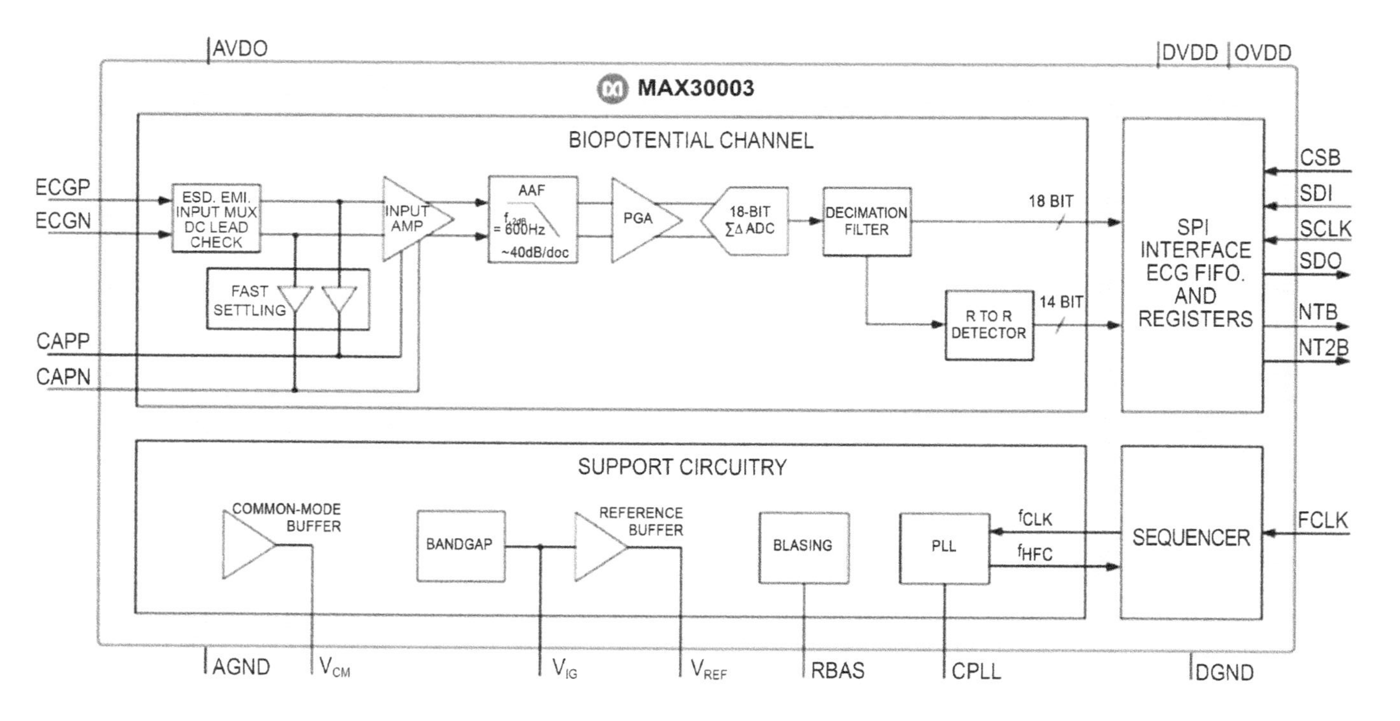

Figure 3.3. *Block diagram of the Maxim Max 30003 integrated circuit. For a color version of this figure, see www.iste.co.uk/paret/smartpatches.zip*

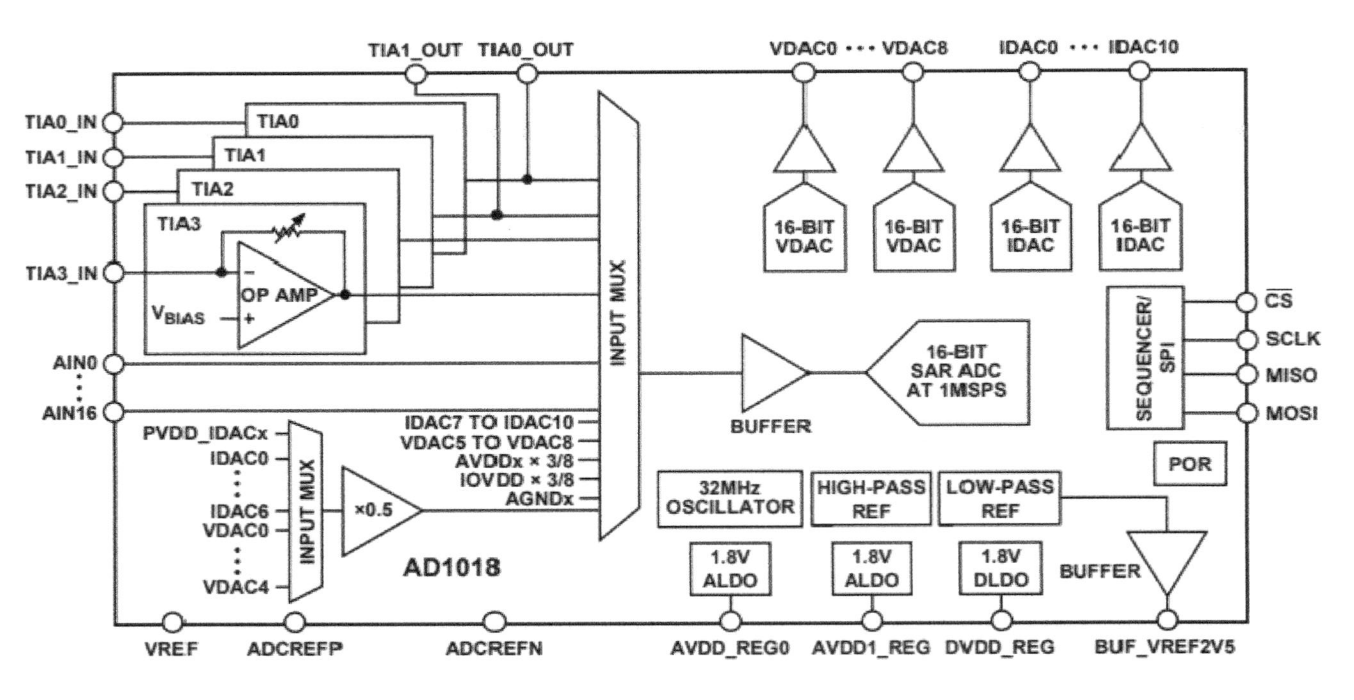

Figure 3.4. *Block diagram of the integrated circuit from an analog device*

Cardiac activity (the activity of the electrical currents accompanying the contractions of/in the heart) is recorded by electrodes on the surface of a person's skin, in an exam which is painless, non-invasive, pain free, and takes only a few minutes. It can be conducted at a doctor's surgery, at a hospital, at home, walking down the street, etc. It is complex to interpret, requiring a clinician's experience. An EKG can be used to monitor the cardiovascular system, notably for detecting arrhythmia and preventing myocardial infarction.

Immediately, this raises the question of the position of the electrodes and the design of sensors and/or fabrics or clothing which must, at all times, ensure good contact (if possible, without adhesives) with the skin, including hair and sweat (see Paret and Crégo (2018)).

The first important points that need to be taken into consideration in reference to an EKG are the input of the initial electrical signals from the probes. These are analog signals, around a millivolt in strength, and are usually amplified with LNAs in conventional differential mode (see Figure 3.5).

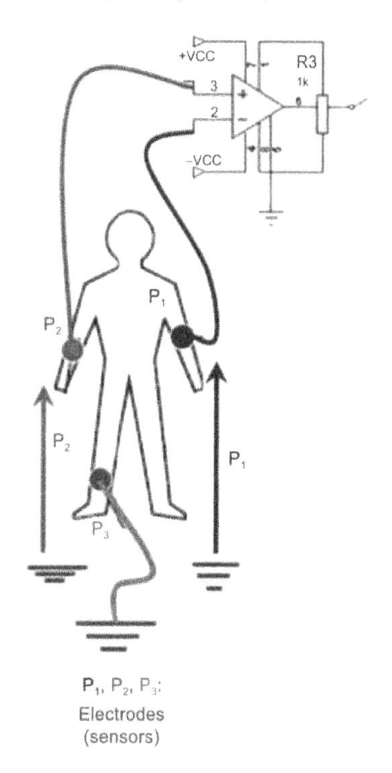

Figure 3.5. *Example of electrode positions for an EKG. For a color version of this figure, see www.iste.co.uk/paret/smartpatches.zip*

Once we have obtained the raw analog data, to obtain the results and ensure they are readable, the AFE needs to process the signal – for example, using:

– An analog and/or digital "low-pass" filter to eliminate high-frequency signals, due to muscle activity in muscles other than the heart muscles, and possible interference from electrical devices in the subject's immediate vicinity.

– Filtration of low frequencies due to respiration to reduce undulations of the baseline to their absolute minimum.

– Improvement of the quality of the signal obtained by "averaging" multiple complex readings. Though this function causes artifacts in the event of irregular heart rate or certain extrasystoles, this technique is particularly widely used in devices suited to effort trials, where many artifacts on the plot are caused by the patient's movements, which is always the case with sports accessories and clothing.

– Common-mode rejection ratio (CMRR) of the amplified signal in differential mode, because there are a number of factors which can interfere with signal quality. As the body moves, the mere motion of the patient's clothes against the body causes interference with the signal. As a general rule, such motion artifacts are observed either in the signals captured by the two measuring electrodes, or are common to those signals, so the common-mode rejection ratio needs to be as high as possible. In addition, the more cumbersome the electronic components associated with the sensor, the more likely it is that the device will move a little when in use, creating further motion artifacts, which must be rejected in order to maintain the quality of the EKG signal.

NOTE.– Numerous devices are combined with either inbuilt software or external software to interpret the plots. Despite the best intentions, and in view of certain constraints (sometimes technical, but primarily commercial), this software is not 100% reliable, and is no substitute for a medical professional's judgement.

Where in smart apparel should the measuring sensors be placed? The case of an EKG

Let us return, now, to the question of information inputs, their effects and implementation, and the positions of the patch sensors on the skin and/or in smart apparel. For this purpose, in order to obtain an accurate view of all the heart's electrical activity, we typically define its electric axis and 12 auxiliary axes distributed 30° apart, meaning 12 analog derivations in total. Of course, the more probes/patches are placed along these axes, the better the result of the EKG obtained will be. This is often where wellness, healthcare and the medical domain begin to intersect!

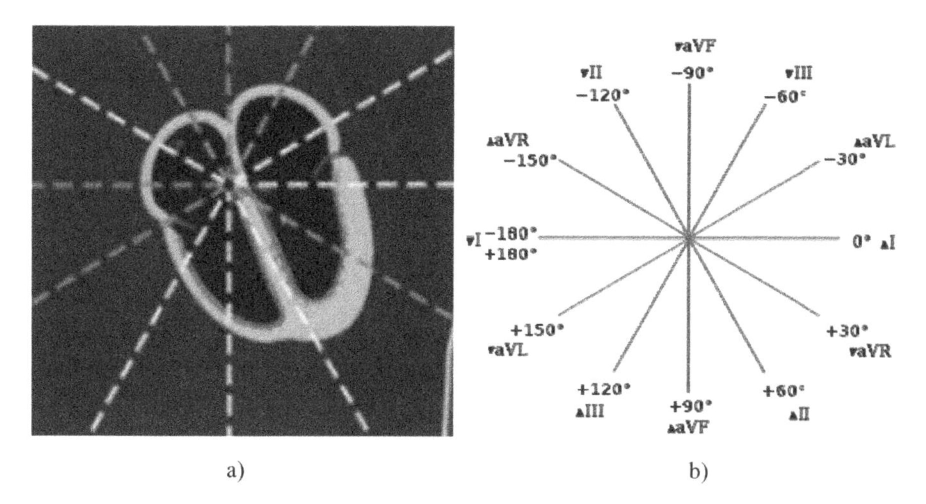

a) b)

Figure 3.6. *a) Electrical axis and b) 12 auxiliary axes of the heart. For a color version of this figure, see www.iste.co.uk/paret/smartpatches.zip*

These 12 (standardized) derivations provide us with a three-dimensional view of the heart's electrical activity, and in fact, an EKG contains (Paret and Crégo 2018) (see Figure 3.6):

– six frontal derivations (I, II, III, aVR, aVL and aVF);

– six precordial derivations (V1-V6).

Let us return to the question of the applications for smart apparel (wellness, sport, PPE, medical devices, etc.), the positions of the sensors in patches, so-called "double-skin" clothing, etc. How, how many, for what purpose, etc.? And where are they to be positioned? (See the examples of Chronolife and BioSerenity in the Conclusion). Before going any further, though, let us once again define a new term: a *lead* is the potential difference between two electrodes. Each lead provides unique information about the heart's activity.

Einthoven's theory holds that the heart is at the center of an equilateral triangle formed between the right arm, left arm and left leg (RA, LA and LL – see Figure 3.7).

With three probes, and the three frontal derivations I, II, and III (so three leads), it is possible to calculate the value of all six frontal derivations. Indeed, in electrics, Kirchhoff's node law and Thévenin's theorem, applied to this triangle, mean that if we know two of the values, we can find the third. Thus, in fact, two probes/patches are sufficient, which can make matters easier, or reduce the number of sensors (patches) to be attached to the skin or installed in a garment… or almost. With the leads, potential differences refire to measurements between two points (the bipolar measurements of I, II and III). For the sake of ease of measurement, often, it is preferable to measure each point RA, LA and LL in relation to a reference point – often the earth. Thus, another electrode, referred to as the "indifferent" electrode, is placed at the bottom of the right leg: the RLD – right-leg drive – electrode (see Figure 3.8).

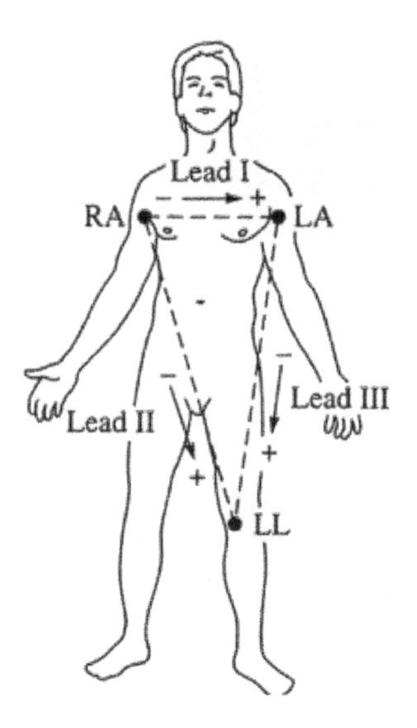

Figure 3.7. *Examples of definitions of leads*

In total, then, we have four frontal electrodes to place (with patches or integrated into a garment, of course): VR, on the right arm; VL, on the left arm; VF, on the left foot; and Vindif on the right leg, giving the three bipolar measurements I, II, III and three unipolar measurements (with respect to Vindif), aVR, aVL and aVF. For example, based on the values of I and II, we obtain:

$$III = II - I$$

$$aVF = II - I/2$$

$$aVR = - I/2 - II/2$$

$$aVL = I - II/2$$

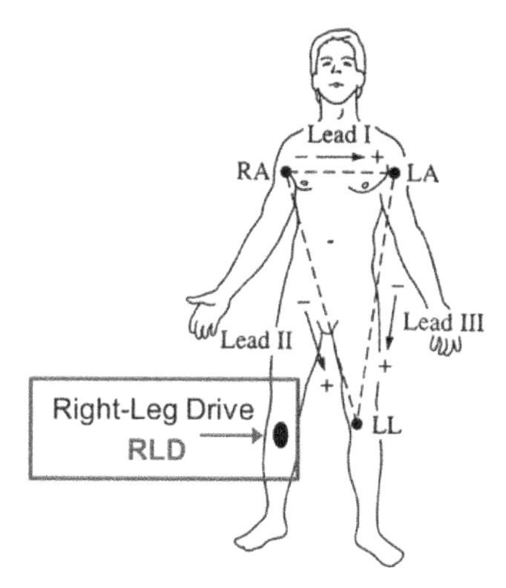

Figure 3.8. *Examples of definitions of leads with the earth reference point. For a color version of this figure, see www.iste.co.uk/paret/smartpatches.zip*

These equations illustrate why, in reality, electrocardiograms often record no more than two derivations, and simply calculate the remaining four. The same is true of the six precordial derivations measured on the chest to obtain V1, V2, V3, V4, V5 and V6 (see Figure 3.9).

The table in Figure 3.10 summarizes these facts. More specifically, in relation to smart apparel, we can define the number of sensors which need to be included in a garment, and make them communicate with one another… perhaps by means of IBC (see Figure 3.11).

We will take a detailed look at the possibilities and concrete solutions in the chapter on IBC patches in a network.

Talking specifically about what happens in the real world, the suppliers of professional medical devices built into jackets, belts, etc., – which can be used to monitor certain conditions (e.g. epilepsy, etc.) – include between 30 and 50 sensors (which is costly). Of course, these sensors are built to be washable a certain number of times, which limits the lifespan of the garment, and on this point, it becomes difficult to make any return on investment (see the Conclusion).

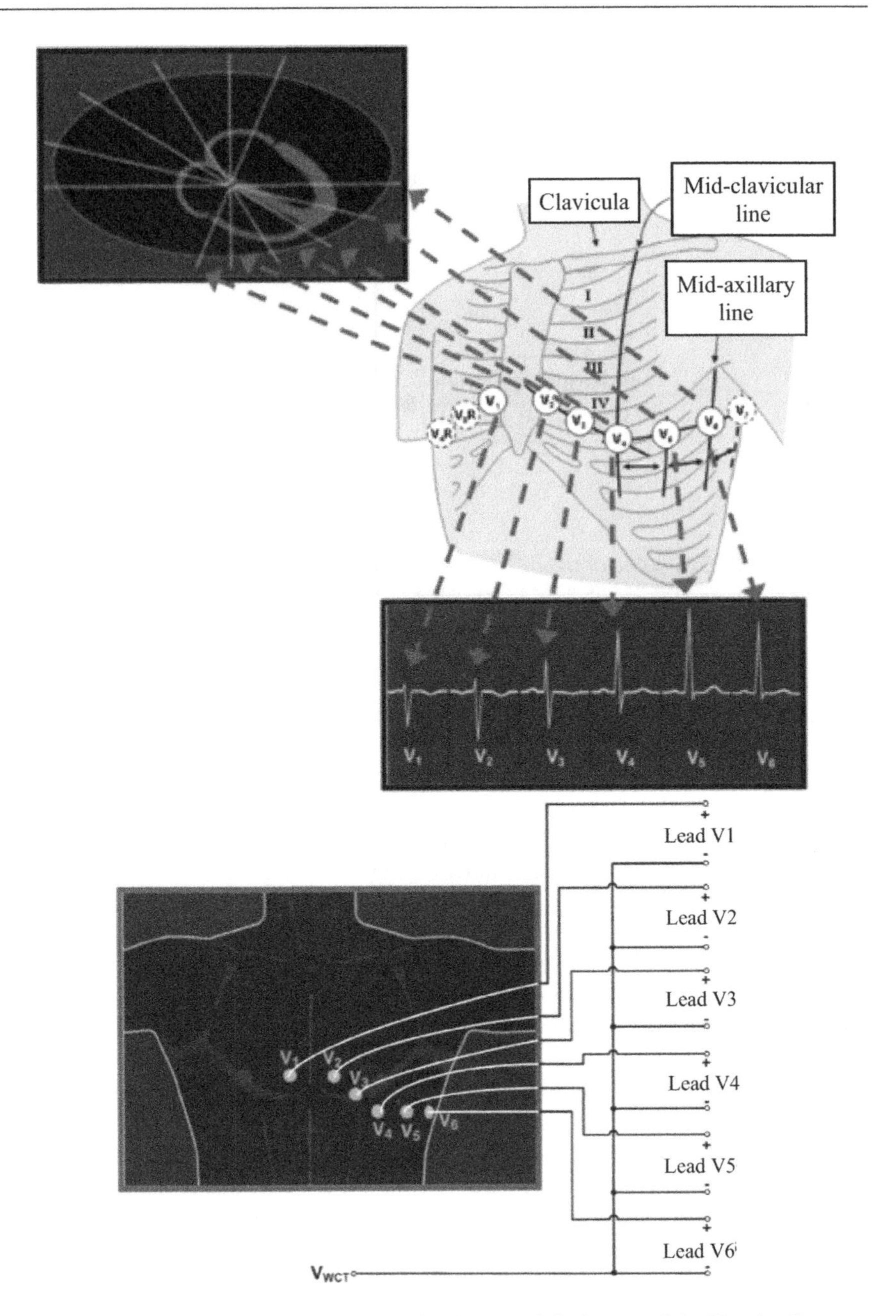

Figure 3.9. *Examples of precordial derivations and their associated leads. For a color version of this figure, see www.iste.co.uk/paret/smartpatches.zip*

Lead	Electrode Formula	Measured or Calculated	Calculation Formula
Primary Limb Leads			
I	LA - RA	Both	Lead II - Lead III
II	LL - RA	Both	Lead I + Lead III
III	LL - LA	Both	Lead II - Lead I
Augmented Leads			
aVR	RA - (LL + LA)/2	Both	−(Lead I + Lead II)/2
aVL	LA - (LL + RA)/2	Both	(Lead I - Lead III)/2
aVF	LL - (LA + RA)/2	Both	(Lead II + Lead III)/2
Chest Leads			
V1	V1 - WCT	Measured	-
V2	V2 - WCT	Measured	-
V3	V3 - WCT	Measured	-
V4	V4 - WCT	Measured	-
V5	V5 - WCT	Measured	-
V6	V6 - WCT	Measured	-

Figure 3.10. *Relations between frontal derivations, precordial derivations and the associated leads*

Number of Leads	Leads Used	Number of ADC Channels
1	Lead I	1
3	Lead I, Lead II, Lead III	2
6	Lead I, Lead II, Lead III, aVR, aVL, aVF	2
12	Lead I, Lead II, Lead III, aVR, aVL, aVF, V1 – V6	8

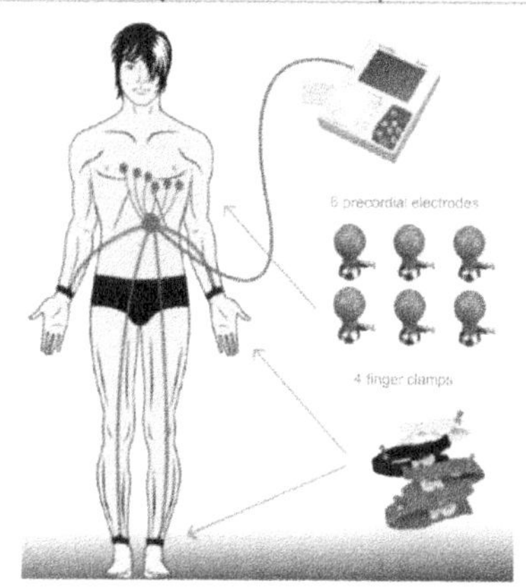

Figure 3.11. *Number of leads which can be used (source: Texas Instruments). For a color version of this figure, see www.iste.co.uk/paret/smartpatches.zip*

3.2.2.3.3. Concrete example

Of the various examples on the market, earlier, we looked at an integrated circuit from Texas Instruments, designed for low-consumption applications in smart apparel in the medical, sports and fitness domains. The application block diagram of this circuit is shown in Figure 3.12. The pins SDO, SDI and SCLK of the SPI bus serve as a link to the external microcontroller, carrying out the functional analysis of the EKG.

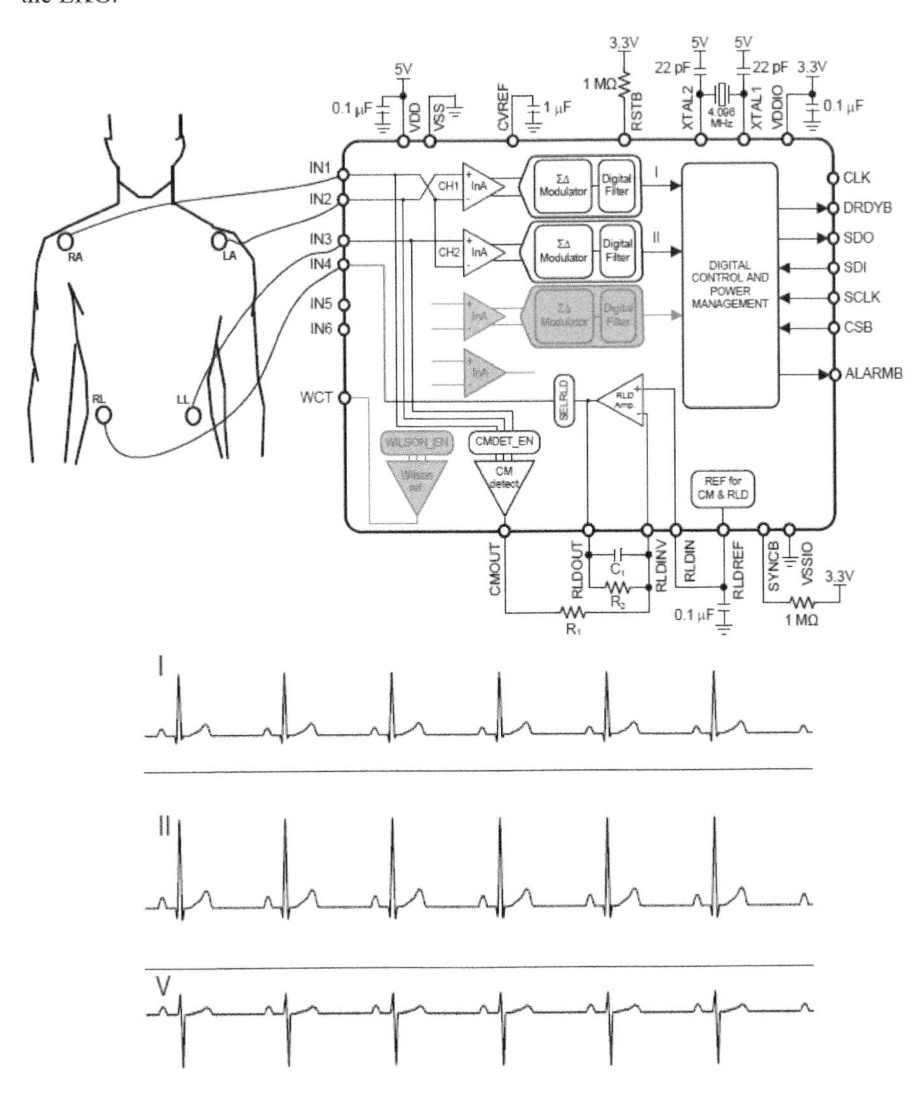

Figure 3.12A. *Application with four leads*

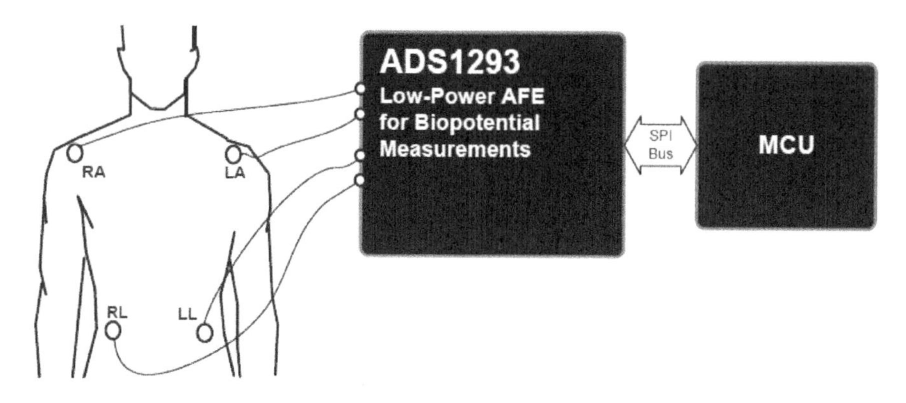

Figure 3.12B. *Application with four leads (cont.)*

We will continue to discuss smart apparel, which is, of course, worn on the body. These types of devices are designed to be used for sports training or leisure activities. In these kinds of smart apparel, the Analog Front-End, the microcontroller and the associated electronics must be capable of:

– measuring respiration cycles;

– taking EKG-type measurements with high-quality graphite rubber electrodes, which can be attached directly on top of hair without the need to shave, be used multiple times, and stand up to transpiration and sweat;

– in parallel, instantaneous and precise geolocation on the x, y, z axes, using an inertial orientation sensor and a GPS, whose precision in x and y is ±20 cm, and whose precision in z is 5-10 cm, to refine the applied strength in order to clear obstacles.

The patch must also have connectivity to a mobile phone, using an NFC or Bluetooth Low Energy (BLE) connection, in order to use the phone's GPS to directly feed the measured data to the user's mobile phone and provide nutrition guidance.

APPLICATION TO SMART APPAREL.–

All of the above indicates that:

– wellness, fitness, sport, PPE, medical = heart, cardiography, etc., but at different levels of finesse of measurement and interpretation;

– specialists are needed for accurate interpretation and understanding of an EKG;

– the algorithms must be “concrete” from a medical point of view;

– only accredited medical devices can be marketed as EKG systems;

– at different levels, the analog sensors being connected to one another need to be flexible, washable, economical, etc., to form an intelligent fabric;

– once again, wellness, fitness and sport are one thing; PPE and medical equipment are a whole different ball game!

3.3. The CPU

Once the electrical signals from the biosensor have been harvested and passed to the AFE, a central processing unit (CPU) in the patch or smart garment uses a dedicated software layer to carry out a suitable algorithmic analysis (an EKG or other process), to extract the parameters needed for a particular application. For this purpose, numerous technical elements need to be taken into account. For example:

– correct calibration, proven by a calibration signal;

– high immunity to parasitic electrical phenomena;

– improper positioning of the electrodes must be investigated and rectified, etc.

3.3.1. *Signal processing*

Now the signals are clean, the integrated circuit of the CPU in the bioprocessor performs a process (perhaps a little more analog processing, followed by digital processing) on the raw signal/digitized information, to refine it, and to record and store analog integrated data following digital conversion.

Bioprocessor components are typically made up of multiple elements (see Part 1):

– analog-digital converter(s) (ADC);

– a microprocessor, with its RAM and EPROM memory;

– radiofrequency communications management;

– a power supply unit (battery, energy harvesting system, etc.) and its management system.

We will begin by looking at analog-to-digital conversion.

3.3.1.1. *A/D conversion (by the AFE or CPU)*

Frequently, to render the input signals and the "raw" analog physical data measured initially from the sensors as truly exploitable (pressure, temperature, EKG signals, etc.), they are converted into digital signals by analog-to-digital converters, of "Σ Δ" types, whose resolutions are increasingly fine (16 or 24 bits) and which should, if possible, consume minimal amounts of energy.

Figure 3.13 shows the orders of magnitude of the data rates associated with analog signals from different sensors, which means that if there are x terminations per sensor, the data rates will increase proportionally.

Types of biosensors	Abbreviation	Number of sensors	Data rates (bytes/s)
12-lead EKG	EKG	20	Uncompressed, 12,000
Peripheral capillary oxygen saturation	SpO_2	1	750-1000
Respiration	RESP	1	50-100
Temperature	TEMP	1	10
Invasive blood pressure	IBP	1	1000-2250
Lung capacity	SPIRO	1	500
Blood CO_2	CO_2	1	250
Total			*15 810*

Figure 3.13. *Biosensors and their respective bitrates*

3.3.1.2. *Digital processing*

It is then necessary, within the CPU (microcontroller), to have a computation unit to apply digital processing, using specific algorithms for each type/functionality of the biosensor, to make corrections and produce "refined, exploitable" data. Consequently, it is almost unavoidable to have small and/or larger local microcontrollers.

3.3.1.2.1. Data refinement

The purely mathematical part of "signal processing" can then commence, with its calculations and equations to refine the signal: filtering of all types (Bessel, Butterworth, etc.), noise reduction, Fourier transforms, wavelet transforms, Hilbert transforms, convolutions and convolution products, etc. It is then possible to extract the specific, usable, “refined”, interpretable signal, from the sensor/biosensor, which can be fed to a management device or a digital network.

3.3.1.2.2. Sensor data fusion

It is not uncommon to require multiple pieces of physical information from multiple sensors in order to deduce the “truly useful, refined data” – the data we actually need – be it a state, a diagnosis, a prediction, etc. For this purpose, we now need to carefully apply certain algorithms to fuse (i.e. mix) different elementary refined data, to extract and obtain the desired final data (see Figure 3.14).

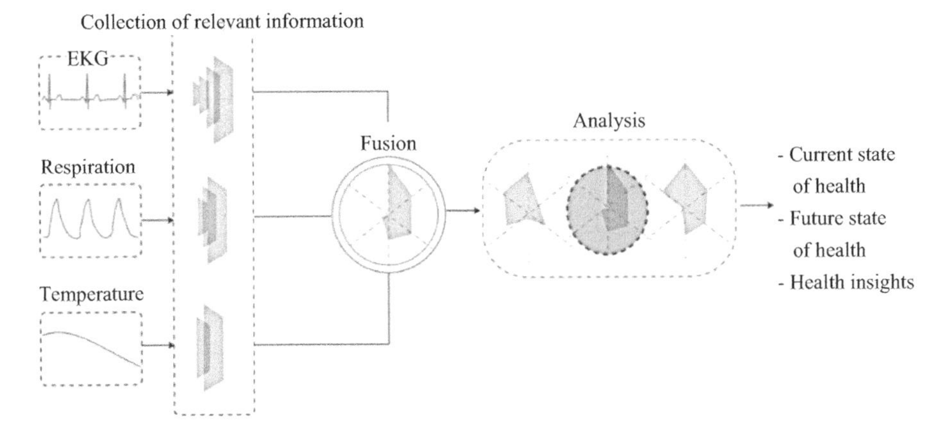

Figure 3.14. *Data fusion from different biosensors*

In this area, too, there are special formulas, deep secrets and numerous technical patents (designed for commercial purposes), which are jealously guarded (for example, in security, medical laboratories, etc.), because of the quality of the data obtained, the swiftness of the process, the level of energy consumption, etc.

Principle of data fusion operation

With data fusion, we are diving straight into the domain of AI (Artificial Intelligence). There are dozens of principles for data fusion, free/open and/or proprietary. Every company proclaims the merits of the solution they themselves have adopted in their field. As this book is not intended to be an encyclopedia, in the next few sections, we have chosen to present a fairly generic example of the data fusion process.

Signal detection may be done by each individual biosensor, and by the software of the fusion system. The raw data are first input for preprocessing, followed by a number of steps, including a certain degree of intelligence – primarily, signal shape recognition.

– preprocessing: the analyzed points which are very unlikely to be "objects" (such as an EKG signal) are filtered and tagged as such;

– segmentation: the clusters of points, or point clouds, which could belong to the same object are identified. The phrase "could belong", here, indicates the concept of probability;

– feature extraction: the individual features are then identified (more detailed shape recognition).

Next come four steps, carried out by the data fusion software, deploying numerous algorithms in parallel and in series:

– Association: the previously identified segments are preliminarily matched with their corresponding object.

– Object update: after and between each sensor analysis, the variations in the objects (in terms of amplitude and movement over time) are updated, based on an interacting multiple model (IMM), because of course, it is important to take account of the patient's movement and simultaneous motion, while the analysis is being performed.

– Track maintenance: the system checks the likelihood of the dataset being processed. This vectorial analysis is based on the orientation (the direction) and the value of the difference of the object (its size) in relation to its previous position (the point of application). If certain segments are determined to belong to the same object (or found not to), a logical process is applied to merge or split them, on the basis of their behavior or their appearance.

– Classification: the principle of *Track before Detect* is then applied. This means that an object is tracked across successive analyses as many times as necessary to clearly identify that object (*final matching*).

Figure 3.15 gives an overview of these processes.

In summary, the main functions and technical characteristics of a data-fusion unit are, commonly, the ability to simultaneously fuse data from 6-8 biosensors, preprocessing of the data (classification of objects/signals, etc.), detection and tracking with other sensors, detailed information about signal amplitudes, movements over time and shapes, object classification and tracking, estimation of uncertainty in the measurements obtained, and also, ranking of the probability of existence for all classified objects, with estimations of the FAR and FRR (False Acceptance Rate and False Rejection Rate), an interface for the incorporation of data obtained with other sources/sensors. This all needs to fit in as small a volume as possible, consuming little or no energy, and dissipating no heat. Nothing new there, then!

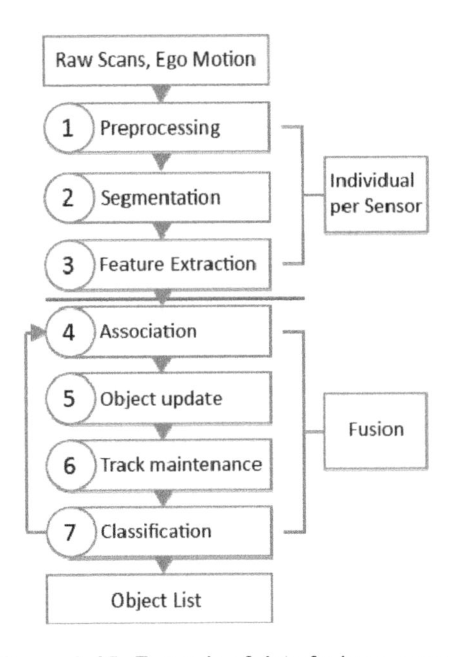

Figure 3.15. *Example of data fusion process*

Let us now use two examples to briefly illustrate what is meant by data refinement and data fusion.

EXAMPLE 3.1.– Simple – when measuring pressure, in order to have a clear idea of its exact value, expressed in relation to its original frame of reference, it is always important to correlate, correct and fuse the readings with the value of the ambient temperature. Thus, we no longer need only one sensor, but two: one measuring pressure, the other temperature, and a computation unit able to apply a particular algorithm to correct the measured pressure as a function of the ambient temperature.

EXAMPLE 3.2.– A little more complicated, in a different field – So as not to burden the system unnecessarily, to avoid producing and consuming information/data that contribute nothing to the system, and thus at the same time, save on the energy consumption of the patch/smart garment, it is sensible to keep, work on and send, over a digital network, only those data that are deemed relevant, and avoid duplicate information, etc. Here, again, the initial data must be preprocessed locally by software.

In these two examples, which are highly representative of situations encountered in daily life, we need to seek out the most suitable microcontroller, for a not-excessive price, which has sufficient computational power, whose computation activities will not "burden" the system's energy consumption, and which will not impact the system's dimensions, weight, volume, esthetic appearance, etc., depending on the dimensions of the cell/battery, which may or may not be used locally.

3.3.1.3. *Data interpretation and formulation*

Once the major preliminary steps of refinement and fusion of the initial raw data have been performed, then (often in the form of software embedded on a microprocessor), the "smart" part of the patch comes into play, with signal analysis and interpretation, processing with various specific algorithms, complex, patented, secret, proprietary to one company or another, etc. We then obtain new data which "represent", "interpret" or "formulate" certain values of physical or biological properties, making the "dumb" electrical signal give us the information we need from it (diagnosis, prediction, etc.) and transforming the patch into a marketable "xxx sensor", giving it data to which specific names can be attached – for example, "pH value", "SpO_2 value", "cardiac arrhythmia sensor", "EKG sensor", etc.

It should also be noted that to read and interpret these parameters generally requires great skill, which can only be acquired through regular practice, and merely serves as one of a set of tools, giving arguments to medical personnel to support their diagnosis, or giving information to patients/users. Thus, generally, the data are interpreted by professionals. In parallel, on the market, there are software packages that come with certain integrated circuits, and their application libraries, which can support diagnostics, but under no circumstances should they replace doctors or other specialists.

3.3.1.4. *Data confidentiality: boundaries between signals, data and personal data*

Generally, in the applications of patches or smart apparel of interest to us here, those analog signals are measured, transformed into digital data, fused and interpreted, becoming data that are directly connected to an individual. From then on, it is possible to clearly define whether or not the data qualify as "personal data", and whether they fall within the regulatory and legal remit of the GDPR, sensitive data or personal data, likely to be subject to very stringent considerations, because they are biometric, behavioral or health-related. That is where things become complicated.

Indeed, often, the application for a patch and/or smart garment is used in combination with a smartphone, so that the user can view the results of the analyses. Thus, questions relating to the interpretation and formulation of these data carry very significant consequences.

– Has the processing of the personal data been performed entirely inside the patch, in which case, is the smartphone app only there to carry out certain ancillary computations on preprocessed data, and to display the results?

– Does the patch only transmit refined data, which have not been interpreted or formulated, and are they formulated by the smartphone app or by an external reader/terminal?

The problem of managing privacy, liability and implementing the GDPR, when the data are not supplied or processed by the same people, then arises, either in relation to the patch or to the smartphone. To avoid legal proceedings, every time, all of these questions require an in-depth analysis of the problem to determine where legal responsibilities lie. Thus, trust no-one, and exercise constant vigilance! Otherwise, beware of the hefty fines mentioned in Chapter 1.

3.3.1.5. *Data security*

We have already mentioned the specific point of "cryptography and end-to-end security" in Chapter 1, where we introduced the concept of a *secure element*. Remember that a system cannot be said to be secure without the use of a genuinely secure element, in charge of all cryptographic processing of the data (Paret and Huon 2017). Obviously, presented in this manner, the addition of a secure element may seem cumbersome for a patch application. In fact, all of a bioprocessor's functions (AFE + CPU + cryptography/secure element + management of NFC and IBC connections) can easily be resolved by using monochip integrated circuits, with minimal power consumption, small dimensions and large volumes, such as those used in conventional contactless cards. Indeed, these functions have a great deal in common with the analog, digital and security-related functions required by the patch, and the chips already have the majority of the communication capabilities at 13.56 MHz for uses in NFC (Near-Field Communication), and can easily be adapted at the same frequency for OBC (Out-of-Body Communications) and IBC (Intra-Body Communications). An example is given in Figure 3.16.

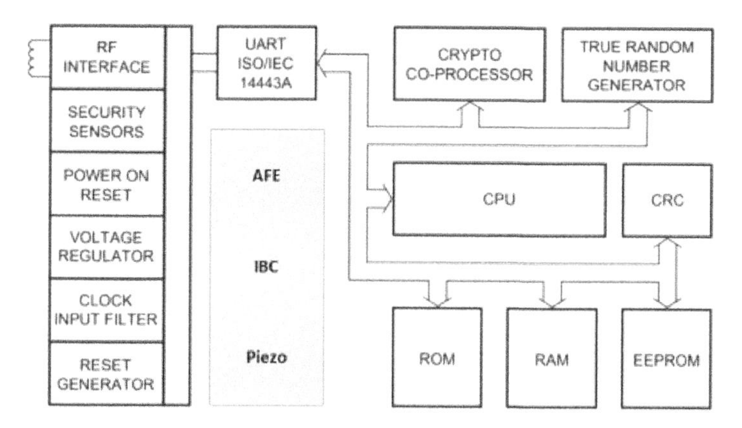

Figure 3.16. *Example of a circuit for a contactless chip card (MIFARE DESFire) suitable for a patch. For a color version of this figure, see www.iste.co.uk/paret/smartpatches.zip*

3.3.1.6. *Storage and cloud access*

Through processing the data (raw, and then refined), it is possible to automatically trigger alerts. By connecting to a dedicated portal, healthcare providers and patients will be able to access (a portion of) those data, and view a report that allows them to take the necessary measures. The data are also available for other systems through the access API, meaning that the patch can be integrated into other remote healthcare systems.

4

Power to the Patch

In this chapter, we examine the ways in which to supply and/or harvest and manage the energy needed for a patch to work.

4.1. Problems surrounding power supply to a patch

Power supply is a crucially important point in relation to all connected smart apparel. One of the main problems relating to patch systems is power supply and power management, to allow them to work properly.

4.1.1. *Choice of means of energy supply to a patch*

The terms “power supply” and “energy consumption” tacitly include the terms “range” and “life span”, which are often one of the key points in sales, purchasing and application of products.

Range is directly related to the availability of an energy source $W = U\ Q$, expressed in joules (and its technology), and voltage U, expressed in volts. In principle, in view of the previous equation, the energy source must be quantified in terms of the quantity of electricity Q – expressed in coulombs – that it contains, and is able to supply to the device (conventionally presented as the product of a current I in amperes, multiplied by time t in seconds, $Q = I\ t$), but often also expressed differently in “amperes × hour”, Ah.

Let us present this problem differently: for the same number of coulombs Q (i.e. for the same volume/dimension or cost of the energy source), how are the two variables I and t distributed? In other words, to maximize the operating time t, how can we reduce, optimize, etc., the consumption I in mA, in μA, in nA, etc.? Of course, everybody has been working on just this problem for decades, and there is no end in sight.

Various technological methods can be employed to supply energy to a system/patch – using a cell, a battery, a rechargeable accumulator, an energy-harvesting device, or indeed certain combinations of all of the above. In addition, as the devices in question are increasingly being miniaturized, and are therefore often difficult to access, it is possible that the energy source will not be easily replaceable, or easily rechargeable, depending on how the technology is implemented (consider the case of implantation in the human body, for example), or that the overall size of the device will not be sufficient to store a large amount of energy. Also, the typical applications of smart apparel require powering the patch, reading stored data and, if possible, passing the data to a server. Thus, we need to be able to supply the patches with energy and, simultaneously (or alternatingly), facilitate wireless transmissions of the data over short (or sometimes medium) distances, whilst minimizing energy consumption as far as possible. In view of this breakdown of the device's required functions, we may suppose that an external relay will be used, by a reader such as a smartphone. In addition, at the time of writing, such devices are able to communicate over short ranges using Bluetooth, Wi-Fi and NFC.

In our case, an NFC-type communication solution offers a range of well-known advantages – in particular, due to the fact that the *interrogator* (= the reader = the smartphone) is capable of transmitting and radiating high-frequency energy at 13.56 MHz, in order to remotely power the *target* (the patch). With this technology, we have the possibility, during the phase of interrogation and exchange between the smartphone and the patch, of harvesting the energy transmitted by the smartphone to recharge either a battery or a supercapacitor inside the patch. In the case of a supercapacitor made from graphene, it will recharge very quickly indeed (see Chapter 2).

4.1.2. *Estimating a patch's energy balance*

Whatever technological solutions are chosen for energy supply, however slight the energy consumption of the electronic components of the patch, it is always necessary to examine and quantify, in detail, every phase of the application, step by step.

To estimate the energy balance of a patch, we shall use the coulomb (symbol: *C*), which is the true unit of electrical charge. But why the coulomb? Though it may seem complicated, the reason is actually a very simple one.

4.1.2.1. *The "coulomb-Ah"*

To estimate the lifespan/range of the energy source in the application, we must first of all quantify the operation of the application for each micro-slice of time (for example, μs by μs), calculating all the quantities of electricity consumed by the different elements and the different functions of the smart garment.

During the phases of activity of the application (measuring, data processing, etc.), the patch also goes through numerous other activities (communications, sending of data, etc.). Each phase consumes a certain amount of current i over the course of a time window dt, which requires a certain quantity of coulombs in the form $dq = i\ dt$ (this can also be expressed as $dq = C\ dV$). Then, we need to find the sum of all these dq values over all the active phases of time, taking account of the cyclic temporal relationships δ of operation of each of the components/elements (phases of waking, activity, rest, normal or deep sleep mode, sleep-mode verification and how this is done, full shutdowns, resumption of certain functions, the presence of partial networks, etc.), in the hope that, at the end of the process, we can find the total by calculating the integral sum "$\Sigma q = Q = \int dq = \int \Sigma(i.dt)$" of the partial consumptions $i.dt$ of the different elements and work cycles.

In addition, by definition of the technical specifications, a smart-apparel project requires a clearly defined or expected temporal functional range (in hours, days, etc.), which must be compatible with the energy range, representing the integral sum of the partial consumptions, presented in the previous paragraph – again, expressed in coulomb.

To conclude, we need to find the average of that quantity Q in coulombs per second, and finally, if need be, express it in amp-hours Ah. We will then be able to make estimations on the basis of the (nominal) specifications of the energy source, representing a "certain energy capacity", in "mAh", of a cell or battery or supercapacitor. From this, we derive the smart garment's range, and its lifespan in days, months or even years.

This lengthy, painstaking process is, unfortunately, the only way to avoid lying, or inventing a value "off the top of your head", for the lifespan on the documentation for your product, which could sometimes end with some highly dissatisfied customers.

We shall now present an example of how to estimate the energy balance of the different phases of the activity of a patch.

4.1.2.2. *Example of how to estimate the range of a patch*

For illustrative purposes, let us consider a realistic example of an ordinary application.

4.1.2.2.1. Phases of activity in the patch

The expressed aim is to run an "insulated autonomous patch", performing short sequences of data measurement every minute, over a period of between 48 and 72 hours (see Figures 4.1 to 4.3), which can be interrogated by RF (for example, at high frequency with NFC). In short, we need to calculate, in great detail, the levels of phases of consumption of the system, and quantify them, in particular, because the more time the system spends in sleep mode or inactive, the less energy it will consume. The same principle applies, of course, in the field of smart apparel.

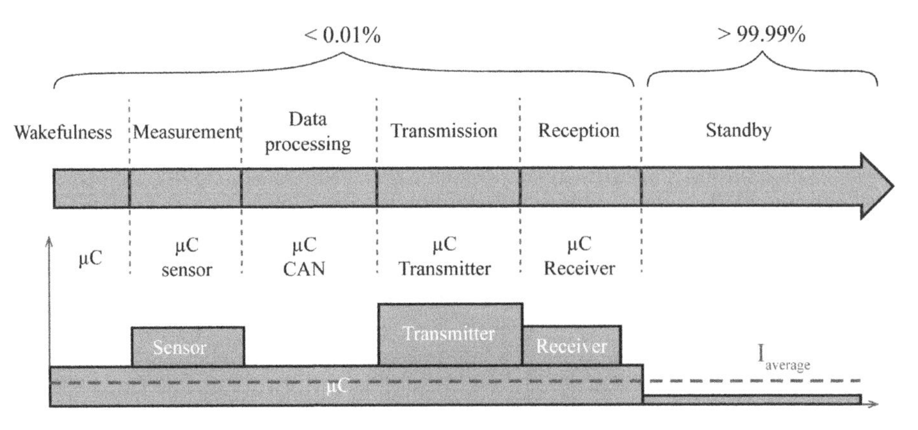

Figure 4.1. *Operational sequences. For a color version of this figure, see www.iste.co.uk/paret/smartpatches.zip*

Figure 4.2 shows some examples of the quantities of electricity consumed during the different phases of operation of a smart garment. We need to add up all these values in order to guide the choice of which type of battery to use.

Suppose, for example, that for each minute, when the measurements are being taken, there are consumption peaks of approximately 140 μA, lasting 30 ms. This represents a need, in terms of capacity, of 1.68 μA.h, and all operational modes (depending on the solutions adopted for the integrated circuits) produce a maximum average consumption per second of approximately 10 μA, which is 10 μC per second, so a requirement of (10 μA × (60 × 60) × (24)) = 864,000 μcoulombs/day, or put differently, 10 μA × 24 = 240 μA.h/day.

A daily source equivalent to 250 μA.h may, in theory, seem sufficient. However, considering systematic recharging of around 80%, we could actually say that we need a microbattery with a capacity of 310-320 μA.h/day. For the desired autonomy of 48 hours = 2 days, we would therefore need a source/microbattery with a capacity of around 700 μA.h.

				Consumption	
Elements				Standby	Active
Microphone		Console (run @ 2 MHz)			500 μA
		Activity time			200 ms
		Q = i t	Coulomb		100 10^-6
		Duty cycle	0.2/1		20 μC
Sensor		Console		0.15μA	140 μA
		Acquisition time			30 ms
		Q = i t	Coulomb		4.2 10^-6
		Duty cycle	0.03/60		2.1 nC
e2prom	via I2C	Console			10 μA
		Write time			100 ms
		Q = i t	Coulomb		1 10^-6
		Duty cycle	0.1/1		0.1 μC
Clock		Console			250 nA
		Write time			
		Q = i t	Coulomb		0.25 μC
		Duty cycle	1/1		
Total		Σ Q per day	Coulomb		1.758 C/day
		(20 + 0 + 0.1 + 0.25) × 3600 × 24			

Figure 4.2. *Example of quantities of electricity consumed during the different phases of the operation of the patch*

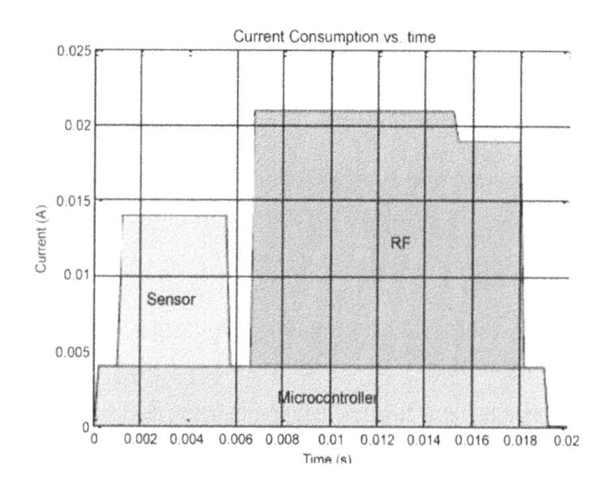

Figure 4.3. *Operational sequences of an IoT system (source: Renaud Briand from Toulouse/Bayonne). For a color version of this figure, see www.iste.co.uk/paret/smartpatches.zip*

4.1.2.2.2. Value of the range of the energy source

In the case of energy-harvesting devices, in which a capacitor C is charged at a voltage V:

$$\text{range} = t = \Sigma q/I = CV/I\ t, \text{ in seconds}$$

EXAMPLE.– Remembering that 1 year = 365 × 24 × 60 × 60 = 31,536,000 seconds, with a capacity of 100 μF charged at 1V, so $Q = CV = 1\times10^{-4}$ coulombs, in order to achieve a 1-year battery life, based on the duty cycles, the average consumed level of current must be no more than:

$$I\_average = 1\times10^{-4}/31.53\times10^{+6} = 3.17\times10^{-12} = 3.17\ \text{pA}$$

4.1.2.3. *Charging time*

In the case of applications supported by energy-harvesting technology, given that the patch's local energy source is also recharged by HF communication in NFC, this only happens during time windows devoted to interrogation by the smartphone. Considering these windows are relatively short (between 1 and 3 minutes each day or 48 hours, at most, including the reading of the results), it is necessary and important to also pay attention to the speed and duration of charging time, which the local energy source for the patch will use. For example, either 80% recharging in 15 minutes for batteries, or perhaps 95% in 2 seconds with graphene-based supercapacitors.

Now that we have estimated all these levels of energy, where does the energy come from, and where will it be stored?

4.1.3. *Choice of battery, supercapacitor and energy harvesting*

Let us briefly summarize the typical technological problems encountered with different types of solutions for local energy supply to patches.

4.1.3.1. *Electric batteries*

Today, batteries represent the dominant energy source for electronic devices. They are found in the form of:

– "button cells", which have limited lifespans and need to be changed regularly. These are single-use batteries, and are not widely recycled;

– disposable alkaline batteries, which require fifty times more energy than they are capable of producing over the course of their lifespan;

– "Li-ion batteries". Depending on the model, cells and batteries may contain nickel, lithium, zinc, cadmium and mercury. Some of these resources are extremely rare, and the process of their extraction generates a particularly large quantity of pollution and is geopolitically unstable. They can be very problematic to use in view of their limited energy-storage capacity, the fact that they need to be recharged regularly, their environmental toxicity, their size and their limited lifespan.

If the choice is made to use batteries which need recharging for a patch, then exact answers must be found to a great number of questions: is the patch disposable? Are the cells single-use? Are the cells to be changed for preventive purposes? Can we easily access and change the cells? Are they easy to install? Who will change the cells: the individual user, or maintenance personnel? If an intervention is required, is the person formally trained and accredited, or is it a stand-in? How much will the intervention cost? For each contract, which battery manufacturer will guarantee their products' battery life? Who signs to guarantee the battery life in real-world application? Are the cells installed directly at the end of the production chain? Have measures been taken in relation to possible lengthy storage in a warehouse? Are the batteries activated only at the time of first use, after having been stored for an unknown period of time? If so, who is to activate them? What will the cost be? And so on, and so forth…

In short, all of these arguments weigh heavily in favor of the smart garment being batteryless, operating instead in a energy harvesting mode.

4.1.3.2. *Batteries… and graphene*

The ability to exchange ions means that graphene-based batteries have unparalleled performances, in terms of energy capacity, fast charging and low loss of charging capacity even after multiple charge/discharge cycles. The physical and electrical characteristics of graphene make it an ideal material for the design of electrodes for batteries.

EXAMPLE 1.– Perforated graphene sheets, doped with silicon, replace traditional graphite anodes, increasing capacity by 3–10 times in comparison to that of conventional batteries.

EXAMPLE 2.– As one example among others, the company Nawa produces carbon nanotubes (rolled up graphene sheets, with capture elements between 1000 and 10,000 times finer than those of conventional supercapacitors) which, when aligned on an aluminum substrate, provide a very large and accessible active surface, which allows charge particles to lodge or dislodge in large quantities, quickly, during charging and discharging.

4.1.3.3. *Graphene-based supercapacitors and ultracapacitors*

Another option is the use of graphene-based electrodes in the design of supercapacitors, also known as “ultracapacitors” to store and manage energy. These components have an ever-increasing energy-storage capacity, and can be charged more quickly than traditional components. In addition, the structure of graphene means that quantities of energy beyond the reach of other materials can be achieved. In comparison to a Li-ion battery, graphene batteries have extremely low internal resistance (~1/10 of a milli-ohm), rapid charging time of approximately a second, experience less heating, have a power density 60 times higher, a longer lifespan, excellent reliability, and very attractive costs.

4.1.3.3.1. Example of graphene supercapacitors

Supercapacitors made with graphene-based techniques are capable of achieving high capacity values (in mF or even in F) in relation to their dimensions/volumes. For example, for a value of 1–5 farads (F), the size is between 1 and 2 cm. This means that, for lower costs, we can obtain levels of power ten times greater than conventional supercapacitors, and a power energy density of 100 kW/kg and an energy density from 8 to 10 Wh/kg, with the aim being to achieve 25 Wh/kg in time.

Of course, we always seek a high value (and thus a large number of coulombs and long range), but in principle, this increases the time required to recharge the capacitor, due to the increase of the time constant of the charging time $\theta = RC$. The curve in Figure 4.5 gives some idea of the values of the serial equivalent resistances of supercapacitors as a function of their capacities.

4.1.3.3.2. Example of minuscule supercapacitors

Consider the minuscule supercapacitor produced by the Daegu Gyeongbuk Institute of Science and Technology in South Korea. To improve the quantity of energy stored, the electrodes, conventionally made of graphene, are in the form of very fine, interleaved comb-teeth and highly porous, providing more space to allow more electrostatic reactions to occur, and thus, more energy to be stored efficiently. In general, this technology is costly and can only be built using flexible and heat-sensitive substrates. To remedy this problem, the graphene ink is specifically sprayed onto the flexible substrate at a 45° angle and 145°C. The micro-supercapacitor is around 25 μm thick, and can store four time more energy than Li-ion batteries, meaning it can be used in flexible electronic patches that can be stuck to the skin, without any degradation of their performances, or to power medical monitoring sensors and remote diagnostic devices. The device has a shorter charging time than that needed for rechargeable batteries, but on the other hand, it is able to store less energy.

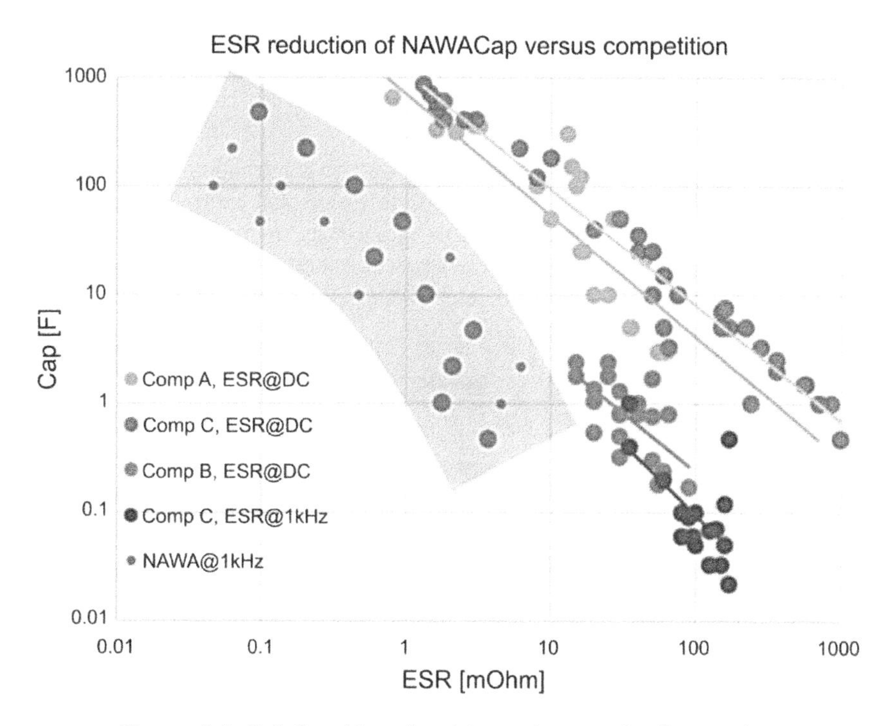

Figure 4.4. *Relationships of resistance to capacity. For a color version of this figure, see www.iste.co.uk/paret/smartpatches.zip*

4.1.3.3.3. Examples of supercapacitors/ultracapacitors

Returning to the values required in our previous example, for a range of ~500 µA.h, which is ~1,500,000 µcoulombs, assuming an operating voltage of 1.5 V, this gives a minimum capacity of (Q = CV), so C = 1,500,000 µcoulombs/ 1.5 V = 1 F = 1000 mF.

4.2. Energy harvesting

Here, we come to another major consideration in relation to patches for smart apparel. How can we supply the patches with the energy to feed their electrical and/or electronic components, without the presence of batteries, or at least, with an extremely small and lightweight battery, which can be recharged by alternative means other than by plugging into a PowerPoint? This is the question we shall examine in these next sections, after a necessary review of certain elements of fundamental physics and nuggets of semiconductor theory.

4.2.1. *General*

As usual, let us begin with a few definitions.

Energy harvesting is typically defined as being "the conversion of energy already present in the surrounding environment into usable electrical energy". There can be a variety of energy sources energy available in our near environment in day-to-day life. For example, light energy, solar, wind power, vibration energy, sound energy, kinetic, thermal, chemical, electrical, electromagnetic (sources/radiated fields in RFID, NFC, etc.), etc. (see Figure 4.5).

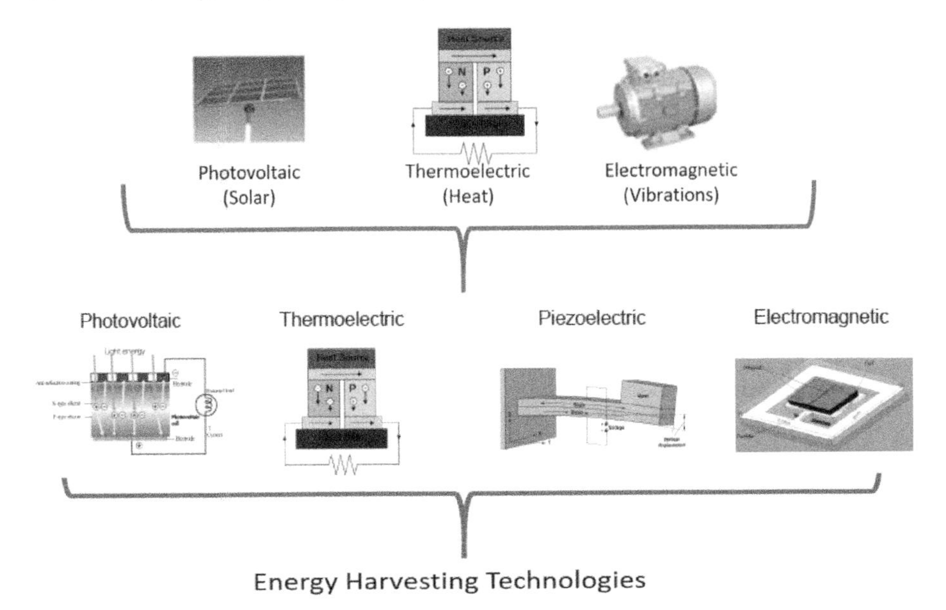

Figure 4.5. *Examples of energy-harvesting technologies. For a color version of this figure, see www.iste.co.uk/paret/smartpatches.zip*

In comparison to the energy stored in standard storage elements such as power cells/batteries, our environment represents a relatively inexhaustible energy source (unless otherwise proven). Energy harvesting is a solution which consists of gathering energy from the patch's environment and turning it into usable electricity.

Rather than being characterized by energy densities, energy-harvesting methods are described in terms of power density (see Figures 4.6–4.8).

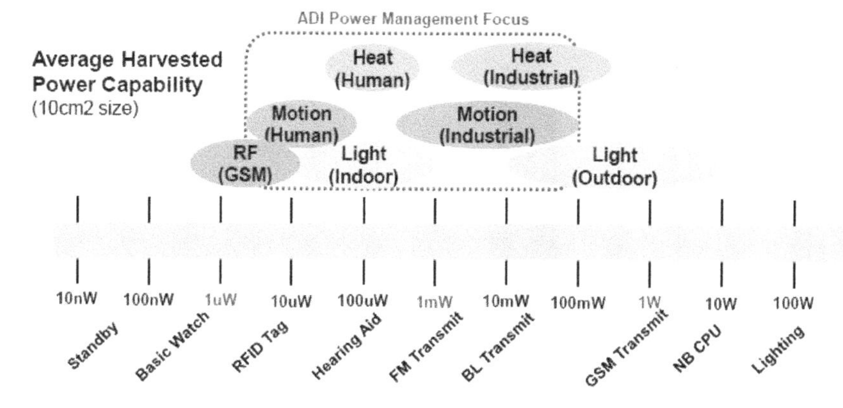

Figure 4.6. *Examples of energy harvesting technologies. For a color version of this figure, see www.iste.co.uk/paret/smartpatches.zip*

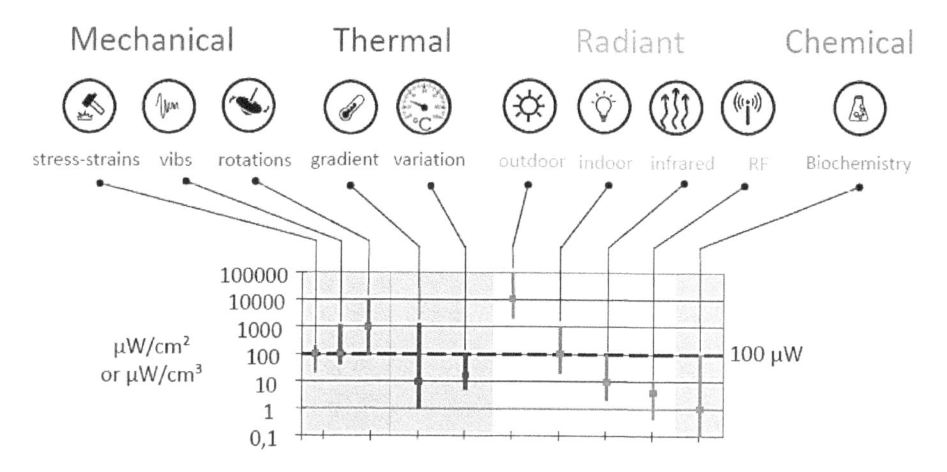

Figure 4.7. *Examples of technological principles of energy harvesting. For a color version of this figure, see www.iste.co.uk/paret/smartpatches.zip*

In our field, the purpose of these technologies is to supply energy to autonomous sensors or a network of autonomous sensors, to power and/or monitor wireless transmissions to a receiver, and power connected objects using radiofrequency communications (mobile and wireless technologies).

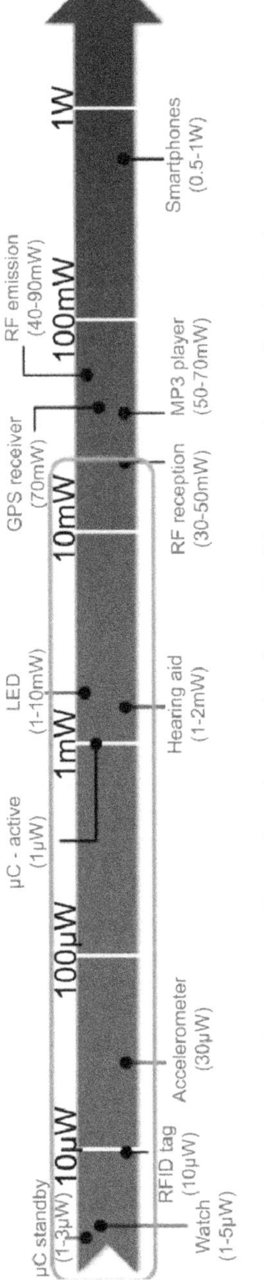

Figure 4.8. *Examples of power levels required by energy-harvesting technology. For a color version of this figure, see www.iste.co.uk/paret/smartpatches.zip*

4.2.1.1. *Energy requirements of patches for smart apparel*

In spite of efforts to limit the amount of energy required by the functions of devices such as biomedical microsensors, etc., wireless structures and environments include numerous functions, and require the systems to be supplied with a certain amount of energy. Unfortunately, because of the volume constraints on products/electronic patches, the available stored energy is limited, often resulting in the devices having rather short autonomous ranges. Thus, we seek, at once, to increase the energy density present in the stores (batteries, etc.), knowing that in any case, their lifespans are short and finite, and to prolong the device's lifespan by effectively managing its power supply, minimizing energy losses that drain the battery needlessly. A sustainable energy supply, independent of the quantity of energy initially stored, is an attractive option, and is increasingly in demand in applications such as healthcare, biomedicine, wellness, leisure, devices for implants, etc. For this reason, a renewable energy store, which can automatically replace the energy consumed, has a place in a large number of applications, particularly, in those of patches, connected and smart apparel. The main question, though, is: how can we harvest that energy from the surrounding environment, and store it locally?

For several decades now, sensors, transducers and micro-electromechanical systems (MEMS) have been capable of extracting or recovering energy from vibrations, thermal gradients, exposure to light, the presence of electromagnetic waves, etc. However, the energy extracted from the environment is generally slight, and unpredictable, because the processes draw energy from external sources in the form of short, intermittent bursts, in discontinuous, irregular and random "peaks", in infinitesimal quantities, and then store it. Therefore, to transfer this energy into refillable reserves (for example, rechargeable Li-ion batteries, capacitors or other electrochemical reservoirs), it is important to develop particular low-loss electronic transfer systems, from the energy source to the peripheral devices/reservoirs. Thus, the energy harvested from the environment, necessary for the local system to operate can be delivered, added/injected into the system, which decreases or even eliminates the need to replace the battery or external recharging cycles.

Of the various energy-harvesting systems, current technologies are able to harvest energy, with relative ease, from solar, wind, vibrations, thermal energy, sound, kinematic, chemical, electrical energy, dynamos, or radiated electromagnetic waves and fields (RFID, NFC, Wi-Fi, etc.). These harvesting systems are generally capable of supplying very small amounts of electrical energy (i.e. electrical current) to electronic circuits operating at very low consumption. Thus, they become systems which can function without a battery (batteryless). Let us briefly examine the six most usual possibilities.

4.2.1.2. *Examples of energy harvesting*

A wireless system, fed by energy harvesting, must take account of various factors, such as the means of energy generation, any conditioning necessary before the energy can be used as electricity, the energy storage solution and the chosen data-transfer protocol. Figure 4.9 illustrates the quantities of energy available for different sources.

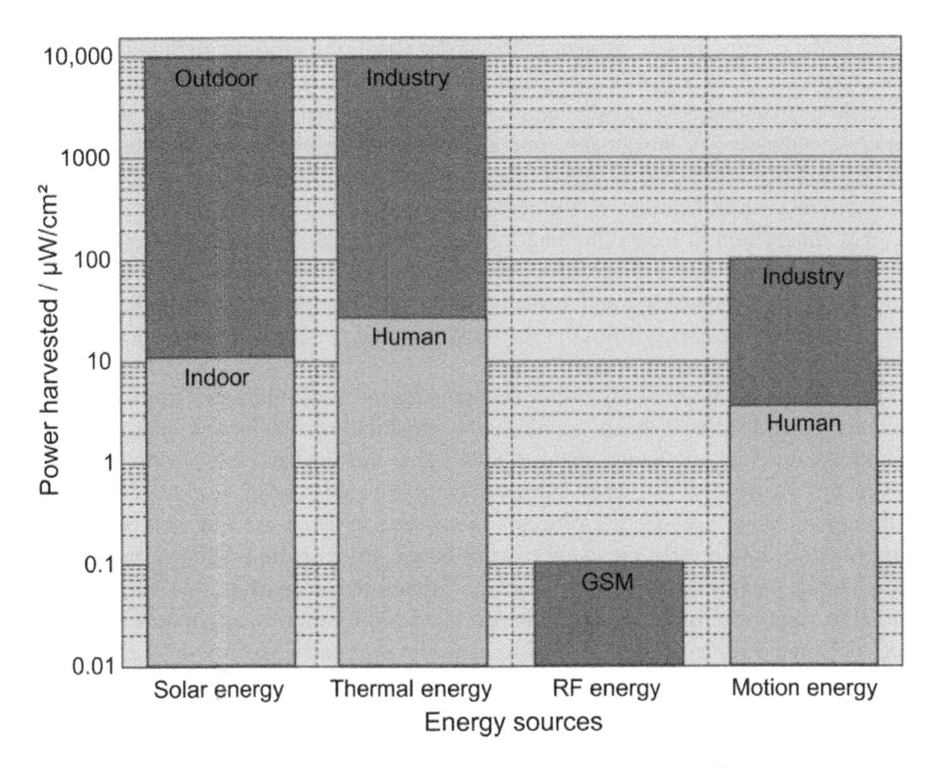

Figure 4.9. *Examples of energy harvesting technologies. For a color version of this figure, see www.iste.co.uk/paret/smartpatches.zip*

4.2.1.3. *Harvesting of light energy: converters and photovoltaic cells*

Photovoltaic cells generate electricity in the presence of light. The quantity of energy generated depends on, amongst other factors, the number of cells used and the light levels available. PV cells provide an excellent power density when directly exposed to sunlight. For reasons of space and cost, it is useful to work with a small number of cells. The consequence, however, is a reduction in the amount of energy available. Maximum Power Point Tracking (MPTT) and the use of a suitable storage element help to optimize energy production when significant variations in luminosity need to be taken into account.

Generally, each cell is made up of a classic semiconductor p–n junction, forming a diode which is inversely polarized. The energy carried by the photons present in the light gives rise to the depletion of electron–hole pairs in the junction, and the resulting electrical field immediately separates the electrons on the N+ side and the holes on the P– side. The positive and negative charges accumulate, developing an open circuit with voltage. When a charge is applied to the terminals of the cell, the excess electrons flow through it from the N+ side and recombine with the holes on the P– side, generating an external current that is directly proportional to the light intensity. Thus, photovoltaic cells can generate enough power to supply the microsystems in smart apparel (e.g. textiles for the fashion industry). This mature technology is compatible with the standard specifications of integrated circuits, and the conversion of photovoltaic energy provides high levels of power in comparison to other energy-conversion mechanisms.

For the applications in smart apparel we are discussing here, light may serve as an energy source, but the availability of output power is heavily dependent on the intended application (indoors or outdoors), and on weather conditions, and the light intensity can change greatly depending on how long the smart garment is exposed to that energy source.

4.2.1.4. *Thermal energy harvesting: thermo-electric converters*

4.2.1.4.1. Thermocouple

Temperature gradients ($\partial T/\partial x$, partial differential of temperature with respect to distance) can be found everywhere in nature, and it is possible to exploit these gradients in order to produce electricity.

The most elementary thermo-electric generator is a thermocouple, made of two different materials, connected by a strip of metal.

The thermal gradients $\partial T/\partial x$ between the bottom and the top of the device's "legs" are converted directly into electrical energy and an electric current, by the Seebeck effect (thermo-electric effect). The temperature differences present between the opposite ends of a conductive material lead to heat flux within that material, and therefore, the mobile charges (free carriers) diffuse from regions with high concentrations C to regions with low concentrations. As the mobile carriers diffuse, in accordance with Fick's law, the concentration gradient $\partial C/\partial x$ due to $\partial T/\partial x$ gives rise to an electric field through the material, making it more difficult for charge transporters to diffuse, and ultimately leading to equilibrium. At equilibrium, the mobile carriers due to the electrical field travelling towards the junction with higher temperature cancel out the number of hot mobile carriers travelling towards the colder junction, and consequently, no voltage is established.

A thermocouple configuration is most appropriate to produce electricity by electrically joining semiconductive materials of types P and N to the hot end. Generally, the P and N legs of a thermocouple are heavily doped in order to deliver better results, combining good electrical conduction with good thermal resistance (in order to maintain the temperature gradient). The heat flux transports the mobile carriers of the dominant charge in each material (electrons in type N and electron holes in type P) to the low-temperature junction, by respectively ionizing each starting electrode with an opposing charge and establishing a differential voltage across the low-temperature electrodes. As the charge carriers leave the hot end, they leave behind ionized molecules which, instead of attracting the inverse flow of the material itself, attract the opposite type of charge carriers in the material. Ultimately, the voltage and levels of power generated are proportional to the temperature difference and to the type of material, with the relation being determined by a so-called Seebeck coefficient.

The choices of materials and thermal gradients are crucial in producing the desired levels of power for applications in smart apparel. In addition, because of the small mechanical dimensions of the microsystems encountered in smart apparel, temperature differences greater than 10°C are rare, and unfortunately, very low levels of voltage are recovered by the thermocouple. Because of these very slight gradients, the thermal energy harvested is limited to around 15 $\mu W/cm^3$, and this technology is only capable of around 5% efficiency in converting heat to electricity. Most research in this area focuses on optimizing nanostructured thermo-electric materials, and their geometry, to produce sufficient power and voltage using temperature differences as small as 5-10°C. One of the challenges lies in maintaining the temperature gradient between the hot and cold regions on so small a scale – in particular, when the device is being used on the human body.

4.2.1.4.2. Peltier elements

Peltier elements are used to transform temperature differences into electrical energy. The current they generate is significant, but typically, the voltage is low. Using specific components, it is possible to make DC/DC converters which work with an input voltage of only a few tens of millivolts. Another approach is to use thermo-electric sensors capable of producing sufficient voltage to directly power the electronic circuits.

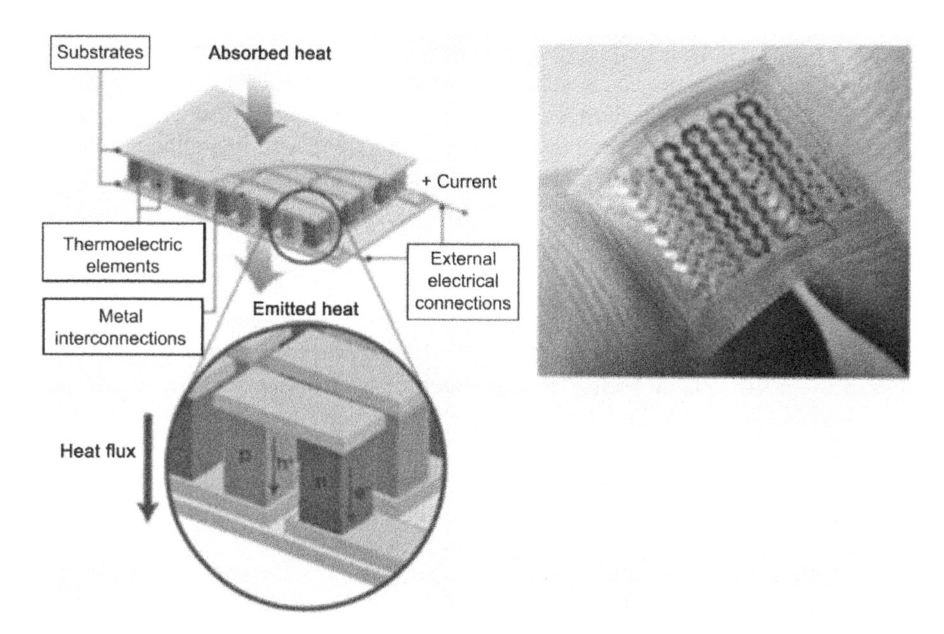

Figure 4.10. *Examples of thermal energy harvesting technologies (source: Webinar, CRESITT Industrie). For a color version of this figure, see www.iste.co.uk/paret/smartpatches.zip*

4.2.1.5. *Vibrational energy harvesting*

4.2.1.5.1. Mechanical

The extraction/harvesting of energy from mechanical vibrations is based on the classic principle of movement of a spring-mounted mass, vibrating in relation to its support, in which the mechanical accelerations produced by vibrations cause the motion of the mass, accompanied by a phenomenon of damped oscillations (a well-known second-order differential system of the mass, spring and damper). These oscillations produce forces of friction and damping exerted against the moving mass (microjoules are lost with each mechanical pulse), which absorb the kinetic energy from the vibration, leading to a reduction and even total stoppage of the oscillations. This conversion uses two mechanisms: direct application of force and use of the forces of inertia acting on the mass. For example, this occurs when an electrical damping force (with a magnetic field) or an electrical field (electrostatic) is imposed on a moving piezoelectric material (see the next section). The vibrational mechanical energy is harnessed and converted into electrical energy, and in this way, energy is harvested from the surrounding environment. One of the most effective means of energy harvesting is to convert mechanical energy from movements, pulses, acoustic pressure or vibrations into electrical energy, using, for example, electromagnetic, electrostatic or piezoelectric transducers. In these cases, wear due to the mechanical movement must be taken into consideration. This vibrational energy is a moderate power source, with power levels ranging between 1 and 200 $\mu W/cm^3$, but once again, heavily dependent on the application.

4.2.1.5.2. Mechanical piezoelectric converters

When a piezoelectric element is in motion as the result of an impulse (a shock), or a continuous action (vibrations produced by walking), it generates alternating current. This voltage can be transformed into direct current, by means of full-wave rectification using a diode bridge and a capacitor. The quantity of energy generated by the external mechanical excitations depends specifically on the amplitude of the vibration or force, the frequency of the source and the efficiency of the transducer. To optimize the yield of resonating energy recuperators, their resonance frequency must be adjusted to that of the mechanical vibration in the environment.

Piezoelectric energy recuperators offer better power density, and have the advantage of being able to be integrated more easily into CMOS (Complementary Metal Oxide Semiconductor) technologies.

Electrostatic vibratory energy harvesting

Vibratory energy harvesting by electrostatic (capacitive) means is generally based on work done against the electrostatic force of a variable capacitor (a *varicap*), dependent on the vibration. Often, the simplest solution is to impose a constant voltage between the terminals of a varicap *C_varicap* – for example, by connecting it to a battery/cell/Super Cap with voltage *V_bat*, notably with a diode, which serves as a voltage clamp between the terminals of the varicap *V_varicap* and that of the battery *V_bat*. As the vibrations cause a mechanical variation in the value of the capacity *C_varicap*, the charge ΔQ (ΔQ = ΔC_varicap × V_varicap_constant) is harvested through a current *i_harvest*, which is channeled to the power cell/energy reservoir at predefined levels of voltage. Thus, the mechanical energy is converted into electricity, and recharges the battery/Super Cap.

4.2.1.5.3. Electromagnetic converters

In the presence of a magnetic field, motion can be transformed into electrical energy (Faraday's law). When generators are subjected to vibrations, this principle can be used to produce several milliwatts of electrical power. Electromagnetic transducers have a high charge impedance, and generate only low levels of voltage (100-300 mV), so other design challenges remain to be overcome. These generators are suitable for the manufacture of hundreds of converters.

Electromagnetic vibratory energy harvesting

Vibratory energy can be harvested by two means: either by electromagnetic means, in accordance with Faraday's law, whereby voltage is generated at the terminals of a coil when it oscillates mechanically through a magnetic field (or vice versa); or, for example, the stress or strain of a piezoelectric material, leading to the separation of charges across the device, causing a voltage drop that is proportional to the applied stress.

These two energy-harvesting methods generate unsteady, alternating current which, consequently, requires the presence of a rectification circuit to obtain direct current, and to condition the extracted power. In principle, this conditioning of harvested power leads to losses of additional power, thus reducing the effectiveness of all the power-supply mechanisms. Finally, these two methods require magnetic and piezoelectric materials which are not easy to integrate.

> APPLICATION OF VIBRATORY ENERGY HARVESTING IN SMART APPAREL.–
>
> The kinetic energy of vibrations generated by human motion (pressure applied manually to a surface, pressure exerted by the heel on an insole while the wearer is walking or running, etc.) is used in the concept of vibratory energy harvesting or piezoelectric sensors/generators. The performances of these sensors are within the frequency ranges of the vibration sources: 50–300 Hz and acceleration ranges: 0.2–1 G.

4.2.1.6. *Energy harvesting from electromagnetic radio waves*

An increasingly popular type of energy harvesting is harnessing the energy in the radiated waves all around us (RFID, Wi-Fi, NFC, BTLE, GSM, etc.). Generally speaking, these waves are within the frequency band 10 MHz–3 or 5 GHz. The propagation of alternating electrical fields E and magnetic fields H in radio-frequency electromagnetic waves creates a "Poynting vector" (equal to the vectorial product of the vectors $E \times H$), transporting power per unit surface expressed in watt/m^2. Obviously, this alternating power, harvested by a wide/multiband antenna (a rectifying antenna, or rectenna for short) is very low, and needs to be rectified and then converted to direct current by a voltage multiplier, through a charge pump, so that it can be stored for local use in a gold/super cap capacitor.

As nothing is perfect, all this circuitry comes with certain losses and a certain yield, which does not make things easier, but there is enough energy left to be able to work.

APPLICATION OF ENERGY HARVESTING FROM ELECTROMAGNETIC WAVES IN SMART APPAREL.–

The antenna of the patch system can also be used for RFID purposes, and the useful load in that case is a small SigFox/LoRa transmitter which occasionally sends messages (generally around 100 messages per day) to link to remote systems (examples of PPE and medical garments).

4.2.1.6.1. Summary of energy-harvesting technologies

ENERGY ON APPLICATIONS

- Typical Sensor Node's power consumption
- 1-5uW: Micro standby mode power consumption
- 300uW-1mW: Micro active mode power consumption
- 50mW: transmission power peak
- 50-500uJoule: typical accumulated Energy for a complete measurement including transmission

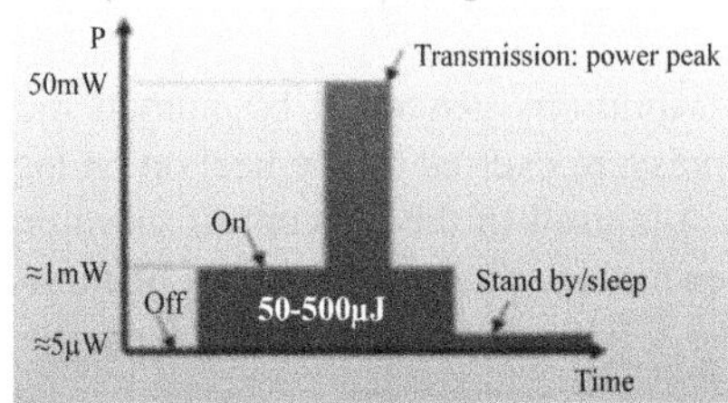

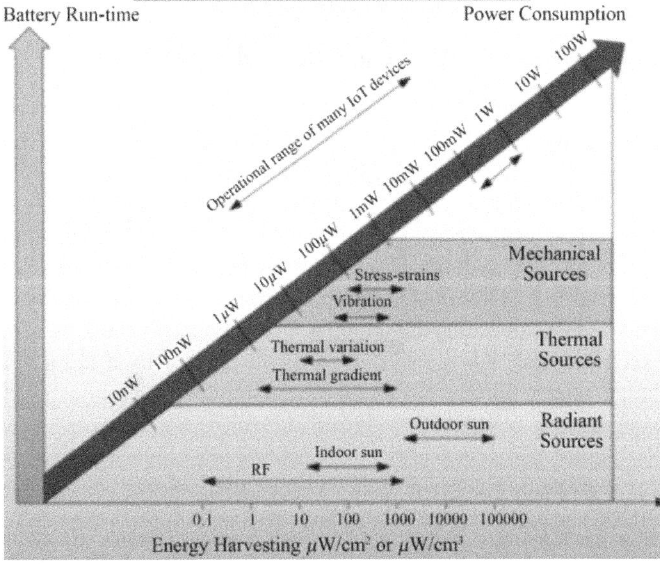

Figure 4.11. *Examples of energy-harvesting technologies (source: Renasas). For a color version of this figure, see www.iste.co.uk/paret/smartpatches.zip*

We have now come to the end of this brief introduction to energy harvesting, which is likely to be an area of growth in the coming years, due to the rise in power of miniature components with MEMS on board, built on the same substrates as the silicon chips used for signal refinement.

4.2.2. *Existing technologies for smart apparel*

At present, many component manufacturers (often the same as for AFEs) offer specific autonomous integrated circuits to use energy harvesting (Maxim, Analog device, STm, NXP, Melexis, e-peas, etc.). These circuits are all made in much the same way, but each of them has their own specific features. In general:

– All of them support the recharging of a battery and/or a capacitor or supercapacitor, but can only use one source at a time, because the optimization options are different and require different types of electronic circuitry and management software for each source, when necessary or specific to sources with an alternating output signal (piezoelectric, RF, etc.); at input, the circuits have a system to rectify the source alternating current into direct current.

– All have a boost system (DC/DC switched-mode power supply), operating at a few MHz, the aim of which is to boost the direct current to a higher level of direct current to drive the function of the integrated circuit. The system is controlled by an MPTT device.

– All have a buck charge converter, which is adaptable to the type of charging element (see Figures 4.12 and 4.13).

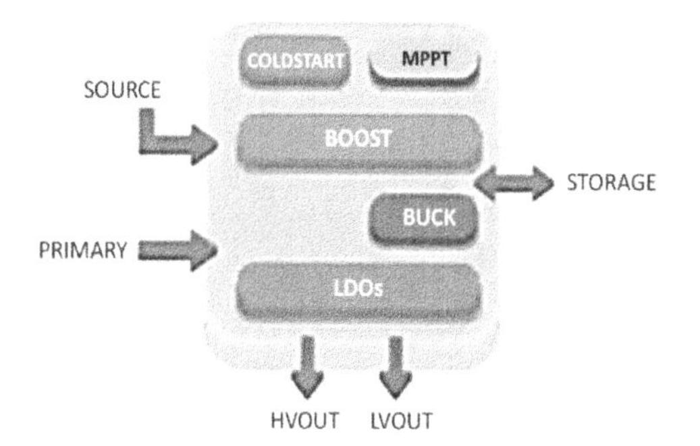

- Boost converter with MPPT tracking
- Buck converter to supply the digital core
- Linear converters (LDO) to provide up to 2 independent regulated voltages
- **Storage element protection** (through CFGx) configuration pins)
- No software - configuration made with I/O

Figure 4.12. *Examples of energy harvesting technologies. For a color version of this figure, see www.iste.co.uk/paret/smartpatches.zip*

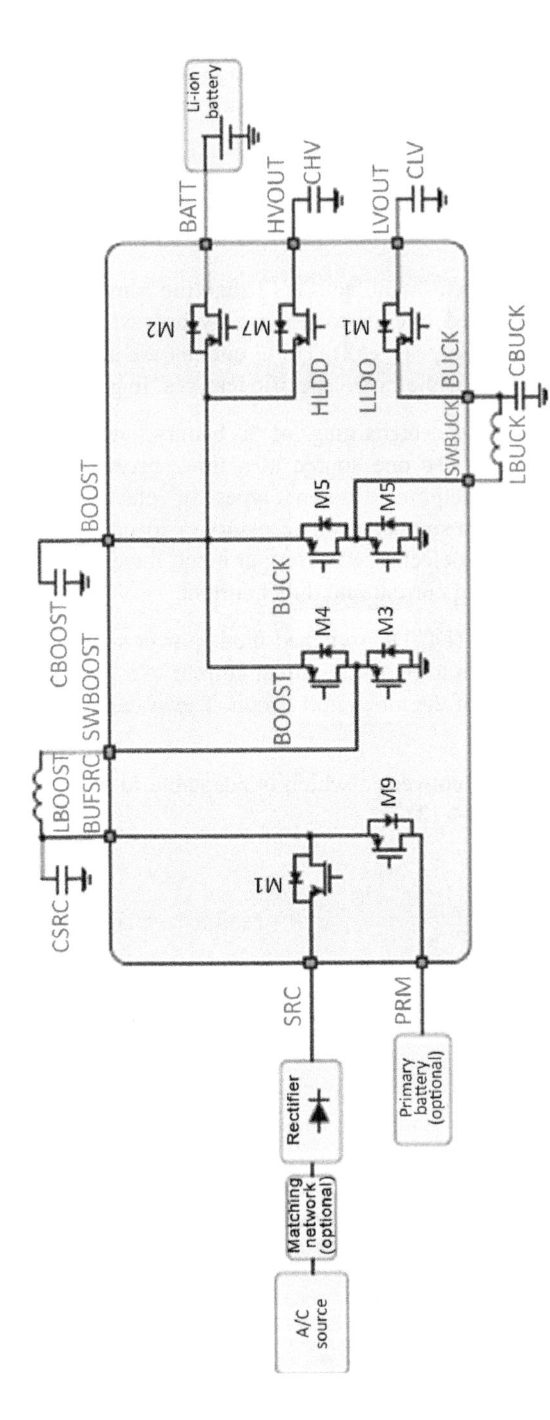

Figure 4.13. *Examples of energy-harvesting technologies, AEM30940 circuit from e-peas. For a color version of this figure, see www.iste.co.uk/paret/smartpatches.zip*

4.2.2.1. *Case of dual-cell supercapacitors*

When using a dual-cell supercapacitor, to avoid damaging it by excessive voltage on one cell, a balun (balanced–unbalanced) circuit allows users to balance the internal voltage. The configuration with the "balun circuit deactivated" must be used if a battery, a capacitor or a single-cell supercapacitor is connected to BATT. If the balun is linked to the node between the cells of a supercapacitor, the balun circuit compensates for any imbalance between the two cells which could lead to overloading of one of the two cells. The balun circuit ensures that BAL remains near to Vbatt/2. This configuration must be used if a dual-cell supercapacitor is connected to BATT.

NOTE.– At present, only a handful of suppliers such as NXP offer circuits which directly integrate energy harvesting with a biosensor, and this represents a real potential solution in the case of patches for smart apparel.

4.3. Example of energy harvesting for smart apparel

This section describes a very interesting case of energy harvesting applicable to patches used in smart apparel.

As previously indicated, the energy contained in electromagnetic wave radiation can be exploited. This energy may come from various transmitter stations in the area (radar, radio, television, etc.), local RFID interrogators or NFC initiators (for example, smartphones, tablets, etc.). The components used for energy harvesting must, of course, be suited to the frequency of the transmission sources. Then, the RF energy can be extracted from the "air interface" link by a receiver antenna, then converted into usable direct current and finally stored in a (super)capacitor in the form $Q = CV$ and energy $= \frac{1}{2} CV^2$. However, the level of power that can be recovered is limited, because the RF energy propagated decreases very rapidly as the distance from the source increases. The low voltage generated through such energy harvesting often means that a voltage multiplier must be used.

4.3.1. *Energy harvesting via an HF (high-frequency) NFC link*

NFC (Near-Field Communication) technology, which is very widely used all over the world, particularly in mobile telephony, offers numerous benefits corresponding to the intended fields of application for patches in smart apparel, in particular, interrogation and simultaneous power supply, to allow the patches to function as sensors of biometric parameters, for possibilities of two-way communication with the patch, and possible supply of energy and remote power to the target (in our case, the patch) by the interrogator (the smartphone).

The quantity of energy harvested by an HF NFC link[1] at 13.56 MHz is perfectly sufficient in conventional NFC communication. In this case, in the few hundredths of a millisecond that it takes to interrogate the patch in HF to recover the desired biometric data with an NFC interrogator (a specific module or a mobile phone), it is possible to simultaneously recharge the supercapacitor integrated into the patch, to provide the patch with 2–3 days' operating range, maximum.

4.3.1.1. *Communication distance with the patch*

A very important comment must be made in relation to NFC technology and its use with a patch for smart apparel. In accordance with the norm NFC2–EN 21 481 and the standards of the "NFC Forum" (used in mobile phones), there are various modes of NFC (types 1 to 5), supporting a variety of standardized communication protocols, in particular:

– *NFC type 2*: standardized by ISO 14443 (contactless cards–proximity cards), with a high communication throughput and an operating distance of a few centimeters.

– *NFC type 5*: standardized by ISO 15693 (contactless cards–vicinity cards), with a slower throughput, capable of function over distances of a few tens of centimeters. The shorthand "NFC type V" (for vicinity) is sometimes used.

NOTE.– In the context of non-time-sensitive applications for patches in smart apparel, communication throughput rates are not overly fast, and NFC Forum Type 5, subject to ISO 15693, tends to be the chosen technology.

Both types 2 and 5 are able to function with a response by the target (the patch) modulating the amplitude of the wave transmitted by interrogator (the smartphone), either by Passive Load Modulation (PLM), or Active Load Modulation (ALM). For its part, ALM, which is also used in ISO 14443-2, offers many benefits. In particular, these relate to *applications of (very) small antennas* (many communicating things in the IoT), but also to applications of antennas *in hostile environments*, with significant loading effects (for example: the presence of metal, readers such as mobile phones, PPE, etc.). Of course, ALM can also be used with ISO 15693 and its derivative, NFC Forum Type 5.

4.3.1.2. *Limiting factors on distance in HF NFC*

To illustrate these problems and these techniques, consider an example. In HF at 13.56 MHz, conventional interrogators compliant with ISO/IEC 14 443, using the classic mechanical format of ID1 cards (85 × 54 mm), and with passive load modulation, have an operating range between 5 and 10 cm. The main factors

1 See the author's 1000+ pages of publications (Paret 2012, 2016a, 2016b).

limiting this communication range are: the capacity of the interrogator to supply the patch with sufficient power to allow it to function at a reasonable distance; the quality of reception of the transmitted data by the interrogator; the possibility of receiving data from the patch, which requires sufficient magnetic coupling between the interrogator and the antennas in the patch (coupling factor k/mutual inductance M). The feasible distance decreases considerably when very small antennas are used on the patch to communicate with a mobile phone. Finally, as the distance range for interrogators is already small, the presence of batteries and other layers of metal in the environment can rapidly cause problems, so the patch can no longer be read by an external interrogator.

4.3.1.3. *Problem solving*

The problem of low available power can be solved very easily. In this case, we simply need to supply the contactless patch with energy using a local power source (the patch then becomes *battery assisted*), but this is a solution which we hope to avoid.

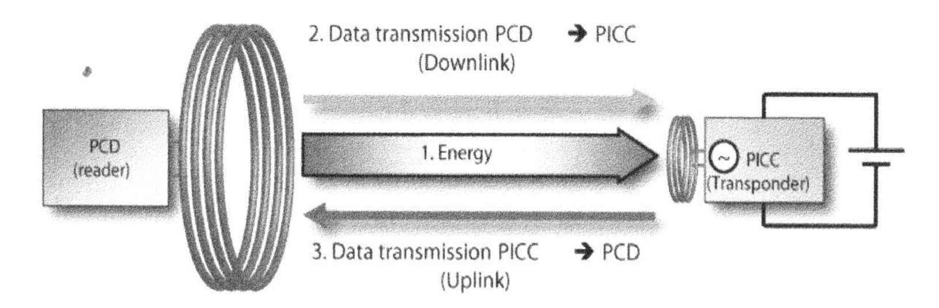

Figure 4.14. *Examples of energy-harvesting technologies. For a color version of this figure, see www.iste.co.uk/paret/smartpatches.zip*

The problem of data transmission from the patch to an interrogator is more difficult to resolve. Even with a patch which has a supplementary local energy source (battery assisted), the conventional technique of backscattering with passive load modulation is not a good solution, because it offers only a marginal improvement on a passive patch, unless the magnetic coupling is improved.

4.3.1.4. *Active Load Modulation techniques*

To transmit the return signal from the patch to the interrogator, another solution is to use a different method to "actively" (that is, for the patch to transmit) a return signal with the same spectral characteristics as those of a backscattered signal with

PLM (see Figure 4.15). This is the method used in the case of patches with small antennas (classes 5 and 6 under ISO 14443 and ISO 15693, and other NFC tags), battery powered or otherwise.

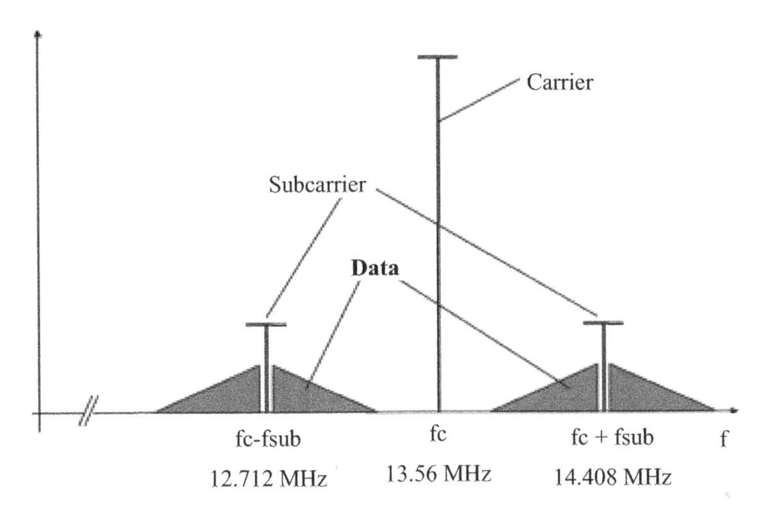

Figure 4.15. *Examples of NFC technologies in energy harvesting*

4.3.1.4.1. Operational principle behind ALM

If ISO 14443 or 15693 are used, in the overall frequency spectrum resulting from the passive load retromodulation, only during the return from the patch to the interrogator, two additional spectral lines appear, symmetrically positioned around the carrier frequency of 13.56 MHz. During the patch's response, these new signals are offset from the interrogator's carrier signal frequency by the frequency of the subcarrier (+ and –848 kHz), with their respective modulation side bands situated on both sides of these two subcarrier frequencies. The data transmitted from the patch to the interrogator are exclusively contained in these side-band modulations. If the energy contained in these side bands is too low to allow for good reception of the return signal, the solution is to generate the two spectral subcarriers described above, and their modulated side bands, containing the data to be passed to an interrogator, to *transmit* the data from a patch, which then becomes *active* (battery-assisted or otherwise), using *active load modulation*. A signal with these properties is known as *double side-band (DSB) modulation*. For this purpose, a simple ASK modulator is used. Then, the signal is amplified and transmitted by the tag's antenna.

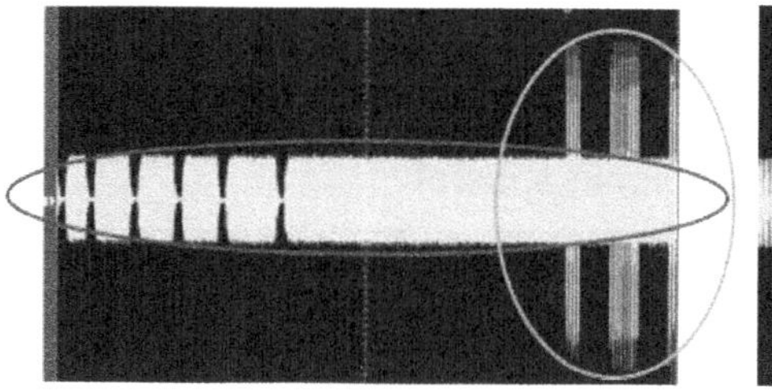
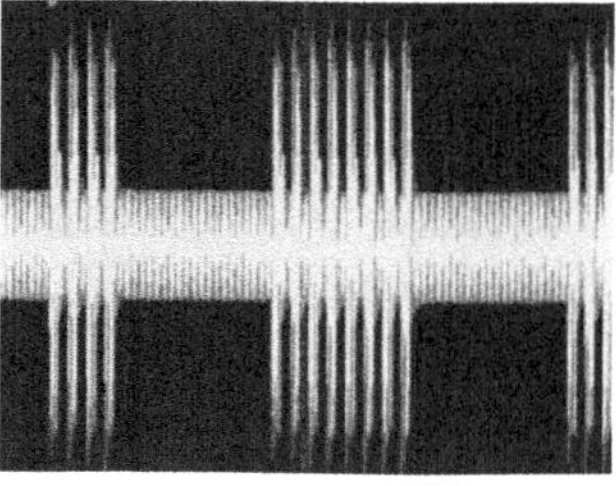

Figure 4.16. *Examples of PLM and ALM (source: Linxens). For a color version of this figure, see www.iste.co.uk/paret/smartpatches.zip*

4.3.1.4.2. Size of a tag antenna versus reading range in ALM

Below are some orders of magnitude of the results obtained:

– Using a conventional ISO 14443 interrogator (with a square antenna measuring 10 × 10 cm), with conventional contactless cards in mechanical format ID1 (antenna measuring around 5 × 8 cm), operating with PLM backscattering, the data can be read over a distance of approximately 7 cm.

– Using the same conventional interrogator and cards with the same type of ID1 antenna, but with an integrated circuit on board, capable of ALM, the reading range becomes approximately 50 cm.

– Still in the same conditions as described above, with ALM, but now with an antenna only 10% of the surface area of a contactless card, it is possible to achieve a communication range of 25 cm.

– Reducing the surface area of the antenna of an interrogator by a factor of 10 halves the reading range, corresponding roughly to the cubic root of the surface ratio between the two antennas. Hence, this technology is particularly well suited to provide an acceptable operating range, even with a very small antenna, on a patch in a smart garment.

4.3.1.4.3. Case of smart apparel with small antennas in hostile environments

This technology is particularly suitable for use in smart apparel, and the patches in smart garments, which need an acceptable operating range even with a very small antenna. For example, a patch in smart garments or smart underclothes:

– With a patch antenna using ALM technology, of the same size as a micro-SD card, in air, it is easily possible to achieve a range of nearly 10 cm.

– Supposing this antenna in micro-SD format is installed in a mobile assembly whose environment is particularly hostile (additional shielding or a significant loading effect), it is still possible to achieve a functional communication range of several centimeters.

4.3.1.4.4. Summary for patches and smart apparel

Medium	Normal interrogator: ISO 15693 or ISO 18000-3	Card/tag				
	Interrogator's antenna	Tag's antenna				
		Mechanical specifications			Type of modulation	
	Dimensions (in cm)		Surface (in mm²)	Format	PLM	ALM
Air	10 × 10	5 × 8	4,000	ID1	30 cm	
Air	10 × 10	5 × 8	4,000	ID1		150 cm
Air	10 × 10	(5 × 8)/10	400	ID1/10		75 cm
Air	10 × 10		130	μ SD	10 cm	
Air	10 × 10	1 × 2 = 2	130	μ SD		20 cm
On metal	10 × 10	1 × 2 = 2	130	μ SD	2 cm	12 cm

Figure 4.17. *Examples with and without ALM with the standards ISO 15693 or ISO 18000-3. For a color version of this figure, see www.iste.co.uk/paret/smartpatches.zip*

The last row in the above table shows the advantage of ISO 15693 – ALM, used for small HF patches, attached or stuck to the skin, or integrated into a garment – this scenario is very common with patches for smart apparel.

4.3.1.4.5. Patch antenna for NFC – type V – ALM

Obviously, the antenna is an integral part of the patch, and must satisfy all constraints to which the patch is subject. In particular, it must be of very small dimensions, be electrically functional, be highly flexible in order to be integrated into a garment, be deformable, ironable, etc. As we have just seen, the ultimate goal is to use the solution ISO 15693, NFC Forum Type V in ALM mode, which allows the dimensions of the antenna to be reduced, whilst the communication range increases, and the energy consumption remains the same.

In principle, an HF NFC patch needs a "loop" antenna with multiple spires. The antenna's dimensions must be compatible with the small dimensions of the patch. Thus, its inductance L will inevitably be low, and in order to achieve a satisfactory value of the quality coefficient Q = Lω/R of the ISO 15693-compliant antenna used in NFC Forum V, a low serial resistance R of the coil is required (Paret 2016b). This constraint can easily be resolved by also creating a small antenna with graphene-based technology. The electrical properties and mechanical robustness of graphene mean that the antenna will be stretchable, washable, and can withstand the rigors of a wash cycle. This helps satisfy many of the required parameters. In business, we are already seeing prototypes of such components emerge (for example, Grapheal in France; see Figure 4.18).

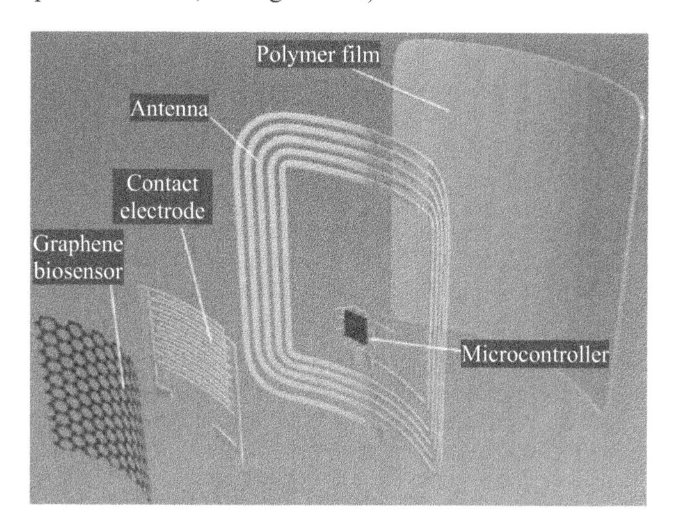

Figure 4.18. *Example of the construction of an NFC antenna for graphene patches (source: Grapheal). For a color version of this figure, see www.iste.co.uk/paret/smartpatches.zip*

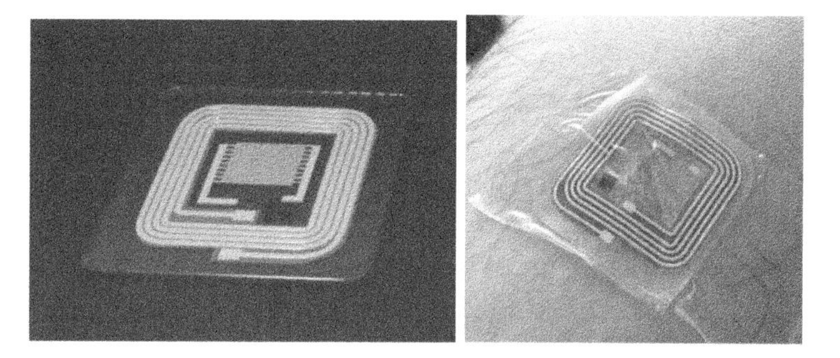

Figure 4.19. *Examples of energy harvesting technologies (source: Grapheal). For a color version of this figure, see www.iste.co.uk/paret/smartpatches.zip*

4.3.2. *Examples of NFC patches with energy harvesting*

In the coming sections, we present two simple examples of patches for smart apparel, using energy harvesting and Out-of-Body Communications (OBC), in NFC:

– Example 1: patches with a specific energy-harvesting device using ISO 15693/ NFC Forum Type 5, connecting to the local biocontroller and its biosensor by means of an I2C.

– Example 2: a simple patch to measure temperature with a monocircuit with energy harvesting, and the biosensor and microcontroller, entirely in keeping with the philosophy of this book, and highly representative of the cases we discuss herein.

4.3.2.1. *Example: energy harvesting via ISO 15693 – Mobile*

Using a mobile phone, OBC and energy harvesting in NFC are based on the use of the NXP – NTA 5332 NTAG 5 boost circuit (see Figures 4.20 and 4.21) – designed around ISO 15693 and the standard/NFC Forum Type V, which includes ALM. Such technologies can be used for very small antennas, with a range of 5 cm, for use in ultra-compact patches for applications in consumer electronics and in industry. In addition, to supplement the patch, this solution requires other circuits to be connected by I2C, acting as biocontroller (analog-to-digital converter) and biosensor, defining the type of sensor desired.

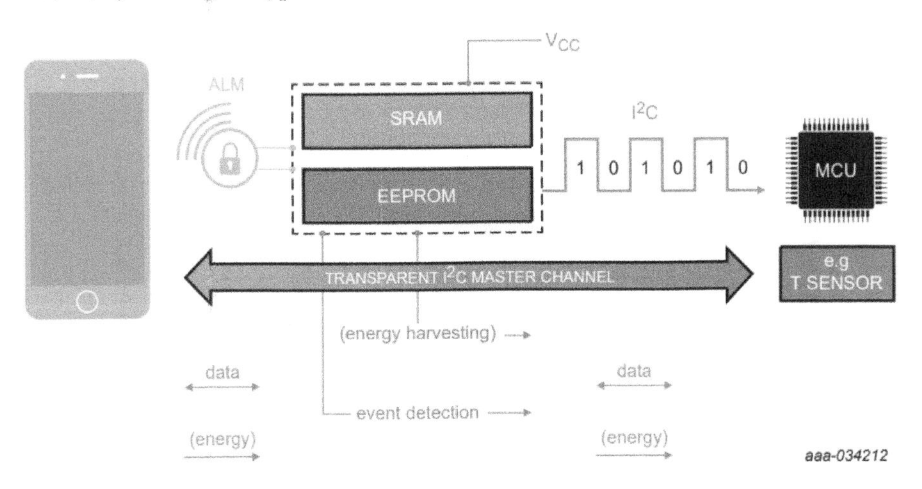

Figure 4.20. *Example of a patch using energy harvesting technology. For a color version of this figure, see www.iste.co.uk/paret/smartpatches.zip*

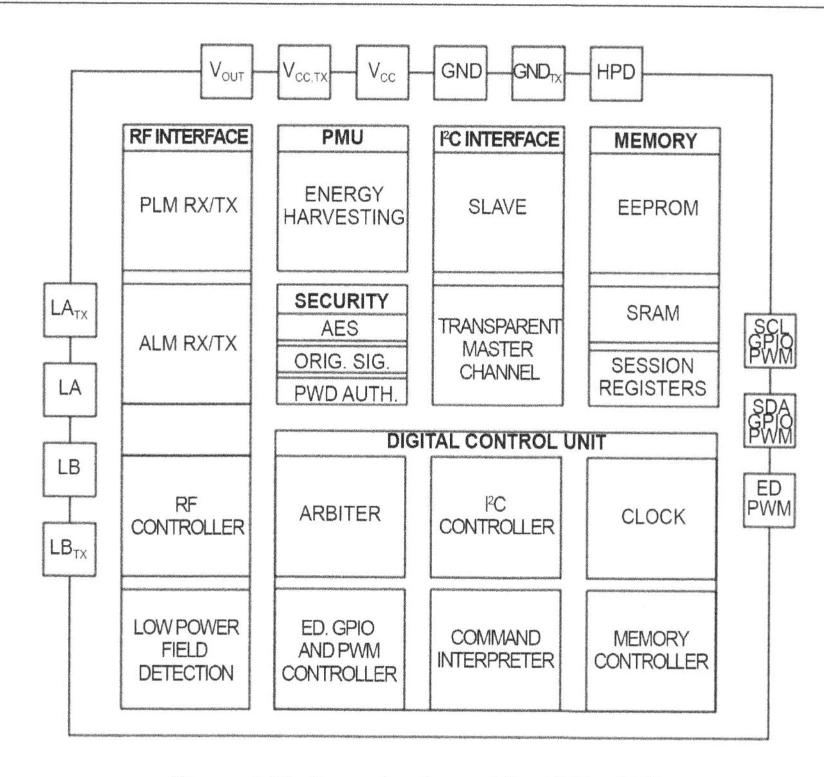

Figure 4.21. *Example of use of the NXP – NTA 5332 NTAG 5 boost circuit*

In this figure, we can see the energy harvesting part, the I2C controller and the terminals L1 and L2 for connection to the antenna.

Similar and/or practically identical approaches are found with the ST 25DVxxK from STm and with the Melexis MLX 90129 circuit.

4.3.2.2. *Example: energy harvesting by ISO 15693 – TAG*

Using a mobile phone, OBC and energy harvesting in NFC are based on the use of the NXP – NTA 533 xxx – designed around ISO 15693 and the standard NFC Forum Type V, which includes ALM. Such technologies can be used for very small antennas, with a range of 5 cm, for use in ultra-compact patches for applications in consumer electronics and in industry. In addition, to supplement the patch, this solution includes the functions of biocontroller (analog-to-digital converter) and biosensor, defining the type of sensor desired.

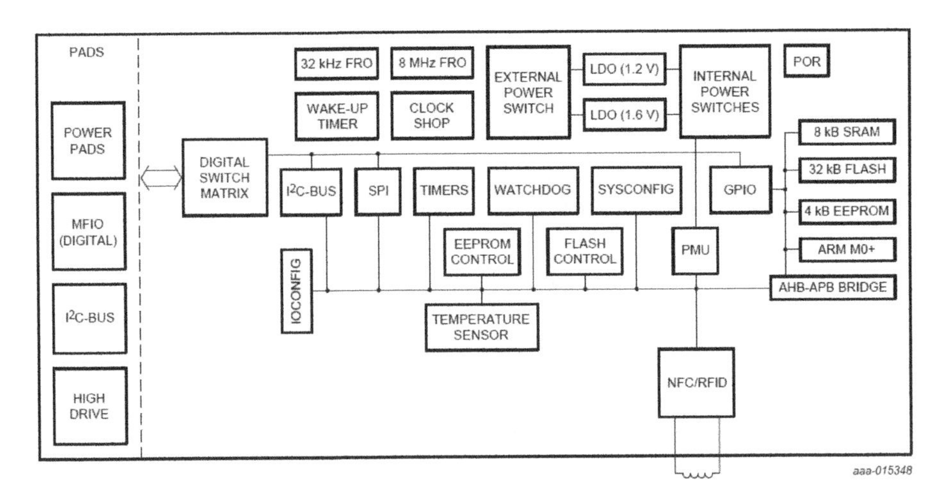

Figure 4.22. *Example of the use of the NXP – NTA circuit*

5

OBC (Out-of-Body Communications) and IBC (Intra-Body Communications) in Smart Apparel

We now turn to a fundamentally important aspect of applications of patches and smart apparel which cannot be overlooked – the silence would be deafening! We speak of the numerous functions relating to connectivity and communications with the outside world and the local environment.

Put briefly, everything that we have mentioned thus far in this book, relating to patches/smart apparel, represents only the visible tip of a colossal iceberg. That tip pertains to data input (by sensors, actuators, etc.) and smart, refined processing of the information measured about a system, an individual, etc. (in the medium term, within the next four or five years, this processing will be performed by the fibers of cloth themselves). The other, submerged part of that iceberg is the sending of those data, by radiofrequency signaling, to centers (smartphones, tablets, servers, server farms, data centers, Big Data centers, etc.), and sophisticated processing for profitable, commercial purposes, dispatching and distribution of data, for pay, through brokers, to end users in business or the general public.

This is then why, from the very outset, we have often drawn a parallel between smart apparel and the IoT. The reason is simple. All data (Big Data) that are collected about a person via all these sensors, and via the Internet, will become Golden Data, which can be turned into cold, hard cash for certain players (insurance companies, pharmaceutical laboratories, for example, for health, etc.). There lies the link between smart apparel and the IoT. Note that really, this is nothing new, and certain companies have been doing it for a number of years (for example, the Big Four).

Obviously, for all this to take place, we need to ensure connectivity (which, in itself, is not easy) between the two parts of the iceberg: the patch and the outside world[1].

5.1. Communications in smart apparel[2]

Now that the information and the measured data are refined, exploitable, present and available in the patch, having been output from the bioprocessor, it becomes necessary to communicate with other patches placed on the same individual or in the same garment, and/or with the world beyond the individual. For example, the same integrated circuit for managing communications could:

– simply exchange their data wirelessly, by NFC, with a smartphone, an NFC watch, etc., and those devices could transmit the data from the patch, via Bluetooth, to the cloud platform where they will be analyzed and stored;

– form an IBC connection with other patches, locally, on the individual;

– the same wireless connection can be used to harvest and/or store the energy needed to power the patch.

To do this, there are two usable RF communication channels: Bluetooth or NFC solutions, sending the data to a relay (a mobile phone) and then to the Internet; a third option is IBC. In this context, though, let us begin with the general definitions of the notions *urbi* and *orbi*[3], which we will have to deal with:

– firstly, for a mixture of communications "around" and "over" the body: Out-of-body Communication (OBC) – see section 5.1.1;

– secondly, for communications "through" or "within" the body: Intra-Body Communication (IBC) – see section 5.1.2.

5.1.1. *OBC*

In the case of smart apparel, remote (*orbi*) communication, referred to hereinafter as *OBC (Out-of-Body Communication)*, as opposed to IBC (Intra-Body Communication), is the most traditional form of communication between a patch

1 This essentially technical chapter is included for the benefit of readers from the world of textiles. To provide a detailed understanding of the issues at stake in relation to connectivity, we present and illustrate a number of types, and characteristic examples, of connectivity in applications for smart apparel.

2 See also Paret and Huon (2017).

3 This terminology comes from Latin: *urbi* means in the city, urban (in other words, local); by contrast, *orbi* means around the globe, the universe, orbit, and around (in other words, remote).

and a specific interface or mobile phone. It uses RF (with a range of between 1 cm and a few meters), with conventional communications protocols, generally managed by the integrated bioprocessor circuit. The two most widely used protocols in the areas of interest to us here are Bluetooth and NFC.

5.1.1.1. *Bluetooth*

This well-known communication protocol serves to connect a patch and a mobile device, serving as an Internet relay.

5.1.1.2. *NFC*

NFC (Near-Field Communication) – operating in HF at 13.56 MHz (mentioned in Chapter 4) and used in telephony/mobile payment (with smartphones) offers a range of benefits for the intended applications of patches, with the notable possibility of two-way communication and remote energy supply to the target by the interrogator, to ensure the patch can capture biometric readings. In addition, the smartphone application provides Internet access and, through its screen, users can view the measurements recorded. From a device such as a smart watch, for example, the readings can be sent via a mobile phone to a dedicated reader. This communication option is the "all-in-one package: cheese course, dessert, coffee and liqueur"!

The amount of HF energy supplied by the NFC interrogator (a smartphone) is sufficient to allow the patches to perform their designated function. Over the course of the few hundredths of a millisecond, or seconds, that the NFC sequence of interrogation and recovery of the desired biometric data from the patch takes, it is possible at the same time to charge, or recharge, the supercapacitor in the patch, by energy harvesting. In France, there are already certain companies with examples of these types of NFC patches on the market (for example, Grapheal).

5.1.2. *IBC*

In smart apparel, local (*urbi*) communication, referred to hereinafter as *IBC (Intra-Body Communication)*, is typically achieved using RF technologies with capacitive methods (and less usually, with galvanic methods), as will be illustrated in this chapter:

– by means of a capacitive connection between the patch inside the garment, serving as a communication relay between a patch and a specific base station or a mobile phone;

– by means of connections between patches (multiple patches, for example, operating in a network), placed on the body, and a local base station on, and in galvanic or capacitive contact with the body (for example, a watch, which can also serve as an *urbi–orbi* relay);

– or by means of direct connection/networking between the patches themselves, arranged on the body.

5.1.3. *Possible means of communication for patches*

Since the very beginning of this book, we have been talking about patches, but what exactly are they? What are their applications? Once again, we need to divide the discussion into two new means of operation: "lone/isolated patches" and "local networks of patches".

5.1.3.1. *"Lone patch" mode*

"Lone patch" mode is a situation in which the user is only wearing one patch, but there is invariably the need for interrogation, to communicate over a certain range, be it short or long.

5.1.3.1.1. Lone mono-biosensor patch

This is the most common case, and the simplest, where a single patch is placed on an individual or on/in a garment, containing only one biosensor – for example, having only one function to perform, with its own energy reserves, and communicating "on request" with an interrogator by NFC or Bluetooth, for instance.

5.1.3.1.2. Lone multi-biosensor patch

The concept of a "lone multi-sensor patch" involves a double biosensor, measuring a combination of body temperature and pH, linked to a piece of management software. The system could even be extended to measure SpO_2, for example. Such a system demonstrates the features and capacities of flexible graphene-on-polymer technology for healthcare/comfort monitoring.

The intended applications require numerous biosensors, of different types. In our case, the envisaged solution combines a range of hardware and software components.

5.1.3.2. *Network of patches*

In other scenarios, it may be necessary to have multiple patches (two, three, etc.) around the body or in the same smart garment, in different places. Those patches may be totally "isolated" from one another, each one working completely independently or in different time slices, or they may need to communicate with one another, and form a "local mini-network" known as a "BAN" (Body-Area Network). In this case, we need to choose the right medium for communication among the patches. These communications may use:

– Hardwired connections: we can discount this solution immediately, as it would be uncomfortable for the users, and fairly primitive.

– Radiofrequency (RF) connections: in principle, this solution may be used. Whether it uses Bluetooth or other technologies, it will require the presence of additional components, which runs counter to our ultimate goal.

– IBC connections: this communication solution, whereby signals are sent through the dermis and/or epidermis of the garment wearers eliminates the problems inherent to the first two solutions listed above, and may, if necessary, solve the communication problems among the various patches. Below, we will examine this solution in great detail, but remain within the bounds of a simple example – for the concrete application of IBC, we can, once again, envisage two types of operation:

- "Galvanic IBC": in this case, the patch needs to be in direct contact with the skin (which means the presence of electrodes, etc.). This solution may be uncomfortable and inconvenient for the user, and we will rarely consider it in this book.

- "Capacitive IBC": in this case, the patch is not in direct contact with the skin, or only makes contact occasionally, and the connection is made by means of capacitive coupling between the patch and the skin. In principle, this is technically more difficult to achieve, but far more advantageous from the point of view of having a non-intrusive device. It is on this type of operation that we will base most of our discussion below.

Figure 5.1 summarizes these possibilities in diagrammatic form.

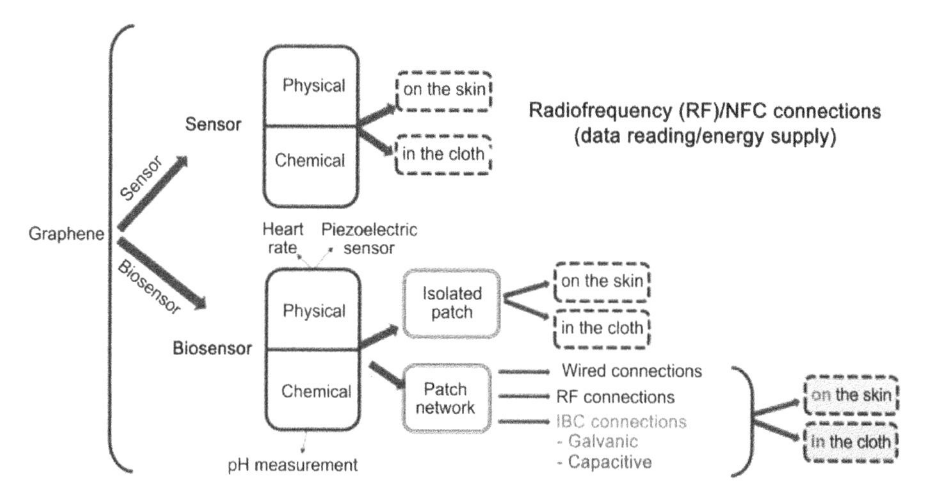

Figure 5.1. *Paradigm trees for graphene-based patches. For a color version of this figure, see www.iste.co.uk/paret/smartpatches.zip*

5.1.3.3. *In summary*

	Urbi – OBC		*Orbi* – IBC	
Type of implementation	On the skin	On/in the garment	On the skin	On/in the garment
"Lone patch" mode	NFC	NFC	NFC	NFC
"Patch network" mode	IBC NCF	IBC NFC	IBC NFC BLE	IBC NFC BLE

Figure 5.2. *Relations between "lone patch"/"patch network" and the functions of OBC and IBC*

5.2. Connectivity and viability of OBC in smart apparel

The recovery of data relating to health, wellness, etc. by means of smart apparel and/or patches equipped with numerous biosensors, operating by OBC, offers a means of collecting patients' health data; processing those data; sending the results obtained to health professionals (e.g. the doctor, the home nurse); raising the alarm in an emergency; facilitating diagnosis in teleconsultation mode; more effectively monitoring patients remotely (telesurveillance); helping patients to correctly follow their own courses of treatment; taking preventative measures in various areas; etc.

Many of the points raised in the list above rely on the patch application, its soundness, cost, profitability, etc. In addition, these points lead us on to other, more pedestrian considerations – in particular, the financial side of patches, also examining complementary and competitive systems referred to as "tele-smart apparel" (these have applications in "tele-medicine", "tele-xxx"). Within the next few years, these will become one of the keystones of this branch of activity. Thus, we must ask ourselves, magnificent as all this is, is it financially viable in relation to competition from other possibilities?

5.2.1. *Viability of OBC connection in smart apparel*

We now come to the point of finance in relation to the market viability of smart apparel and the associated patches.

5.2.1.1. *Financial handling of smart apparel*

5.2.1.1.1. Health insurance

In order for a (connected) smart garment to be covered by health insurance, it must have been prescribed by a doctor, and be on the list of reimbursable products and services (LPPR) as a medical device for individual use. In addition, in order to be able to prescribe a connected device or sell it in a pharmacy, and for it to be covered by health insurance, it must have obtained a "CE label", attesting that it is compliant with the applicable EC regulations. Only some such devices are covered by the social-security system[4]. Increasingly, doctors are recommending mobile apps to their patients, and are also beginning to recommend blood pressure monitors, blood sugar detectors, electronic pedometers, etc. Let us briefly recap some examples of health-related connected objects already on the market: pill dispensers, balances, oximeters, etc. However, in 2019, only 5% of patients were given a prescription for connected objects.

5.2.1.1.2. Health-insurance funds – supplementary insurance

Health-insurance funds and other insurance companies are considering the possibility of reimbursing the cost of such connected objects, because they allow for better monitoring of the patient, and are part of a preventative healthcare approach. Certain formulae used by supplementary health insurance funds already include an annual fee, which covers the purchase of small, connected, medical devices. Very few of these purchases are reimbursed, because aside from demonstrating the added value of the device in comparison to existing comparable products, coverage by health insurance entails the setting of a "responsibility" rate, up to which it will take action.

Such rates are rarely adopted for CE-marked connected smart health devices/apparel, for a range of reasons: the manufacturer has not requested it; the fixed rate is not suitable for them; they do not want to make the investment because the filing of the application requires a €3,220 fee; the application has been denied because the HAS (the French National Authority for Health) or the *Commission nationale d'évaluation des dispositifs médicaux et des technologies de santé* (CNEDiMTS) have failed to recognize that the connected object serves a need or has sufficient added value in comparison to comparable products, considered to be referential, according to scientific data; the application is still being processed: the processing of applications is time-consuming, and it may be difficult to assess the contribution of disruptive connected things, which can slow down the decision-making process.

5.2.1.2. *Teleconsultations and home healthcare staff*

In the interests of exhaustive coverage of the subject, we must also consider the financial cost of a communicating smart garment's application to teleconsultation and home healthcare.

---

4 In particular, these include blood sugar detectors, electrodes, strips and sensors, injector pens, blood clotting measurement devices, continuous positive airway pressure (CPAP) ventilation devices and peak flow meters.

5.2.1.2.1. Teleconsultations

Teleconsultation is billed by the teleconsulting doctor at the same rate as a face-to-face consultation[5]. The means of reimbursement are the same as for a conventional consultation. The rates of coverage by the health insurance provider are the same (70%) or higher if, for example, the teleconsultation relates to a long-standing condition in the context of a healthcare protocol, maternity, etc.

5.2.1.2.2. Home nursing

Similarly, under the French system, there are two types of nursing available at home: AIS (*Actes infirmiers de soins*: Nursing Healthcare Actions): sessions lasting tens of minutes, for assistance with toileting and self-care, for example[6]; and AMI (*Actes médicaux infirmiers*: Nurse Medical Actions): technical actions such as dressing wounds, removing staples, drips[7], etc.

5.2.1.3. *Personal health data… and the sale of them!*

The economic model of the value chain and the business model for these applications show that one of the main aims of the smart apparel project is to collect health-related data so that the user can track the evolution of some of their vital signs in real time… but also to serve the project's financial interest, through the sale of the health data!

To recap, article L.1111-8 of France's Code of Public Health addresses the case of the platforms on which personal health data are hosted, and stipulates that "For the purposes of the activity of prevention, diagnosis, care or social and medico-social monitoring, *any sale of health data, identifiable either directly or indirectly, including with the agreement of the person in question, is prohibited*". Thus, sly as they are, web giants "borrow" data and "rent" them in anonymized form to organizations such as Google with Ascension, a company which works with data from 150 hospitals and 50 elderly homes in 21 US States.

Consider another example. Since 2017, the company Embleema has been marketing technology based on blockchain, designed to allow illness sufferers to share personal information with researchers. The pharmaceutical labs registered on the platform, called "Blockchain santé" (Blockchain Health), can ultimately buy health data directly from the patients. In concrete terms, the patients can commoditize their information, in "anonymized" form, for crypto-tokens (which can

5 In 2020, between €23 and €58.50 depending on the speciality and the sector in which the doctor practises (sector 1, sector 2).

6 An AIS is charged at €2.65 per 10 minutes.

7 For an AMI, the base rate is €3.15.

be cashed in for real money). The types of data involved include: their numbers of daily steps, information about their genomes, or even their medical history[8]. The platform allows the patient to set the price for the sale of their own data, the organization to which they wish to sell the data (labs, healthcare centers or research institutes), and whether or not to allow the use of algorithms. At present, depending on the type of condition, a medical file could sell for between $6000 and $25,000, for the person selling it! "Business is business!" Then, the patients' data are sold on by the companies that buy them, to health professionals such as hospitals, who "anonymize" the data and then sell them in turn to pharmaceutical laboratories. In France alone, this market represents around €200 million. "Again, business is business!"

Other companies use artificial-intelligence solutions in healthcare or, without actually trading them, process data from hospitals and doctors, pertaining to hundreds of thousands of patients.

5.2.1.3.1. Some additional constraints

In addition to the long list of purely technical and regulatory constraints which we listed in Chapter 1, there are certain points specific to the GDPR which must not be forgotten, in relation to the following areas of activity: healthcare, leisure and wellness, which must be assessed on a case-by-case basis; system security and cybersecurity in relation to these data, for example in the implementation of a platform for telesurveillance or harvesting of personal data about the patient by mobile terminals. In drawing up a security policy, we must be vigilant against data theft and diversion in the context of health, wellness or leisure applications. Hygiene constraints must be observed when placing patches directly on wounds, and we must be cautious of the impact of electrosensitivity on the Hub, etc.

In the wake of these generic remarks, and having now built up an awareness of this technical-financial environment, let us return to our examination of OBC communication techniques for smart apparel.

5.3. From the RF-connected world to OBC in smart apparel

We will now discuss the way in which a patch or a smart garment communicates with the outside world (OBC) (via an NFC, BTLE mobile phone, etc.) with deported higher-level applications. This chapter is technical, but this is the price to ensure that the creators, designers, manufacturers, etc., of connected smart apparel are fully

8 In France, today, it is only possible to share data from connected bracelets linked to the system.

aware of what to expect, and what is behind the expression "RF connectivity and its concrete applications to smart apparel".

Smart and connected apparel uses RF communications. This book is not devoted exclusively to RF (for that, see Paret (2008)), but is a book about patches and smart apparel for people working in that area. For that reason, we have decided to include this somewhat more detailed instructional section to go some way towards plugging the gap.

5.3.1. *The absolute fundamentals of RF*

In order to gain a fuller understanding of the rest of this chapter, and of the underpinnings and applications of RF technology in smart apparel, in the following, there is a brief recap of the technical fundamentals that are essential in estimating the possible communication range that we can expect from a device, and the difficulties that we may encounter. For this purpose, we need to have some basic knowledge of a few additional parameters.

5.3.1.1. *Power expressed in W or dBm*

The unit of "transmitted or received power" for an RF wave is normally expressed in watts, but often is also expressed in a derivative unit: decibel-milliwatts (dBm). 0 dBm corresponds to a reference power of 1 mW.

5.3.1.2. *Link balance*

The "link balance" represents the difference (in dB or dBm) between the power levels (EIRP or ERP) of the signal sent by a transmitter and that of the signal received by a receiver. The difference is attributable to transmission losses.

5.3.1.3. *Losses due to transmission and to the environment*

Inevitably, along the path of a signal, it will experience propagation losses due to its journey through the air/atmosphere/the garment, attenuation due to a host of other factors (walls, buildings, etc.), and interference from other signals. Consequently, the level of signal which actually reaches the receiver may be very low indeed. In addition, it is usual to allow safety margins in order to be sure of correct function. The total of these losses is subtracted from the usable link balance, and of course, reduces the true range over which the device is able to communicate.

5.3.1.4. *Regulations*

Of course, the regulations and constraints applying in the field of RF must be respected (see Chapter 1).

APPLICATION TO SMART APPAREL.–

By way of concrete example, in the domain of smart apparel such as PPE for firefighters, airport runway personnel, security agents, and for road traffic and military operations (police, gendarmerie), GIGN, DST, etc., may be in operation over distances of 1-2 km, or even more on their bases (forest fires, runway excursions, etc.) and communications must be preserved.

The same is true for telemedicine for people wearing medical jackets, either in hospitals, or living in isolated villages and using the jacket for medical monitoring.

5.3.2. *Long- or short-range RF connectivity in smart apparel*

Sooner or later, smart garments will need to communicate with the outside world to supply, or receive data from their users, either in the way that we saw in Chapter 2, using sensors (whether or not directly integrated into the material of the garment), or by means of RF connections.

In the context of smart apparel, in order to better understand the resources that these solutions provide, we will discuss various applications in detail. For example:

– The case of communicating devices which often seek out local mini-gateways, which can then be used to send data to an Internet network. Example: a communicating patch using NFC/BLE, seeking to pair with a smartphone serving as a gateway.

– The case in which the smart garment communicates using a contactless connection (NFC, BLE), to signal an action, a presence, the sending of data, etc. (applications: secure verification of access rights, usage rights, etc.). In these situations, the communication ranges in air are relatively small (around 10 cm), or even very small (merely a few centimeters for NFC). On the other hand, data sent back by PPE jackets worn by airport runway maintenance personnel may come from a much greater distance (1-2 km). Figure 5.3 shows, as it stands at the time of writing in 2021, the range of technological possibilities which can easily be used, and indicates their functions and typical positions in the "data throughput/operating range" plot.

To supplement this overview, Figure 5.4 shows the most common wireless connectivity technologies: in green, those for short distances (*Close* and *Near-Field*), in orange, those for medium distances, and in pink, those for long distances (*Far-Field*).

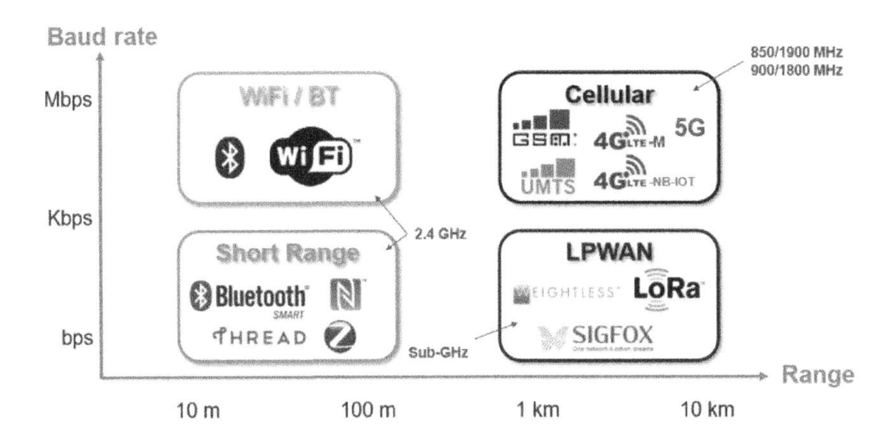

Figure 5.3. *"Baud rate v. range" plot for common connectivity technologies. For a color version of this figure, see www.iste.co.uk/paret/smartpatches.zip*

Network types	Range/Rate	Names	Examples	Carrier frequency in MHz
WPAN WLAN	Short Range	NFC	NFC Forum	13.56
	Short Range Medium Range	Bluetooth	BT LE	2450
		Zigbee		2450
		Thread		2450
		Z Wave		
		Li Fi		Light
		Wi-Fi		2450
MAN WAN	Long Range Low Throughput	LPWAN LTN	SIGFOX, Qowisio	~900
			LoRa, Ingenu	~900
	Long Range High Throughput	LPWAN WB	2G, 3G, 4G	Several GHz
			LTE-M NB-IoT EC-GSM	
	Low Throughput	NB		

Figure 5.4. *The most common connectivity technologies. For a color version of this figure, see www.iste.co.uk/paret/smartpatches.zip*

Let us now give a few details on the subject of communications.

5.3.3. *Short range (SR)*

The left-hand side of Figure 5.3 is referred to as short-range communication – in essence, between 10 and 100 m.

Terms			Order of magnitude of range		
Name	**Sub-terms**	**Data throughput**	**1 to 10 cm**	**1 to 10 m**	**10 to 100 m**
SRLTN	**VSR**	**Near-Field Communication**	NFC		
	SRLTN	**SR low throughput network**		RFID	
SRWB		**SR Wide Band**			Bluetooth
					Zigbee
					Thread

Figure 5.5. *The most common protocols used for SR (Short-Range) technologies*

5.3.3.1. *Very Short Range (VSR) (< 10 cm)*

Very short-range RF connections – with a range of roughly 10 cm maximum – between the "outside world" and smart apparel may include several types: for example, high-frequency (HF) in NFC, ultra high frequency (UHF) in RFID, etc.

5.3.3.1.1. HF (13.56 MHz): very-short-range NFC (1 cm to around 10 cm)

The protocol NFC is well known for its applications and its use in all mobile phones on the market – Android (Samsung, Google, etc.) or iOS (Apple). It offers multiple possibilities for interconnection with the IoT, either directly or indirectly.

APPLICATION TO SMART APPAREL.–

The best-known applications are patches in which NFC is used for connection to the smartphone, in a short-range low-throughput network (SRLTN) (with a range of between a meter and a few meters).

5.3.3.1.2. Short-range UHF and RFID

RFID can operate at various frequencies: at LF at 125 kHz, HF at 13.56 MHz, UHF at 433 MHz and 860–960 MHz, 2.45 GHz and SHF at 5.8 GHz. All of these communication protocols are covered by the ISO 18 000-xx family of standards, and these tags (batteryless patches/smart apparel) can also be supplied with power by means of energy harvesting.

> APPLICATION TO SMART APPAREL.–
>
> Certain applications (PPE, medical and military equipment) use smart apparel, including connections with remote-powered sensors by energy harvesting from the RF wave emitted (often at UHF) by RFID readers.

5.3.3.2. *Short Range Wide Band (SRWB) (tens of meters)*

For certain applications, it is necessary, or even compulsory, for the reporting of an action, a presence, the sending of data to the apparel, to be done over a distance of only a few meters. In this case, to make itself understood to the "outside world", the smart garment can communicate with it (and vice versa) using Bluetooth Low-Energy, for example, at UHF (900 MHz or 2.45 GHz).

5.3.3.2.1. Bluetooth Low Energy (BTLE)

Bluetooth is a short/medium-range wireless communication technology, which is very widespread, being present in billions of mobile phones, tablets and millions of portable computers on the market, etc. Bluetooth Low Energy (BTLE) is defined as the "low-consumption" version of the conventional Bluetooth standard. BTLE v5.x uses different techniques to ensure very low energy consumption, the main characteristics of which are highly useful in patches and connected smart apparel. This power can, of course, be further reduced to optimize the power consumption or reduce interference with other devices, and allows for a typical operating range (in air) of around 50 meters.

> APPLICATION TO SMART APPAREL.–
>
> These types of circuits can be combined with applications for iOS and Android devices with Bluetooth technology. With very low power consumption in standby mode, they are obviously designed for smart apparel with a BLE connection, such as medical devices, paramedic equipment, wellness, fitness, and sensors of all kinds. In addition to feeding back information, smart apparel and PPE for firefighters, and special apparel for the police and armed forces (Soldat 2020) increasingly need to be interconnected in mesh networks – hence, the use of the Zigbee and Thread protocols.

Unfortunately, as this book is not written to be an encyclopedia on the subject, this is where we will cease our discussion of short-range solutions.

5.3.4. *Medium Range (MR)*

Figure 5.6 offers an overview of medium-range solutions.

Correct names		Approximate ranges
Name	**Throughput**	**10 m to 100 m**
MRWB	**MR Wide Band**	Wi-Fi
		BT v5

Figure 5.6. *Most common protocols for Medium-Range technologies*

5.3.5. *Medium Range Wide Band (MRWB) (around 100 meters)*

Multiple standards are available for Medium-Range (MR) operation (over around 100 meters) in wide band. Of these, Wi-Fi is by far the most widely used.

5.3.5.1. *Wi-Fi*

Wi-Fi is a set of radiofrequency communication protocols governed by the IEEE 802.11 family of standards, describing the features of a wireless local-area network (WLAN – Wireless LAN). A Wi-Fi network can be used to connect multiple smart garments apparel in order to establish a link and allow high-throughput data transmission among them. The indoor operating range is up to a few tens of meters (generally between 20 and 50 meters). Thus, it is possible to establish a Wi-Fi network that can connect to the Internet in an area with a high concentration of users (examples include PPE smart apparel: construction sites, airports, hotels, trains, hospitals, etc.). These zones or access points are called Wi-Fi terminals, Wi-Fi access points or hotspots.

APPLICATION TO SMART APPAREL.–

Numerous applications in PPE and professional smart apparel (for firefighters, civil security, medical devices, professional sport trackers, etc.).

5.3.5.2. *Bluetooth v5.0 (BTLE)*

Bluetooth version 5.0 (Bluetooth Low Energy) provides many improvements over previous versions – in particular, increased output power.

APPLICATION TO SMART APPAREL.–

PPE, and certain wellness and sporting equipment.

5.3.6. *Long Range (LR) and Far-Field*

Let us now look at the right-hand side of Figure 5.7, which gives an overview of Long Range – Far-Field solutions.

<table>
<tr><th colspan="3">Correct names</th><th>Approximate range</th></tr>
<tr><th>Name</th><th>Sub-appellations</th><th>Throughputs</th><th>1 m to 10 km</th></tr>
<tr><td>LRWB</td><td></td><td>LR Wide Band</td><td>LTE-M</td></tr>
<tr><td rowspan="3">LRLTN</td><td>LTENB IoT</td><td rowspan="3">LR low-throughput narrow band</td><td>LTE NB IoT</td></tr>
<tr><td>LRLTN
NB
DSSS</td><td>LoRa</td></tr>
<tr><td>LRLTN
NB
FHSS</td><td>SigFox, Qowisio, etc.</td></tr>
</table>

Figure 5.7. *Most common protocols for Long-Range technologies*

Certain requirements for long ranges, low throughputs and low consumption have led to the emergence of new techniques and technologies. These solutions, which deliver very low power consumption and low cost, are often grouped together under the umbrella term LPWAN (Low-Power Wide-Area Networks). Sooner or later, we must establish one-directional and/or bidirectional RF connections to communicate with smart apparel in applications in which they are generally a (very) long way apart (Long Range), and speak little and infrequently (thus, LTN), so perhaps narrow-band (NB) or even ultra-narrow-band (UNB) or another solution, and must consume little energy (low power).

APPLICATION TO SMART APPAREL.–

This solution is highly practical when we need to feed back patient information (to family members, doctors, nursing assistant, family carers, etc.) in long-range applications. In PPE, an outdoor temperature sensor used in a heated parka jacket regularly sends temperature readings to a remote control center. Once it has sent data, it decides to remain actively listening for a few seconds (for example, at most four times per day), so as to be able to receive an instruction to increase or lower the jacket's temperature.

5.3.6.1. *LTE, LTE-M and 5G*

LTE (Long Term Evolution), which is more usually called "4.5G or 5G", is a very high-throughput technology for mobile communications. The specifications of LTE-M NB IoT (Narrow Band-IoT) include IoT "narrow band" (NB) with a reduced bandwidth of 200 kHz and an upload and download speed of around 150 kbit/s.

APPLICATION TO SMART APPAREL.–

The same applications as SigFox or LoRa, with a slightly higher bitrate when necessary. There are applications in PPE or medical equipment.

5.4. Architecture of connected smart apparel chains

Let us now take a look at the overall architecture of a chain for connected smart apparel, communicating over the Internet (medical devices, consumer electronics – PPE), which is complex and in which the smart apparel itself becomes the very small visible tip of an immense submerged iceberg. First, however, we must recap the goal of a smart-apparel application:

– to collect raw or refined data from a range of different sources (e.g. sets of digital or analog sensors) which are remote, and speak little and rarely;

– to assemble, sort, process, etc. all those data centrally in accordance with precise instructions;

– to distribute and feed back the processed data to the relevant end users, in accordance with their needs.

5.4.1. *Technological description of the chain*

Figure 5.8 offers a representative view of the architecture of the overall communication chain, comprising biosensors and smart apparel, smart devices and gateways, back-end datacenters and services, and finally, end users.

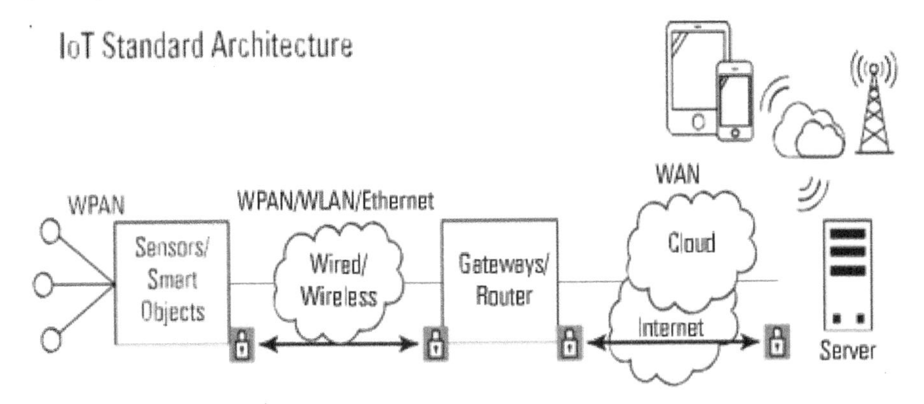

Figure 5.8. *Overview of the architecture of the chain for a smart garment (source: Inside Secure)*

A great many scenarios, sub-scenarios (indeed, an infinite number!) and use cases may be envisaged, and there is no question of looking over all of them in this book. We offer a more detailed examination of the solutions which are general, but can be very easily adapted to readers' own concrete solutions.

Figure 5.9 gives an overview of the general extent of the possibilities for connection with smart apparel, the architecture, a conventional network structure, all possibilities and possible choices of solutions, highlighting the main elements and the four major entities and areas of interest which are as follows: the smart apparel itself, the base stations/gateways, the Cloud Servers and the users, with their control toolkits.

These are not the only options, but the most frequent and most likely.

The chain for connected smart apparel contains a significant ordered succession of events (in the direction of the path taken by the information/data) of the different links in the whole of the chain. This part follows the base signal, the data from the smart apparel.

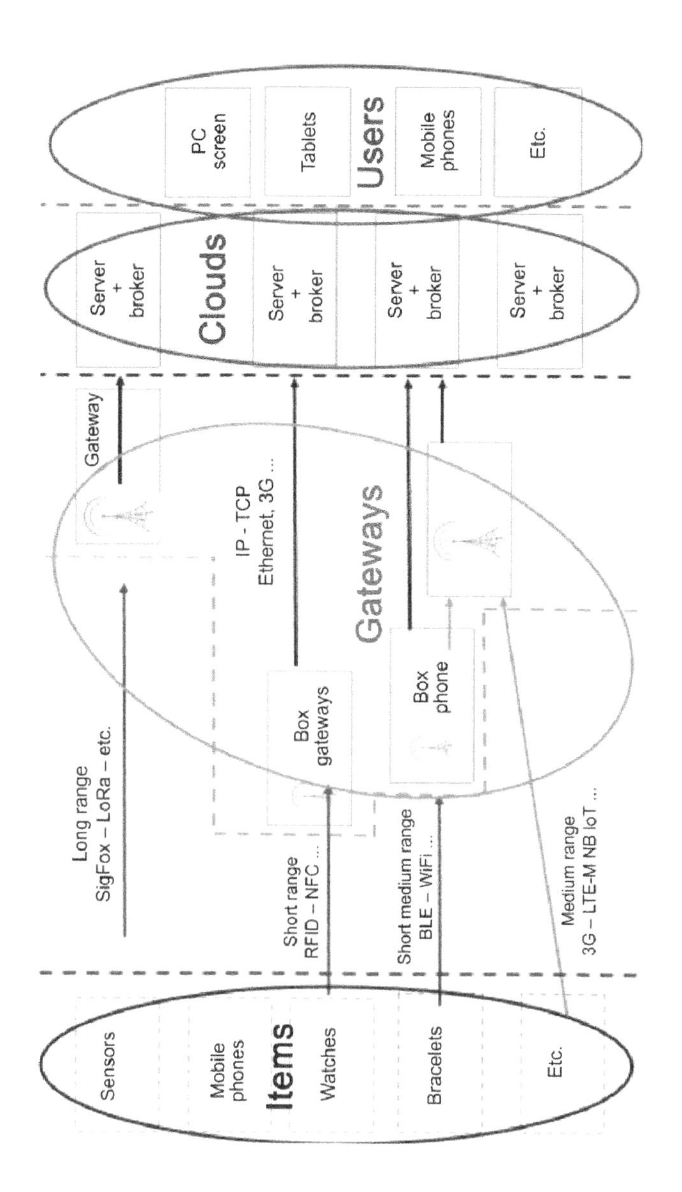

Figure 5.9. *Possibilities and possible choices for solutions. These are not the only ones, but the most commonplace and the most likely. For a color version of this figure, see www.iste.co.uk/paret/smartpatches.zip*

5.4.1.1. *The smart apparel zone*

This is the zone containing the clothing items, patches, wellness equipment, sport, PPE, in which the creation and/or basic information detection and/or preliminary checks are carried out. They are sometimes designed without the need for a direct Internet connection, and normally deported by means of RF connections, either short range (from a few centimeters to a few meters – for example: NFC, BTLE) or long range (several kilometers: Sigfox, Lora).

The description of the connected smart apparel itself can be divided into two distinct main parts: firstly, a part which runs from the garment to the outside world – i.e. the way in which a "user" communicates with the smart apparel (base station, NFC mobile phone, etc.); then a part which deals with the means of communication implemented by the garment to reach a gateway, depending on the distances from it (Short Range, Medium Range or Long Range), and on the types of communication networks we wish to use to reach those gateways (BTLE, Zigbee, Wi-Fi, SIGFOX, LoRA, GSM, LTE-M NB IoT, etc.).

5.4.1.2. *The gateway zone*

The connection from the smart apparel from the network/cloud is generally made via the Internet through an intermediary gateway, which mainly receives messages from the patches/clothes, and whose function is to manage and convert between protocols (networks, transport, sessions, etc.), to serve as a relay, a connection, and then to transfer the data to a collection network, either via a terrestrial system based on Ethernet networks, via a GSM cellular telephone connection, or via any other wireless telecommunication system. Typically, the gateways are connected to a network server using standard IP connections.

5.4.1.3. *The server zone*

In the context of PPE and medical smart apparel applications, the network server and all its potential variants have the main roles of managing the network to ensure it works correctly, using the most appropriate protocols for the exchanges, organizing acknowledgments of the messages, taking measures to eliminate duplicate packets, adapting the datarates to the different peripheral devices, managing the end-to-end security of the data transport, refining the data if necessary, storing the data, and above all, finally, distributing the data to the various users/subscribers through a broker.

Currently, the major providers of computer services, on behalf of end users, and on a paying basis, handle the hardware and software maintenance of the servers, security (e.g. in the area of certain medical garments) and duplication of data on other remote servers, forming a "server farm" or a server cloud.

5.4.1.4. *The cloud zone*

The term cloud, or Cloud Computing, covers all solutions for remote data storage. The data, rather than being stored on hard drives or local memories, are available over the Internet, from remote servers. The various providers of cloud computing have enormous banks of storage servers, commonly called datacenters.

Thanks to gateways, the pertinent data taken from biosensors are sent to the remote network server in a cloud, whose role is to manage the network, take measures to eliminate duplicate packets, organize acknowledgments, adapt the datarates to the chosen solution if necessary, carry out all or part of the tasks of organization, sorting, selection, distribution of data into the cloud using software running on operating systems (OS). Then comes the organization, processing and distribution of returning useful information to users, and finally, dispatching over various communication channels (GSM, Internet, etc.) to the people authorized to receive the data (patients, doctors, hospitals, etc.), on computers, tablets, mobile phones, etc.

Cloud Computing is divided into three main types, generally known as *IaaS* (infrastructure as a service), *PaaS* (platform as a service), and *SaaS* (Software as a Service). SaaS is a model for the commercial exploitation of software, in which the programs are installed on remote servers rather than on the users' machines. Depending on the user's (client's) skill level, they may manage the server themselves or content themselves with using remote applications in SaaS mode – the clients do not pay for a user license for a version, but can freely use an online service or, more usually, pay for a subscription. The main applications of this model are, for example, customer relationship management (CRM) and feeding back information to the clients and/or end users.

These services are designed to help clients/organizations to evolve more quickly, reduce their computing costs, manage their infrastructure without compromising on scalability, security or reliability, and scale up their applications for market. It is in this direction that the field of smart apparel professional applications is developing (i.e. PPE and medical gear), and it is in this space that computer developers calibrate all the ramifications of the range of applications which mark the specific market offer.

5.4.1.5. *The broker zone*

The term broker refers to the intermediary between two parties in a transaction. The function of a broker used and necessary for a smart apparel/cloud structure is designed to support what is a very high upload workload, through the numerous small messages from the smart apparel, and also to serve as a broker, to disentangle and distribute all these data/input to specific outputs for actions. The primary function of a broker, therefore, is to redistribute the information "published" by "sensors" to all "subscribers" to that data (see Figure 5.10).

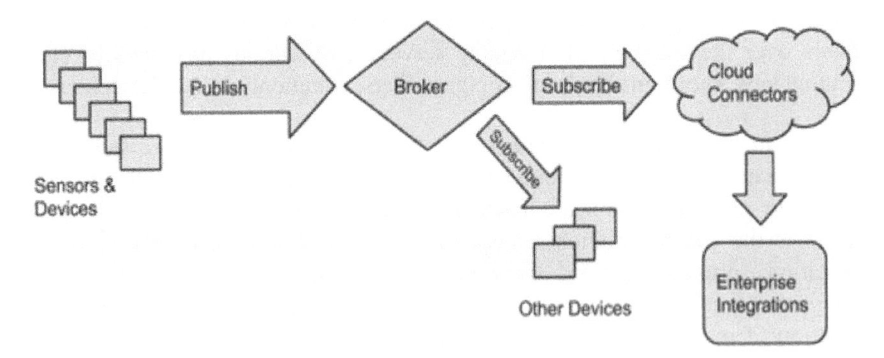

Figure 5.10. *Architecture of a broker (source: iotworld.com)*

5.4.1.6. *The user zone*

The user zone includes the return of signals from the server and the cloud to the end users. This last link in the chain for connected smart apparel is one of the most important, because it is here that the canvas of software applications unfurls, and commercial applications are created (PPE, medical, sport, fitness, etc.), which businesses like yours can then offer clients in return for cold, hard cash! Indeed, after successfully migrating the relevant data to a cloud server, then comes the work of organizing, sorting, selecting, etc., the data stored very briefly in the cloud, organizing the return of the usable information to the users, and organizing the dispatch/distribution of those data over various communication channels (GSM, Internet, etc.) to people authorized to receive them, on computers, tablets, mobile phones, etc., in short any conceivable "Man–Machine Interface" (MMI) – a screen, a PC, a mobile phone, etc.. Of course, the term "authorized persons" here could, in a business context, be replaced by "customers", along with various suppliers, various IoT managers, initial and end users, who have paid on a contractual basis to receive the data, breathing life into the business).

Thus, for a given user, it may seem that with a remote computer, they can control the actions of the smart apparel or view their data, with the whole network described above being almost transparent. It is in this area that MMIs are found, along with the different OSs that can be used (Linux, Microsoft, etc.). With certain architectures, the whole network described above may be almost transparent, and a single remote computer can control the actions of the smart apparel and view or distribute the data to the various suppliers.

APPLICATION TO SMART APPAREL.–

Cloud Computing offers a simple way to access powerful servers, have capacious storage facilities, and draw upon databases and a wide range of applicative services over the Internet, without having to make major initial outlays on hardware or software. In addition, it means there is no need to waste precious time on managing the hardware/software and spend money needlessly on ensuring they work; it replaces a considerable portion of the functional costs and capital expenditure (CAPEX) with variable operating expenditure (OPEX); users only pay for the computing resources that they consume; they no longer need to pay for datacenters and servers without knowing how they will use those resources; they no longer pay for the maintenance of datacenters, nor even cloud computing suppliers, who own and maintain the network-connected hardware required for these applicational services. Thus, the cloud offers essential operations to a business and quick access to flexible computer resources, at a low cost.

5.4.2. *Big Data*

Smart apparel for medical purposes, PPE, wellness and sports is characterized by the generation of vast quantities of data – commonly referred to as Big Data. We then need to process the uploaded information and extract the pertinent information from the mass of data. Thus, three problems arise: how do we eliminate incoherent data? How do we identify which data are actually useful? How do we make the best possible use of those exploitable data?

The mass gathering of information makes it very difficult to identify infrequent signals, which are said to be "weak". However, very often, ISs (information systems/structures) are incapable of dealing with these vast masses of hierarchical data, and must be fed with summarized, pertinent and qualified information. Unlike traditional Big Data approaches, which manipulate "cold" data (consumer behavior history, visits to a store/site, etc.), certain smart apparel (for health, PPE, etc.) work with "hot" or even "red hot" data, which must be processed almost in real time, and require a response within a few seconds (e.g. a patient going into an epileptic fit; we can easily understand the importance of responding instantly, rather than an hour later!).

APPLICATION TO SMART APPAREL.–

In the market for connected smart apparel (PPE, wellness, medical equipment), so-called Fog Computing is on the rise. For applications at a local scale, which require greater fluidity, stability and speed, along with greater security, local clouds known as fogs offer users easier implementation – for example, in local medical monitoring, and on large-scale work sites with laborers wearing personal protective equipment, etc.

5.4.3. *The numerous protocols used*

As stated a little earlier, taking a bird's-eye view, this architecture can be summarized by Figure 5.11, which shows the main conventional protocols that are used, and their respective positions in the overall architecture of applications.

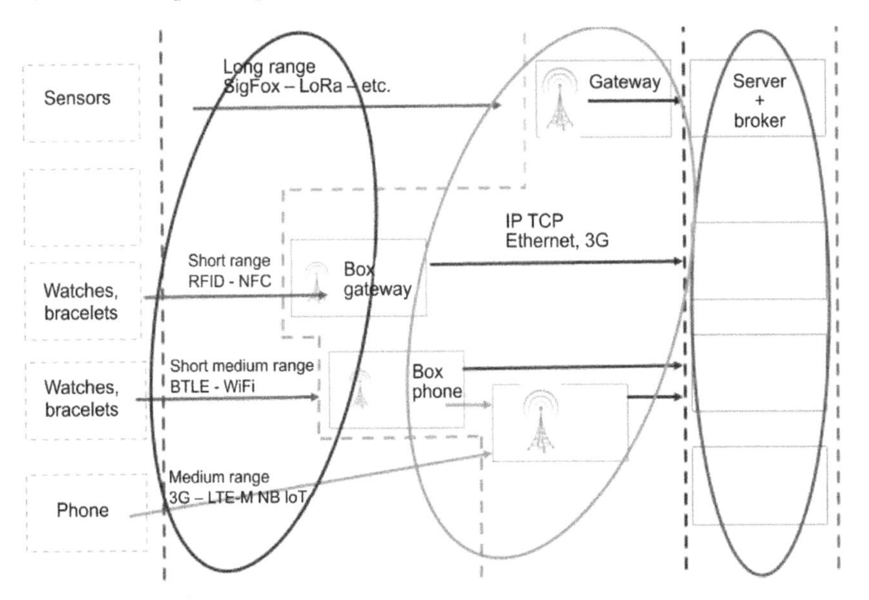

Figure 5.11. *Main connection protocols used. For a color version of this figure, see www.iste.co.uk/paret/smartpatches.zip*

We have now set out the general framework. Let us move on to look at networks of patches.

5.5. OBC and IBC patch networks in smart apparel

New questions, new problems!

It may be that in certain applications, over a certain range, with a certain area (A = Area), when two or more patches (each having single or multiple biosensors) are present on an individual simultaneously (either in the same garment or integrated into the make-up of a garment), we may wish for (or need) them to communicate with one another, forming a mini-network (N = Network). Here, we come to a new adventure, with ANs (Area Networks), to which the next few sections of this book are devoted.

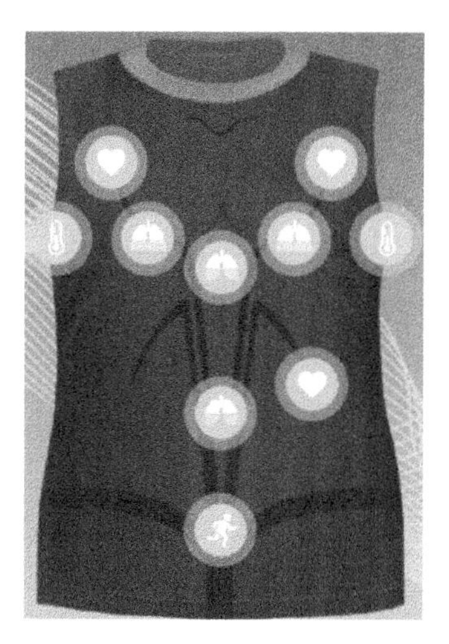

Figure 5.12. *Example of smart apparel with multiple patches and multiple sensors, requiring the patches to form a network (source: Chronolife)*

5.5.1. *The numerous terms in the x AN (x Area Network) family*

Before diving into this section, let us revisit the numerous definitions, which often lead to confusion. Figures 5.13 and 5.14 list the most common "names" and give a rough idea of the usual "ranges" for the main types of x Area Networks.

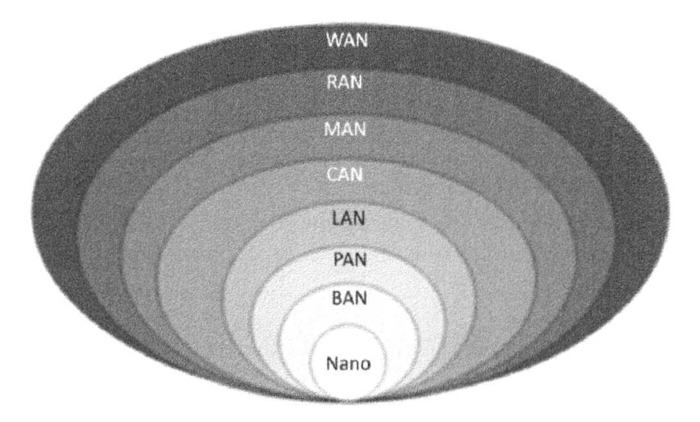

Figure 5.13. *Main types of x Area Networks. For a color version of this figure, see www.iste.co.uk/paret/smartpatches.zip*

Names	Abbreviations x AN	Operating ranges
Nanoscale		
Body	BAN	On or in the body
Near-Field	NFC	10 cm
Personal	PAN	Around the person
Near-me	NAN	Near to the person
Local	LAN	A single room
Home	HAN	A single home
Storage	SAN	
Wireless	WLAN	Wireless, depending on the range of the wave signal
Campus	CAN	Across a campus
Backbone		
Metropolitan	MAN	Across a large city
Municipal wireless	MWN	
Wide	WAN	Large geographic region
Cloud	IAN	
Internet		
Interplanetary Internet		

Figure 5.14. *Definitions of the main types of x Area Networks. For a color version of this figure, see www.iste.co.uk/paret/smartpatches.zip*

In these cases, for reasons of ease, applicative flexibility, financial cost, etc., but also to get around the physical constraints (implantation in hazardous environments, in hard-to-access places, inside the human body, etc.), a wireless connection may be sought. Unfortunately, wireless connections have the disadvantage of being somewhat more energy-hungry to transmit the information. Therefore, we need to find ways of reducing such consumption while still respecting the physical and regulatory constraints.

In the following, we present a brief state of the art, concerning solutions for x PAN, using radiofrequency.

5.5.2. *RF x PAN (x Personal Area Network)*

Three solutions from the IEEE standards offer network structures dedicated to low-energy wireless (W) connections:

– *IEEE 802.15.4* for wireless personal area networks, *WPAN*.

– *IEEE 802.15.431* for low-power wireless personal area networks, *LPWPAN*. This technology is characterized by a range of between a few meters and a few hundred meters, and throughput less than 250 kbit/s. It is optimized for very little use of the medium.

– *IEEE 802.15.6*, which focuses on Body Area Networks, *BAN*.

Figure 5.15 summarizes the main solutions.

<table>
<tr><th>Types of networks</th><th>Range
Network Rate</th><th>Name</th><th>Carrier frequency (in MHz)</th><th>Range</th></tr>
<tr><td rowspan="7">WPAN</td><td>Short Range (SR)</td><td>NFC</td><td>13:56</td><td>1 to 10 cm</td></tr>
<tr><td rowspan="4">Short Range
Low Throughput Network
(SRLTN)</td><td rowspan="4">RFID</td><td>LF at 125 kHz</td><td rowspan="4">1 to 10 m</td></tr>
<tr><td>LF at 450 kHz</td></tr>
<tr><td>HF at 13.56 MHz</td></tr>
<tr><td>UHF between 860 and 930 MHz</td></tr>
<tr><td rowspan="2">Short Range
Medium Range</td><td>Bluetooth
Zigbee Thread
Wi-Fi</td><td>2450</td><td>10 to 100 m</td></tr>
<tr><td>Z Wave</td><td>868 in Europe</td><td>Around 50 m</td></tr>
<tr><td rowspan="2">MAN
WAN</td><td>Long Range
Low Throughput</td><td>LPWAN LTN
(SIGFOX/LoRa)</td><td>Around 900</td><td rowspan="2">1 m to 10 km</td></tr>
<tr><td>Long Range
High Throughput
Low Throughput</td><td>LPWAN
WB
NB</td><td>A few GHz</td></tr>
</table>

Figure 5.15. *AN solutions in the field. For a color version of this figure, see www.iste.co.uk/paret/smartpatches.zip*

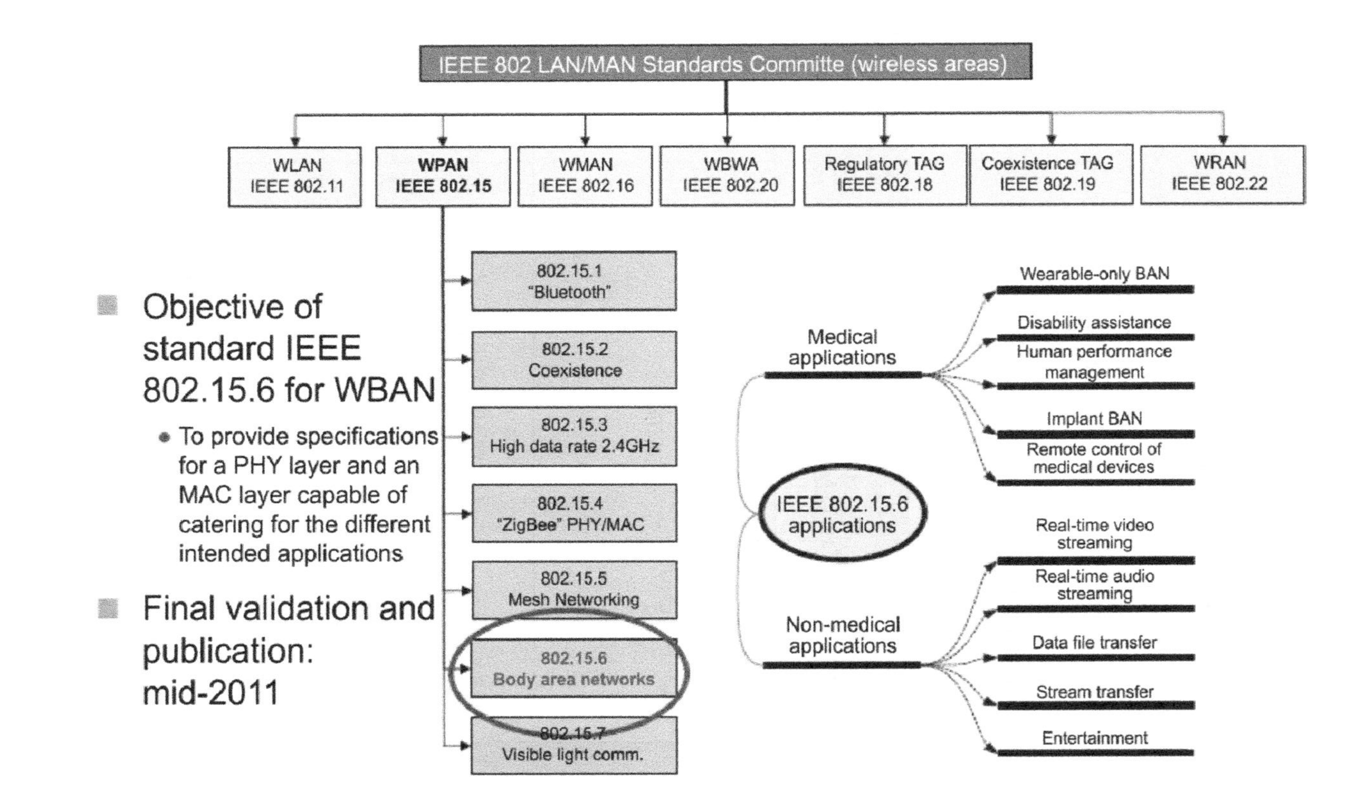

Figure 5.16. *Standards in the IEEE 802.15 family. For a color version of this figure, see www.iste.co.uk/paret/smartpatches.zip*

5.5.3. *From the WPAN (Wireless Personal Area Network) to the WBAN (Wireless Body Area Network)*

Instead of sending information between near peripheral devices via a LAN (or WLAN), there are other wireless RF networking technologies known as "WPANs" (Wireless Personal Area Network), defined by the standards of the IEEE 802.15 family, denoting a type of computer network made up of few hardware elements, designed to cover small areas such as private homes or individual workspaces, and generally implemented in an area of around ten meters. Because of the communication range entailed by the term "personal area network", numerous technologies have infiltrated into PANs and WPANs – e.g. Bluetooth, Zigbee, UWB, etc.

NOTE.– Other names for this type of network are: home networks or private networks, generally used to set up a network of personal devices, such as a laptop computer, a smartphone, a tablet and mobile devices.

5.5.4. *PAN and the IEEE 802.15 family*

Figure 5.16 gives an overview of the IEEE 802 family of standards, and in particular, the WPAN IEEE 802.15.x branch.

Figure 5.17 offers a snapshot, at a specific date, of the overall throughputs, frequencies and ranges of the main wireless standards listed in this section.

<table>
<tr><th>WPAN standard</th><th>Frequency</th><th>Datarate</th><th>Range</th><th>Type</th></tr>
<tr><td colspan="5">802.15.x family</td></tr>
<tr><td>802.15.1
Bluetooth</td><td>2.4 GHz</td><td>3 Mbit/s</td><td>100 m</td><td>PAN</td></tr>
<tr><td rowspan="2">802.15.4
Zigbee</td><td>868/915 MHz</td><td>40 kbit/s</td><td rowspan="2">75 m</td><td rowspan="2">PAN</td></tr>
<tr><td>2.4 GHz</td><td>250 kbit/s</td></tr>
<tr><td rowspan="3">802.15.6
WBAN</td><td>NB</td><td rowspan="3">> 1 Gbit/s</td><td rowspan="3">10 m</td><td rowspan="3">BAN</td></tr>
<tr><td>UWB</td></tr>
<tr><td>HBC 21 MHz</td></tr>
</table>

Figure 5.17. *Datarates, frequencies and ranges of the standards in the IEEE 802.15 family*

Let us now take a look at the 802.15.x WPAN family of standards.

5.5.4.1. *IEEE 802.15.1 – Bluetooth Low Energy (BLE)*

Already mentioned multiple times in this book, *Bluetooth Low Energy* (BLE) technology is designed to wirelessly connect very small devices with limited energy resources.

5.5.4.2. *IEEE 802.15.4 – Zigbee*

Zigbee is a standard which applies to the upper layers of the network and IEEE 802.15.4, and to the physical layers and MAC. This technology, created with the objective of low power consumption in mind, is also used to implement BANs. This standard primarily targets rational energy use in applications relating to home automation, construction and industry, but has also been proposed to meet the requirements of the *Continua Health Alliance* for monitoring in the fields of health and fitness. There are a number of points which have prevented Zigbee's widespread adoption in the field of BANs – in particular, the value of the frequency band used by ZigBee/IEEE 802.15.4 (2.4 GHz), which is considered to be congested by WLAN traffic, and sensitive to interference. Though Zigbee is still used for communications between sensors, the standard IEEE 802.15.6 (dating from February 2012) establishes a new dedicated standard for WBAN communications.

5.5.4.3. *IEEE 802.15.6 – BAN*

We will now continue our tour of the IEEE 802.15 family, with the particular section 802.15.6, which represents a new type of network known as a BAN (Body Area Network), and its wireless version, WBAN.

5.6. BAN

5.6.1. *Definition of a BAN – IEEE 802.15.6*

Let us begin by defining the concept of a BAN, which is very similar to that of a PAN, differing from it only in terms of the range.

The wireless RF network technology of BANs (and WBANs) is defined by the IEEE 802.15.6 standard as a communication standard optimized for low-power devices, which operate *on*, *in* or *around* the human body (but not limited to humans) to serve a wide range of applications (including medical ones), consumer electronics, entertainment technologies and other applications. It involves forming a connection on, around or in the human body, between a set of miniscule devices (biosensors), which can take measurements or actively carry out an action (actuators) in order to monitor the body's vital signs (i.e. biometric data).

NOTE.–

– There are numerous variants which include and use the words *Wireless* and/or *Sensor* – hence the abbreviation WBAN. It should be noted that certain publications indifferently employ the terms WBAN and BAN.

– The Body Sensor Network (BSN), the abbreviation BASN (Body Area Sensor Network) and WBASN (Wireless Body Area Sensor Network) are used to refer to the same thing as a BAN.

Biosensors, which have long battery life and use very low levels of power, can communicate with one another and pass the data collected to a base station located outside of the body using wireless technology, and be connected with a remote service center. For example, the data can then be redistributed in real time to a medical emergency service, a hospital, a clinic or any other users who need them.

In this chapter, in relation to smart apparel, our discussion will now shift to the ranges typical of BANs and shorter.

5.6.2. *History of BANs*

The development of new low-power MAC protocols has always been an important focus of research, and considerable effort has been invested in seeking out and putting forward solutions which could satisfy the needs of WBANs in terms of power consumption. In addition, a number of researchers have advanced the idea that WBAN IEEE 802.15.4 technology could be used for low-throughput applications, but was insufficient for use in applications which require datarates higher than 250 kbit/s.

(W)BAN technology began to be developed around 1995, hinging on the idea of using wireless personal area network technologies to establish communications *on*, *near* and *around* the human body. In 1996, the IBM research labs developed a new technology for PANs, using the conductivity of the human body itself – *in the body* – to transmit data. Thomas G. Zimmerman's publications began to circulate, describing the use of the conductive properties of the human body, in which very weak electrical currents circulate (around a nano-ampere). The first intended applications, therefore, were data exchanges between portable devices, with the user's body serving as a transmission medium (e.g. an anti-ignition/immobilization signal for a car) and automated identification of a person (replacement of payment cards (see the example of Gemplus), and recognition of people for medical purposes).

In 1997, the term '(W)BAN' became part of conversations, and only six years later, the term 'BAN' was officially used to describe systems in which communication takes place entirely within, on or in the immediate vicinity of a human body.

It was in around 2007 that the working group IEEE 802.15, under IEEE 802, set up Task Group 6, whose mission was to focus on technologies for systems of wireless communication which could function on, in the immediate vicinity of, or actually within the human body, and TG6 began work on the specifications for WBAN. Finally, in February 2012, the BAN was defined by the published standard document *IEEE 802.15.6*, as 'a communication standard optimized for low-power devices, which operate *on*, *in* or *around* the human body (but not limited to humans) to serve a wide range of applications (including medical ones), consumer electronics, entertainment technologies and other applications".

IEEE 802.15.6 lays down the specifications for an international standard in wireless communications in the area surrounding the human body for communications with low emissive power, that are short range, extremely reliable and support a wide range of datarates for a variety of applications. It defines the physical layer and the MAC (Medium Access Control) layer, as defined in the OSI model, for BANs in the true sense of the term, and uses the existing frequency bands reserved for ISM (Industrial, Scientific and Medical) applications, and other bandwidths approved by national medical and/or regulatory authorities. In addition, this standard takes account of the effects of the presence of a person (male, female, thin, fat, etc.) on the antennas. In addition, the radiation pattern is shaped to minimize the specific absorption rate (SAR) in the body, and changes in the characteristics when the user moves. Quality of service handling, extremely low power and datarates of up to 10 Mbps are necessary, whilst also ensuring that strict directives on non-interference are met.

Since then, the *IEEE 802.15.6* working group has been charged with designing low-power devices, developing and disseminating them, and supporting the development of (Wireless) Body Area Network applications, the aim being to handle a variety of applications in health monitoring and consumer electronics in real time. The main applications of these types of networks are in the fields of healthcare, first aid, military operations, defense but also entertainment, leisure, sport, ambient intelligence or man–machine interactions, etc. In addition, the increasing demand for biosensors for consumer electronics – in particular in the medical domain and sport/gaming – is helping to drive down the cost of components for BANs.

5.6.3. *BANs in smart apparel*

In a smart garment for healthcare, PPE, sport, work (applications in posture, for example), we may have multiple biosensors at different points within the clothing (jacket, overalls), meaning we need to think about creating a mini-BAN. Let us briefly examine what a BAN requires.

The architecture of a BAN has numerous points in common with IoT architectures, which we have described in detail in two earlier publications[9].

5.6.3.1. *Architectures for the 'communications' in a BAN*

BANs are generally subdivided into three main categories:

– *in-body* (communications with implants and other medical sensors);

– *on-body* (communications between the devices and systems placed on the body);

– *off-body* (communication between the devices and systems which are in proximity to the human body and devices placed on the body).

This means that, in terms of the communication architecture, a BAN is also generally divided into three levels.

The first represents *intra-BAN* communications, referring to exchanges (via RF technology) which take place in the "immediate vicinity" of the human body. Thus, care must be taken to distinguish communications between the sensors themselves and communications from the sensors to a PDA.

These two types of communications are described succinctly in this book. The biosensors (and actuators if they are present) are at the heart of the first functional level, which corresponds to intra-BAN communication.

The second represents *inter-BAN* communications. A BAN rarely operates independently. Inter-BAN communications are those communications between a personal digital assistant and a network access point. This type of communication will be described in this book.

The third represents *out-of-BAN* communications. At this level, service comes into play (e.g. in the healthcare domain, this level represents the telemedicine service run by a hospital). This type of communication will only be described briefly in this

9 *Secure Connected Objects* by Dominique Paret and Jean-Paul Huon, and *Wearables, Textiles and Smart Apparel* by Dominique Paret and Pierre Crégo.

book. Interested readers are invited to refer to the lengthy descriptions given in our book on the IoT.

5.6.3.2. *Topology of a BAN*

The physical topology of a BAN is generally star-shaped, having a single hub at its center. The networks are generally made up of multiple body sensor units (BSUs) and a single body central unit (BCU). Communication frames are exchanges between the hub and the nodes in accordance with the chosen communication protocols.

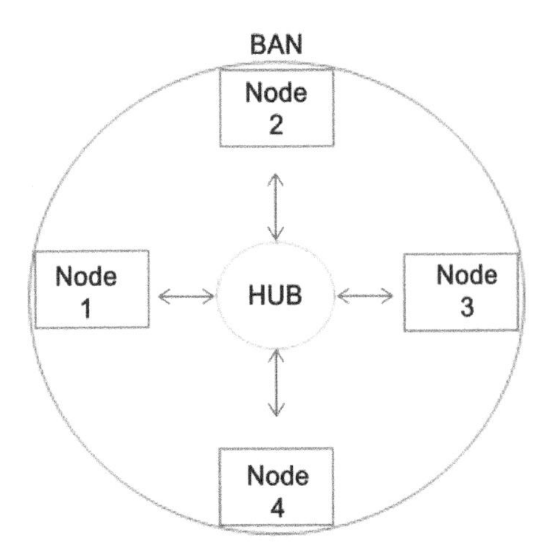

Figure 5.18. *Ordinary topology of a BAN*

BAN devices can be inside the body within implants, project from the body in a fixed position (smart apparel technologies) or accompanied by devices which human beings can carry in different ways, in their pockets, in their hands or in a bag, for example.

A WBAN system may use WPAN wireless technologies as gateways to achieve longer ranges. With gateways, it is possible to connect devices worn on the human body to the Internet. In this way, health professionals can access patients' data using the Internet, irrespective of the patient's position.

5.6.3.3. *Position of BANs in relation to other networks*

Figure 5.19 shows the position of a BAN, in terms of throughput/power consumption, in comparison to the other typical networks found on the market.

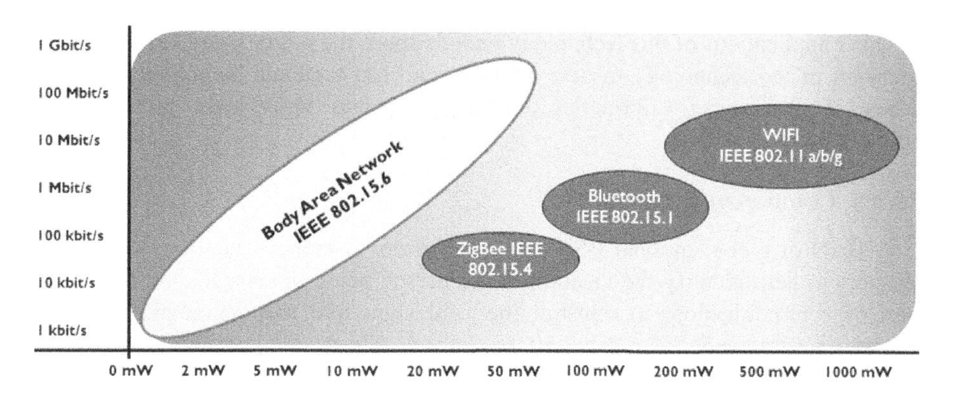

Figure 5.19. *Position of a BAN in terms of datarate and power consumption. For a color version of this figure, see www.iste.co.uk/paret/smartpatches.zip*

5.6.3.4. *Applications*

The rapid rise of physiological biosensors, low-power integrated circuits and wireless communications has paved the way for a new generation of networks of wireless sensors, and BANs seem to have applications mainly in monitoring in the domain of healthcare – in particular, for constant monitoring, alerts and recording of the vital signs of patients suffering from chronic conditions such as diabetes, asthma and heart disease. For example:

– a BAN placed on a patient can alert the hospital, even before they actually suffer a heart attack, by measuring changes in their vital signs;

– a BAN on a diabetic patient allows insulin to be automatically injected by a pump when their insulin level drops.

Body-area networks represent an interdisciplinary domain which allows for low-cost constant health monitoring, with real-time updates of medical files over the Internet. A number of smart physiological sensors can be integrated into a portable wireless BAN, which can be used for computer-assisted rehabilitation or for early detection of medical conditions. Sensors implanted on or in the human body record various physiological changes in order to monitor the patient's state of health, wherever they may be. The information is transmitted wirelessly to an external processing unit. This device instantly transmits all the information in real time to doctors anywhere in the world. In case of emergency, the doctors can immediately alert the patient through the computer system, by sending messages or triggering appropriate alarms. This has led to the concepts of telemedicine and mHealth.

Other applications of this technology include sport, the military and security. The extension of the technology to new domains can also facilitate communication by means of transparent information exchanges among individuals, or between individuals and machines.

5.6.3.5. *Components of a BAN*

A BAN or a conventional BSN requires sensors to monitor vital signs, motion detectors to help identify the location of the person being monitored, and a certain form of communication, to transmit the vital signs and motion readings to the doctors or caregivers. A typical kit to set up a body-area network comprises biosensors such as EKG sensors, SpO_2 sensors, a blood pressure sensor, an EEG sensor, etc., and a microprocessor, a transceiver, and a power-supply unit.

5.6.3.6. *Technical challenges posed by a BAN*

The problems connected to the use of this technology include certain performances, which we will discuss below.

5.6.3.6.1. Quality and data management

Once the BAN has managed to collect large volumes of data, they can play a key role in the process of patient care. It is essential to ensure the quality of those high-level data so that the decisions made are based on the best possible information, and there is a need to manage and maintain these datasets, which are extremely important.

5.6.3.6.2. Data consistency management

The data hosted on multiple wireless mobile devices must be collected and analyzed transparently. In BANs, the essential datasets for the patients may be fragmented across a number of network nodes or portable computers. If a doctor does not have access, from their mobile device, to all known information, then the quality of care provided to the patients may suffer.

5.6.3.6.3. Validation of biosensors

The omnipresent detection devices are subject to constraints inherent to the communication and the hardware, including unreliable connections (be they hardwired or wireless), interference and limited power stores. This may lead to the sending of corrupted data to the end users. It is of prime importance – particularly in healthcare – that all readings from the biosensors are validated. This helps reduce false alarms and identify possible weaknesses in the hardware and software design.

5.6.3.6.4. Consistency of performance

WBANs must perform consistently. The measurements taken by a sensor must be accurate and calibrated, even when the WBAN is switched off and then on again. The wireless connections must be robust and function in environments with many users.

5.6.3.6.5. Security

Considerable efforts are needed to make WBAN transmission secure and accurate. We must ensure that the patient's "secure" data come only from the dedicated WBAN system for each patient, that they are not mixed with other patients' data, and that there is secure and limited access to those data. Privacy, authentication, integrity and data freshness, along with availability and secure management, are security requirements in the field of WBANs.

5.6.3.6.6. Interoperability

WBAN systems must provide homogeneous data transfer between standards such as Bluetooth, ZigBee, etc., in order to promote the exchange of information, be evolutive to ensure efficient migration between networks and provide uninterrupted connectivity.

5.6.3.6.7. Invasion of privacy

If applications go beyond "secure" medical usage, WBAN technology may be felt to be a potential threat to individual freedoms. Social acceptance is essential when we are seeking a larger-scale application of this technology.

5.6.3.6.8. Interference

The wireless link used for on-body sensors reduces interference and increases the coexistence of sensor nodes with other network peripheral devices available in the environment.

5.6.3.6.9. Non-intrusiveness

The WBAN must be portable, lightweight and unintrusive. It must not alter or encumber the users' day-to-day activities. Ultimately, the technology must be transparent to the user, meaning that it needs to be able to carry out its assigned monitoring tasks without the user being aware of it.

5.6.4. *Physical layer of a BAN*

The primary function of the PHYsical layer is to transform the "Physical Layer Service Data Units" (PSDUs) into "Physical Layer Protocol Data Units" (PPDU). For this purpose, the standard IEEE 802.15.6 introduces three different physical layers for WBANs, based on the specific requirements of the applications:

– the physical layer *NB PHY*, which is optional, for narrow-band (NB) communications;

– the physical layer *UWB PHY*, which is compulsory in ultra-wide-band (UWB) communications;

– the physical layer *HBC PHY*, which is compulsory in human-body communications (HBC).

In order to send binary data over a wireless medium, we must use a digital/analog modulator.

In the above three standards, the PPDU structures vary (see Figure 5.20), as do the types of modulation. This alters the performance, throughput and range of the WBAN.

5.6.4.1. *Narrowband (NB) Physical layer*

The narrowband physical layer NB PHY is designed to handle communication from sensors worn on or implanted in the human body. It works primarily by activating and deactivating the radio transceiver, CCA (Clear Channel Assessment) and transmission and reception of data over the networks. It also provides a means of converting PSDUs into PPDUs. PPDU structures in the NB PHY comprise three main components:

– Physical Layer Convergence Protocol (PLCP) Preamble: the preamble is added to assist in packet detection by the receiver (synchronization and carrier offset recovery), and the header includes the necessary information for the receiver to decoder the PSDU;

– PLCP header;

– PSDU.

The whole NB communication system handles the sending and receiving of the data at the frequencies presented in Figure 5.21, in which 230 channels have been defined across seven operating frequencies. Due to the high power of NB communication, this option is generally used for a BAN.

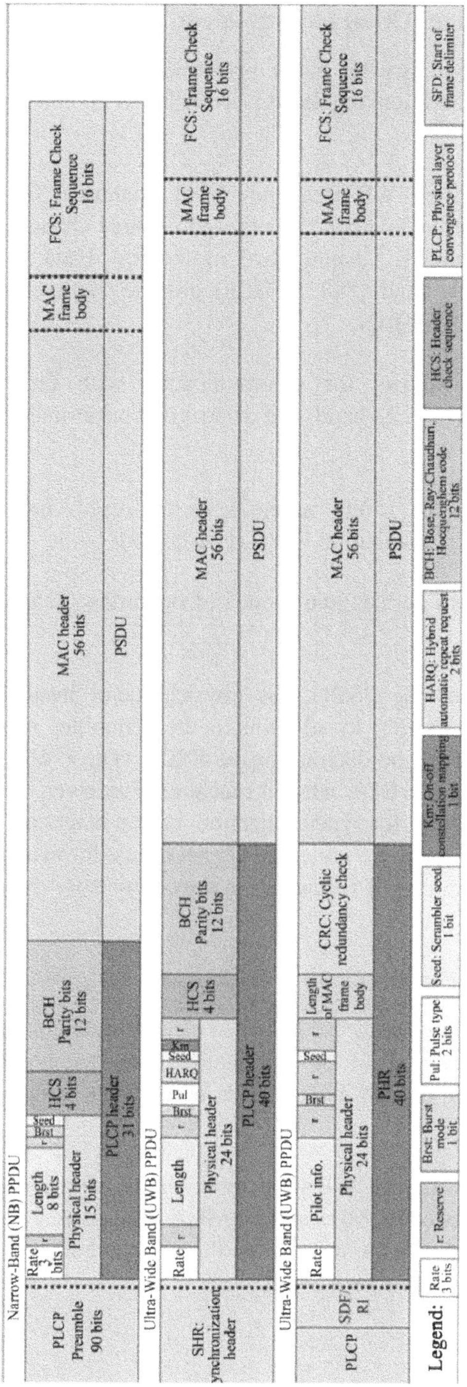

Figure 5.20. *PPDU structures for the three standards. For a color version of this figure, see www.iste.co.uk/paret/smartpatches.zip*

5.6.4.2. *Ultra Wide Band (UWB) physical layer*

UWB offers the possibility for better performances, little complexity and low power consumption. In addition, this band is available in all countries, barring a few exceptions.

In IEEE 802.15.6, the UWB frequency band, spanning 3.1 to 10.6 GHz, is divided into two categories: low band, composed of three channels, and high band, composed of eight channels. Channel two in the low band (3993.6 GHz) and channel six in the high band (7987.2 GHz) are the main compulsory channel frequencies used in UWB WBAN.

In addition, UWB allows for communication with the medical implant communication systems (MICS) band, and offers good performances with extremely low power.

Two different types of UWB technologies are used: impulse radio UWB (IR-UWB) and frequency modulation UWB (FM-UWB).

The specification also defines two modes of operation: default mode and high QoS mode.

Each PPDU includes the PSDU, the physical layer header (PHR) and the synchronization header (SHR). In addition to the scrambler initializer, the PHR provides information about the datarate in the PSDU. The UWB system uses data from the PHR to decode the PSDU when it reaches the receiver. The SHR is divided into the preamble (necessary for synchronization, packet detection and carrier offset recovery), and the start frame delimiter, SFD (necessary for synchronization of the frames). The nominal bitrate is 1 Mbps, but in the compulsory channel, it is reduced to 0.4882 Mbps.

5.6.4.3. *Human Body Communication (HBC)... or the premise of IBC*

HBC is the first technology to use so-called electric field coupling (EFC), and is therefore sometimes referred to by that name. This technology contains two categories, which we will discuss in detail later on: galvanic coupling and capacitive coupling.

The standard IEEE 802.15.6 WBAN provides a specification for HBC, using a range of frequencies between 5 MHz and 50 MHz. Capacitive coupling is primarily used in HBC to send and receive information through the human body. The datarate in capacitive coupling is higher than 2 Mbps.

5.6.4.3.1. Physical layer HBC

Transmitters (in our case, patches), which are exclusively digital, have no electrodes, no antenna, and no bulky RF module, which makes them easy to transport, and means their power consumption is low. The frequency centers solely on 21 MHz.

5.6.4.4. *Summary of NB, UWB and HBC frequency bands*

Figure 5.21 summarizes the frequency bands and bandwidths for the different physical layers in accordance with IEEE 802.15.6.

<table>
<tr><th>IEEE 802.15.6</th><th colspan="2">Frequency (MHz)</th><th>Band (MHz)</th><th>Band</th><th>Modulation</th></tr>
<tr><td rowspan="7">Narrowband (NB)</td><td>402-405</td><td>10 channels</td><td>300</td><td>MICS</td><td>M-DPSK</td></tr>
<tr><td>420-450</td><td>12 channels</td><td>300</td><td rowspan="2">WMTS</td><td>GMSK</td></tr>
<tr><td>863-870</td><td>14 channels</td><td>400</td><td rowspan="5">M-DPSK</td></tr>
<tr><td>902-928</td><td>60 channels</td><td>500</td><td rowspan="4">ISM</td></tr>
<tr><td>956-956</td><td>16 channels</td><td>400</td></tr>
<tr><td>2360-2400</td><td>39 channels</td><td>1</td></tr>
<tr><td>2400-2438</td><td>79 channels</td><td>0.5-1</td></tr>
<tr><td rowspan="2">Ultra-Wideband (UWB)</td><td colspan="2">3200-4700
3494.4
3993.6
4492.8</td><td rowspan="2">11 channels
(central
frequency)
499.2 MHz.</td><td rowspan="2"></td><td rowspan="2"></td></tr>
<tr><td colspan="2">6200-10,300
6489.6
6988.8
7488.0
7987.2
8486.4
8985.6
9484.8
9984.0</td></tr>
<tr><td>Human Body (HBC)</td><td colspan="2">16
27</td><td>4 MHz is used in South Korea, the USA and Japan.
The 27 MHz band is valid in most European countries.</td><td></td><td></td></tr>
</table>

Figure 5.21. *Frequency bands and bandwidths for the different physical layers according to IEEE 802.15.6*

5.6.5. *MAC (Medium Access Control) layer*

The MAC layer is responsible for preparing for the transmission and receipt of frames. In this layer, a node must send a connection request frame to a hub, and the hub must send a connection assignment frame in order for the node to be connected to the hub.

In numerous applications, a WBAN is required to operate for months or more without any intervention, whereas others may require a lifespan of only a few tens of hours, given the nature of the applications. In each context, the most important characteristics of the MAC layer protocol are security and energy efficiency.

5.6.5.1. *Security management and collision management*

To facilitate and secure exchanges between the hub and the nodes, the IEEE 802.15.6 standard provides for a key-exchange mechanism.

When a collision takes place due to the fact that multiple data packets are transmitted simultaneously, this layer helps optimize network communication, which plays an important role in the energy consumption of the whole system. For example, in terms of managing response times, frame acknowledgements, collision management and inactivity:

– the colliding packets are then retransmitted, which entails additional energy consumption;

– the second source of energy wastage is listening mode, because each node actively listens on a free channel to receive data;

– the third source of wastage is over-listening – that is, listening to packets that are addressed to other nodes;

– the final source of wastage is excessive transmission of control frames, meaning that control data are added to the usable payload of the messages.

The MAC protocols conventionally used are CSMA/CA (Carrier Sense Multiple Access/Collision Avoidance) and TDMA (Time Division Multiple Access).

5.6.5.2. *Other important points*

The development of BANs, in implantation inside the body (MICS) and outside the body, is favored, thanks to the data link layer, and the frequency bands used (ISM or UWB).

This protocol must take account of the electrical properties of the human body, morphological diversities and the different types of nodes. It must also be able to deal with changes in network topology, the position of the human body, and the density of the nodes. Processing must be quick and reliable. This protocol needs to be highly evolutive and highly adaptable to changes in the network, time of passage, throughput and bandwidth used. Thus, it is energy-efficient and flexible to the different operating techniques, which is necessary in order to reduce listening time, packet collisions and problems of packet control in general. The following are different types of protocols that are in place or have been proposed for BANs: IEEE 802.15.4; H-MAC: Heartbeat-Driven MAC Protocol; DTDMA: Reservation-Based Dynamic TDMA Protocol; Frequency-based DTDMA; Preamble-based TDMA; and BodyMAC Protocol.

Power-saving mechanisms play an important role in the performance of a good data link layer. They include LPL (Low-Power Listening), SC (Scheduled-Contention), and TDMA (Time Division Multiple Access), which are continuously analyzed/discussed in the context of BANs. However, TDMA is considered the most reliable and energy-efficient protocol.

5.6.6. *Fields of application for BANs*

Applications in the domains of smart apparel relate mainly to healthcare, wellness, medium- and high-level sports, PPE, monitoring posture, fatigue, hazardous places, measuring radiation, virtual-reality games, etc.

5.6.6.1. *The health and medical domain*

The main use of BANs is in medical and health-related fields, to such an extent that some commentators present this type of network as being devoted exclusively to healthcare. Indeed, the increase in healthcare costs and life expectancy in developed countries is a driver of innovation in healthcare, and one such innovation is real-time monitoring of patients with sensors connected in a WBAN. This way, we can monitor a person's vital signs and, if necessary, alert the emergency services. It is even possible to inject medication using an actuator close to the body.

The use of a BAN and its microsensors also opens the door to new endoscopy techniques with a "capsule", which consists of a micro-camera the size of a pill, originally designed to record images of the digestive tract for subsequent diagnosis. The use of transmission techniques based on UWB now means that we can view the images in real time. In the future, microrobots could explore other parts of the human body using similar technologies.

5.6.6.1.1. "Virtual Doctor Server" applications

In the medical domain, we can also envisage using BANs in the concept of a "virtual doctor". An external server, known as the VDS (Virtual Doctor Server) includes a set of applications that can be used to provide healthcare and first aid. This server may have the following uses: recording the patient's history; providing guidance to the patient or the healthcare personnel on the basis of the vital signs recorded by sensors; calling the emergency services if necessary, allowing the patient to be guided through administering first aid to themselves; and where the patient has lost consciousness, guiding a rescuer through administering first aid.

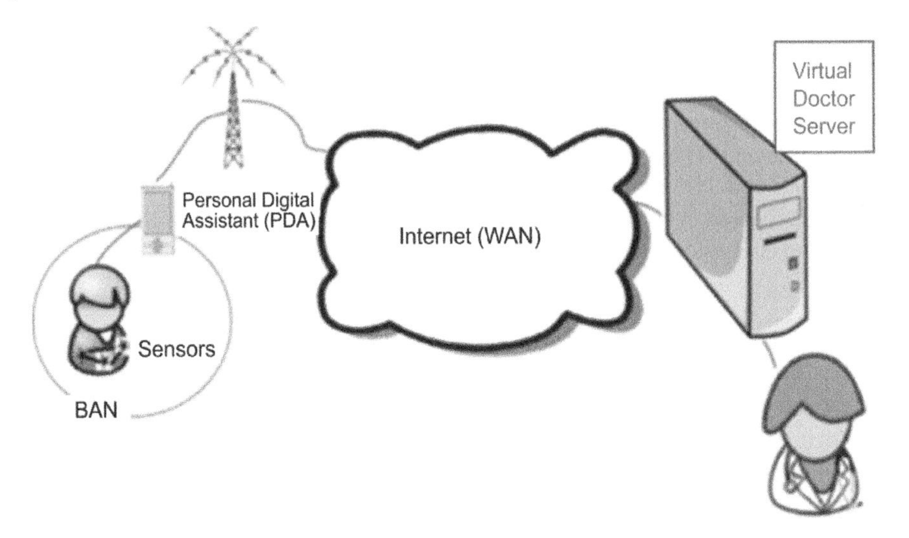

Figure 5.22. *Examples of a "virtual doctor"*

5.6.6.1.2. Applications for the emergency services

The use of BANs in the emergency services could save the lives of personnel. For firefighters, sensors collect and transmit information about the environment (levels of methane, oxygen, carbon monoxide, air temperature and humidity), the wearer's state of health (body temperature, heart rate, blood oxygen saturation) or the status of their equipment (levels of air in the tanks). Firefighters in difficulty or in danger could be saved thanks to GPS location systems.

5.6.6.1.3. Issues and prospects

The aging population and rising cost of healthcare in developed countries are leading those countries to adopt new solutions, including WBANs, for patient monitoring.

A 10-year report from the Directorate of Hospitalization and Organization of Healthcare (attached to France's Ministry of Health and Sports) looked at the role that telemedicine may play in the organization of healthcare. The report underlines the following observation: for several decades now, life expectancy has been increasing by three months per year, and such longer life brings with it an increase in chronic health conditions (diabetes, hypertension, cardiac and renal weaknesses, cancer, etc.).

Since the report's publication, numerous research projects in the field of smart apparel have come to fruition (examples in France include BioSerenity, Chronolife, etc.), and have been taken up at the European level. These garments, which include accelerometers and magnetometers, are used for the remote monitoring of elderly patients in their homes, alerting healthcare staff in the event of a fall, which is a major cause of death among this segment of the population. Numerous European projects are aimed at developing "biomedical smart apparel", including sensors integrated into the fabric of the garments, and capable of transmitting the data wirelessly to a computer or a specialized hub. Such apparel would allow for constant monitoring of patients' vital signs. Tools like this would help improve quality of life for the elderly, the ill and the infirm. The aforementioned report indicates that "these people could have continuous, personalized and transparent medical monitoring thanks to smart apparel, including sensors, which they would continuously wear, and which would, in real time, via a BAN (Body Area Network), communicate all the medical data to their doctor, their reference hospital or a care home".

From a financial point of view, the global market in smart apparel could be worth billions of dollars. Indeed, with our aging population, such devices could become as indispensable as mobile phones are today.

5.6.6.2. *Applications in sport*

In the field of medium- and high-level sport, BANs are used to measure athletes' performances and activity in real time. The different sensors record data such as heart rate, analyze movements, temperature flows, respiration, and geographic location via a GPS. For example: in golf, in addition to a sensor in the club handle, players are connected to a gyroscope, accelerometer and magnetometer, and can analyze the quality of their swing in order to achieve better results.

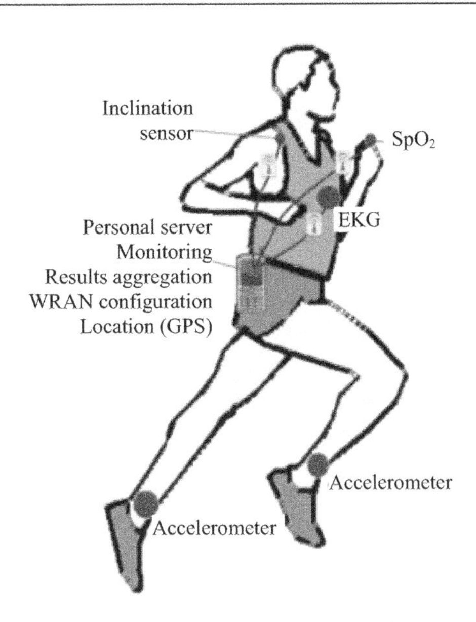

Figure 5.23. *Example of applications in sport. For a color version of this figure, see www.iste.co.uk/paret/smartpatches.zip*

5.6.6.3. *Domain of PPE and defense*

Technology such as this has a range of uses in PPE and in defense. The stress levels, effectiveness or hardship experienced by a laborer or a fighter pilot, for example, are monitored via a sensor network including: an electroencephalogram, an electrocardiogram, an electromyogram, sensors to measure respiration, blood pressure, pulse, skin conductance, etc. For laborers or the armed forces, the transmission and analysis of data collected through sensor networks worn by active personnel are used to monitor the impact of extreme heat, intense cold, high altitudes, professional activities, monitor posture, fatigue, measure radiation in hazardous environments, etc., physical training, deployment operations and nutritional factors on personnel's health and performance. Thus, each laborer/soldier becomes a network platform, interconnected, with target location systems, a night-vision system, and 3D audio systems to identify and locate noises in the surrounding area.

5.6.7. *Main challenges in the design of BAN patches*

In these different fields of application, the main problems and challenges in the design of systems and patches are as follows:

– power supply to the sensors (see Chapter 4);

– lifetime/operating life of garments powered by a power cell/non-battery technology;

– size and weight of the sensors (see Chapter 4);

– range of wireless communication and transmission characteristics of the sensors (see Chapter 4);

– the position and accuracy of the biosensor(s) and how they are arranged in relation to one another;

– cost of highly refined sensors, small in volume and at low cost, which can be manufactured in large quantities from a textile point of view, and woven, glued, printed, etc., directly into the fiber;

– usage (washing, ironing);

– global economic considerations;

– the configuration of a homogeneous system;

– automatic data transfer;

– the user interface (which must be simple and intuitive to use);

– constant transmission, notably for accelerometers or gyroscopes;

– compliance with the GDPR and social issues surrounding WBANs: in view of the health related communication between the sensors and servers, all WBAN and Internet communications must be encrypted to protect the users' privacy. A legal regulation is needed to regulate access to nominative data;

– etc.

5.7. IBC

5.7.1. *From the BAN to IBC*

This new major section of this book dedicated specifically to BANs (Intra-Body Communications – IBC) may, to some readers, seem anecdotal. It is divided into several parts which, once again, reflect the direct relationship with e/smart textiles. To begin with, however, to clarify the situation, let us now briefly recap where we are up to in our discussion:

– in Part 1, we looked at smart textiles and smart apparel (and the various constraints that come with them, etc.), and went into great detail about the technologies likely to be used (graphene, etc.) for these mono/multi biosensors involved in potential intelligence for apparel, and satisfying these applications;

– in Part 2, we turned our attention to how such devices are powered and how these biosensors communicate with the outside world;

– we will now look at how to make them communicate with one another if necessary, through the body as a medium, in contactless mode, but through simple capacitive coupling. This will be the goal of this section on IBC.

By way of a general introduction, note that, for certain readers, this new section may be considered simply to be ahead of its time:

– simple "bio" patches stuck to the skin are already out there, being used;

– simple patches integrated (stuck on, stitched in, etc.) into garments are in the process of emerging;

– multibiosensor patches are also beginning to emerge;

– today, it is a little too early to see industrial-scale direct communications between patches stuck to the skin – this is galvanic or capacitive IBC, however, it is on the way – that is our next focus in this chapter;

– in tomorrow's world, the different patches, not stuck to the skin, will be integrated into a single garment, and will be able to communicate with one another, by means of capacitive IBC.

For their part, the following subsections touch on the vast world that is Intra-Body Communications, media coverage in the mainstream press and in specialist publications, etc., and the concrete reality of defining, designing, producing, concretely rolling out and scaling up a product to industrial levels, and in particular, successfully selling it!

The goal of this part, therefore, is to give readers an introduction to the techniques of IBC and how they work. So as not to confuse the terms which often come up under Intra-Body Communications and smart apparel, the first subsection gives a brief overview of the vocabulary and definitions that we are dealing with.

5.7.1.1. *Definition of IBC*

Let us start with the origin of the term Intra-Body Communications – communications through the body: *Intra* = in, within; *body* = the (human) body; and *Communications* = communications.

The elements of Intra-Body Communications are, in fact, a specific subset of connected things, which may or may not have Internet connection. Smart Intra-Body Communications are Intra-Body Communications endowed with "intelligence". If that intelligence allows them to connect to other elements, they are often part of connected objects.

NOTE.– An official memo from the ISO simply states that a "*smart device* is a element which has electronic connectivity and/or embedded computational power".

Simply put, IBC is a technology:

– for short-range wireless connectivity;

– using the capacity of certain features of the human body as an electrical communication channel for signal transport;

– consisting of a method of data transmission offering network structuring possibilities;

– providing intuitive communication, simple and securable between two compatible elements;

– in which connections and communications between IBC elements take place by simple proximity detection when they are in direct contact or a few centimeters away from a human body.

Having given this general introduction, let us now take a look at how the technology came to be.

5.7.2. *Genesis of IBC*

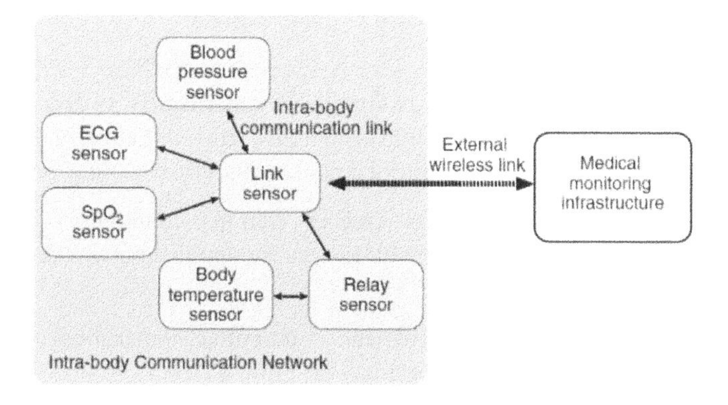

Figure 5.24. *Principle of Intra-Body Communication*

The original basic idea of IBC came about as a result of a vision of a wireless biomedical monitoring system using multiple (2, 3, etc.) sensors, *on or very near to the body, or even implanted within it*, with the aim of monitoring vital functions and transferring the data through the human body to a central monitoring unit or to other elements (see Figure 5.24). These technologies, which offer greater freedom, comfort and opportunities in clinical monitoring are used, in particular, for at-risk patients.

5.7.2.1. *History*

The implementation of the concept and technique of body-area networks and IBC dates from the 1980s-1990s (see the patents listed below). It then appeared to die down for a long time, and re-emerged in 2015 – 35 years later, with new ideas about applications. We will now briefly summarize the various steps in the evolution of the technology.

IBC techniques in general have already been discussed in great detail, over the course of many years, falling under a variety of names:

– HBC (Human Body Communication). It was in the context and around the concept of HBC, devised by the IEEE, that other communications systems have developed.

– BCC (Body Coupling Communication). For its part, the study of BCC began in the early 1990s, and numerous articles were published under the banner of initiatives as a "portable computing group".

– The IST Ambient Networks Project (AmI).

Thus, what we are dealing with here is by no means new! The following are a few particularly salient facts in the history of the discipline:

– the seminal reference is Personal Area Networks (PAN): Near-Field Intra-Body Communication, published by Thomas Guthrie Zimmerman B.S., Humanities and Engineering, Massachusetts Institute of Technology, IBM, in February 1980, with a few projects and variants, and in industrial domains other than healthcare;

– patents were filed by Thomas Geisler (Philips, Hamburg – 1990) for contactless inductive and capacitive input devices for applications in the automotive industry;

– the documents produced by Siemens Automotive Regensburg Continental Automotive are also particularly important;

– the Catrène "eGo" project, headed by Alain Rhélimi from Gemalto, is geared toward applications in identification and chip cards;

– the ECMA 401 standard was ratified in December 2011.

5.7.2.1.1. In France

In 2008, CEA-Leti launched Projet BANET[10], financed by the national research agency, the ANR, aiming to bridge the gaps and remedy the weaknesses in wireless BANs, by:

– providing knowledge of the propagation channel, taking account of the role of the configurations and design of antennas interacting with the body, including the constraints on the system's use (mobility, technologies and arrangements of antennas, apparel, environment, path loss, etc.);

– providing a justification for the choice of an air interface and a medium-access protocol capable of covering most needs in terms of availability of the radio link, throughput, consumption, lifespan and quality of service;

– providing a study of the coexistence of a BAN in an environment comprising other BANs and other wireless systems, and of the benefits offered by cooperation between BANs measuring diverse bodily parameters;

– taking account of the essential rules of the use of communicating objects/radio implants around/in the human body, in terms of radio emissions, and of biological effects, and reducing these systems' energy consumption as far as possible, in order to maximize their ease of use, battery life, optimize the size of the terminal nodes, which could ultimately lead to a network that is autonomous, thanks to energy harvesting (self-powering).

5.7.2.2. *Objectives and specific challenges*

The objectives and specific day-to-day challenges of an IBC network of wireless biosensors for patient monitoring are as follows:

– the presence of the sensors must not cause the patient discomfort;

– the transmission of the signal must not interfere with the functions of the human body;

– the real-time requirements must be satisfied in order to manage emergency situations;

– each sensor must operate over a long period of time, with or without a small battery;

– a certain number of sensors must be able to be handled without interfering with one another;

---

10 Partners in the BANET project: CEA-Leti, Decathlon (Oxylane Group), Ela Medical, de Sorin Group, ENSTA ParisTech, France Télécom-Orange, CITI, INSA public laboratory, Movea, CEA start-up, UPEMLV-ESYCOM and UPMC-L2E.

– a sensor must be placed on the body (e.g. in a dressing), under the skin or implanted (e.g. swallowed in pill form);

– the existing regulations must be strictly complied with.

Thus, there is nothing fundamentally new in this book, apart from the new fields of application and the new technologies that are applied.

5.7.2.3. *Comparison with other technologies in terms of comfort*

Let us give a brief rundown of what IBC offers in comparison with the technologies commonly used today. IBC technology allows:

– long-term monitoring of physiological signals, which requires sensors distributed over a wide area of the body;

– the obtention of a comparable signal-to-noise ratio, giving good diagnostic quality for monitoring respiration and cardiac output (as specified in IEC60601-2-47) by multi-lead electrocardiography (EKG) and impedance cardiography, which requires electrodes to be placed near the neck and at the end of the sternum;

– comfort and acceptability of devices in application scenarios for sports and healthcare;

– a stretchy support for the biosensor (graphene, for example), offering good adherence of sensors to the skin, with stabler contact impedance for the electrodes;

– the ability to use NFC and BTLE connections.

5.7.2.4. *The main areas of application of IBC*

In the context of the applications we have been considering since the start of this book, two major branches of IBC solutions are available, depending on the nature of the biometric measurements desired and the nature of the biosensors involved in taking those measurements:

– Firstly, in applications (not directly linked to smart apparel) in which the patches with their biosensors need to be in direct galvanic physical contact with the skin (for example, a dressing-type patch directly applied/stuck to an open wound, measuring pH to detect diabetes-related problems).

– Secondly, in other cases of applications, in which the patches and their biosensors do not necessarily need to be in direct contact, constantly or periodically, with the skin, but only in the vicinity (for example, measurements of heart rate, rises in body temperature, measurements linked to transpiration, etc.). In these cases, the patches and their biosensors may be integrated directly into the clothing (glued on, sewn in, woven in, etc.), and thus only operate with capacitive coupling between the patches and the skin (direct applications to smart apparel).

In these cases, where multiple patches are distributed across the body or in the garment, we are dealing with applications that require the creation of BANs of patches arranged on the body or on a particular part of the body. It is possible, in certain applications where multiple patches (*each of which may have multiple biosensors*) are present on an individual (*or are present in the same garment or integrated as part of a garment*), that we may wish to (*or need to*) have them communicate, or communicate with one another. Here, we come to a new field of exploration, which is the very focus of this book.

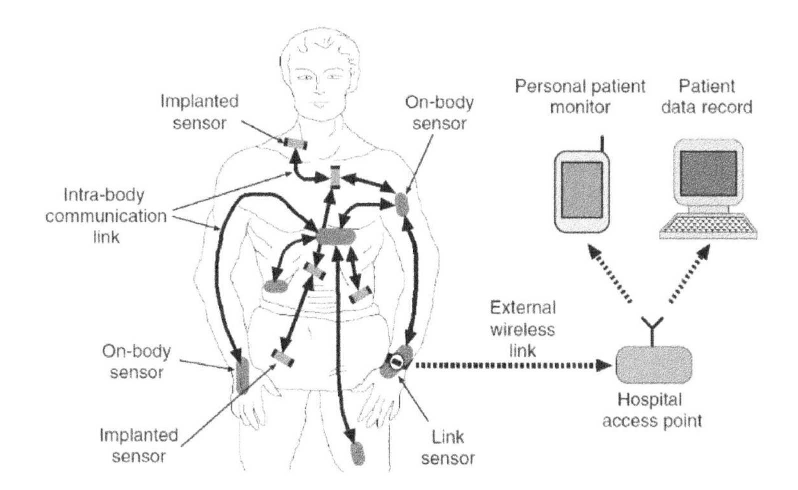

Figure 5.25. *Examples of graphene patches directly adhered to the human body*

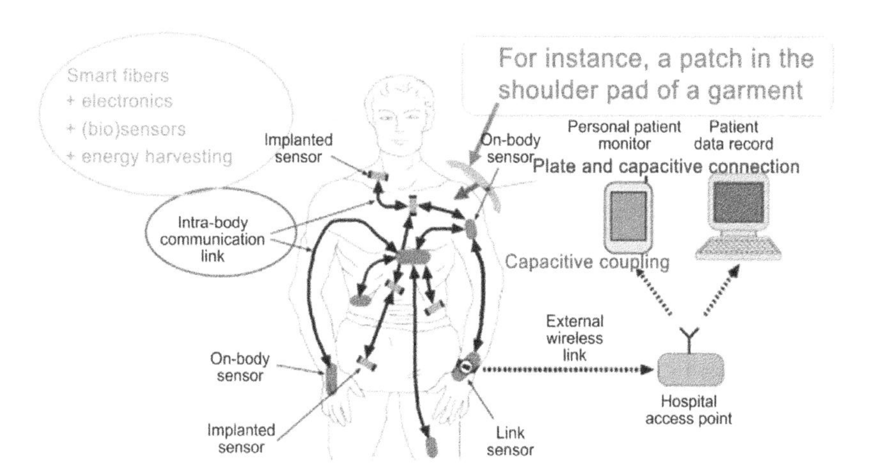

Figure 5.26. *Examples of graphene patches directly integrated into a garment. For a color version of this figure, see www.iste.co.uk/paret/smartpatches.zip*

In these cases, for convenience's sake, for reasons of cost, but also to get around certain physical constraints (implementation in dangerous environments, in places that are difficult to access, in the human body, etc.), for reasons of flexibility in application, etc., a wireless connection may be sought. Unfortunately, in transmitting the information, wireless connections have the disadvantage of being highly energy-hungry, so we must find ways of reducing that energy consumption whilst respecting the physical and regulatory constraints.

In the case of particular applications, in order to improve the efficiency of a system equipped with multiple patches and ensure wireless communications between multiple biosensors on the body or in the clothing, two main types of connections – BANs through the skin and IBC – can be envisaged, to handle electronic data transfer (physical, biological, physiological, positional, motion, stress and emotion in response to effort, etc.) between intelligent elements (patches) worn by the person and the outside world (a cloud server), from the person to a hardware reader, or person to person.

Figure 5.27 offers a brief overview of the solutions which can be considered for these multiple variants and options.

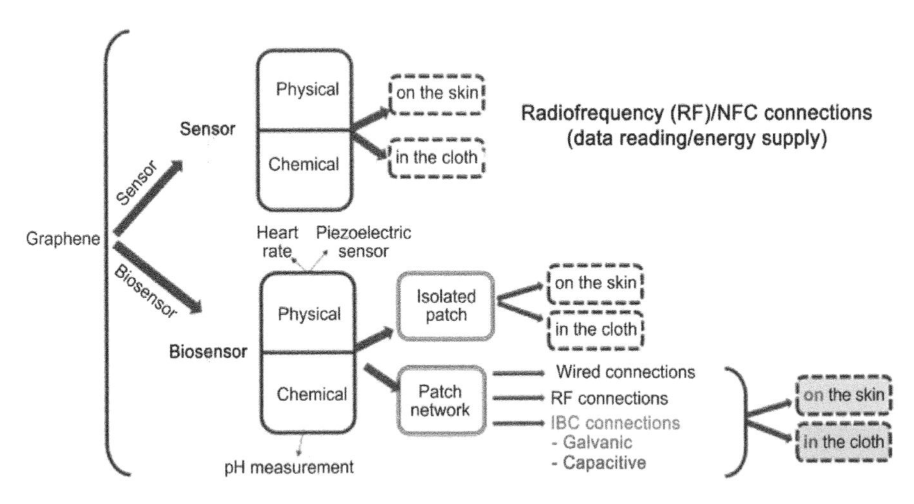

Figure 5.27. *Examples of paradigms for the applications of graphene patches. For a color version of this figure, see www.iste.co.uk/paret/smartpatches.zip*

5.7.3. *The major principles of IBC*

Over the past 20 years, a vast number of articles, theses and books have been published on the theme of IBC, so there is no need for us to repeat their contributions here[11]. Similarly, over the past three decades, numerous portable electronic devices, attached to the body, have been developed (for example, various body implants, medical devices to monitor health parameters, and technologies for sporting activities, which monitor athletes' performances). In this book, therefore, we will restrict ourselves to recapping the most fundamental aspects of this work, and drawing the direct connections between those aspects and the subject of our research: connections/communications between patches equipped with biosensors, which may be present in smart apparel.

5.7.3.1. *The body as a communication channel*

The study of the body as a "communication channel" and IBC technology for wireless communication, serving to transmit/supply data from the human body to devices (e.g. medical devices, sports equipment, mobile phones, etc.) partly involves the study of the surface, or the near surface, of the body.

5.7.3.1.1. Anatomy of the skin

As Figure 5.28 shows, this surface, the skin, is composed of three layers: the epidermis, the dermis and the subcutaneous layer.

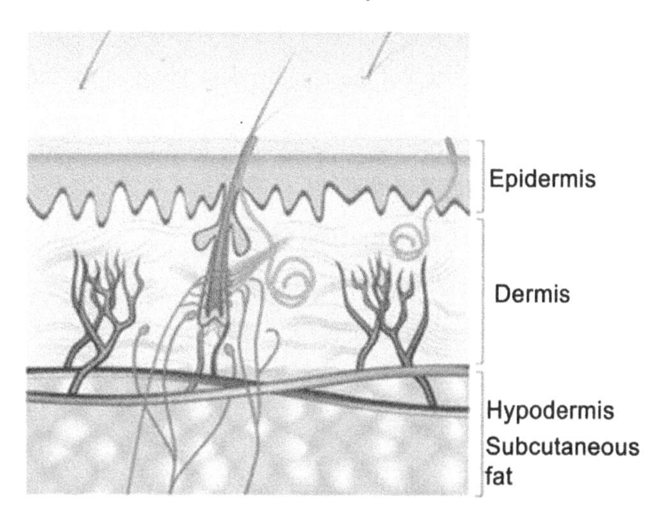

Figure 5.28. *The structure of the surface of the skin. For a color version of this figure, see www.iste.co.uk/paret/smartpatches.zip*

11 Taking only the people that we could have cited in the References, even if duplicates were eliminated, we would have at least a 350-page book!

– The epidermis can be considered as the barrier between the body and the outside world.

– The dermis includes the body's temperature-regulating apparatus, such as blood vessels, sweat glands, sebaceous glands and the hair-raising muscles. Blood vessels contract to conserve heat and dilate to dissipate heat. Sweat glands cool the skin, sebaceous glands prevent the skin from drying out, and the hair-raising muscles contract to help retain heat.

– Amongst other things, the subcutaneous tissue includes blood vessels. The blood carries heat throughout the body[12].

We will come back to this point.

Returning now to a technical perspective following this biological introduction, the study of the different parameters characterizing the "body" as a medium and the communication channel consists of studying:

– the possible means of coupling with the body;

– the body's intrinsic electrical qualities (resistivity, etc.);

– the different paths that signals may take through the body;

– the qualities (attenuation, etc.) of these paths, depending on the individuals, the movements of the individuals, etc.;

– the minimum and maximum values of the signals to be transported (currents, fields, etc.) and how they are regulated;

– the influences of the signaling frequencies used;

– the types of signal modulations to be used;

– the means of signal processing to be used;

– the signaling rates...

– ... and a wide variety of other points!

5.7.3.2. *The different possible types of coupling with the body*

IBC uses the human body as a medium for the transmission of the signal/modulated electrical field *on*, *in* or *through* the user's body, using four conventional means of coupling, recapped below.

---

12 We could have also discussed *fat IBC* – that is, IBC which passes through fat rather than through the epidermis.

5.7.3.2.1. Galvanic coupling

With galvanic coupling, the signal is controlled by the circulation of an applied current. The human body can be considered a waveguide, and the propagation of the current through the body does not require any return path outside of the body. Consequently, this method is less sensitive to noise.

5.7.3.2.2. Capacitive coupling

With capacitive coupling, the electrical signal is applied to an electrode of the transceiver in order to create an electric field. This approach mainly aims to maximize the coupling between the transceiver and the human body, reducing interference due to the ambient noise.

5.7.3.2.3. Inductive coupling

With inductive coupling, electromagnetic coupling is used to provide a means of communication for implanted devices (*in* or *on* the body), with an external coil being placed near to the patient, coupled with a coil implanted *on* or *under* the surface of the skin. Power transfer takes place thanks to inductive coupling between the transmitting and receiving coils. In this case, the implant is powered by the harvested magnetic field.

5.7.3.2.4. RF coupling

"RF-*based*" IBC is one of the alternative techniques which can increase bandwidth and facilitate two-way data communication (capacitive, galvanic, etc.).

In IBC, the two methods conventionally used to couple the signals to the body are galvanic and capacitive. The other two approaches, inductive coupling and RF coupling, are less frequently employed.

Let us recap the details of the galvanic and capacitive types of coupling used in IBC.

5.7.3.3. *IBC using galvanic coupling*

As shown in Figure 5.29, in galvanic coupling mode, both electrodes for each of the IBC devices *are in direct contact with the human body.*

A single path for the differential signal transmitted is established by the flow or a current (of a few picoamperes), which penetrates human tissue. Owing to this characteristic, galvanic coupling is often held up as a viable alternative for communication between implanted sensors. In this case, the patch is affixed and makes direct contact with the skin, and IBC takes place by means of galvanic coupling between the skin and the patch. However, this is beyond the aim and scope of this book.

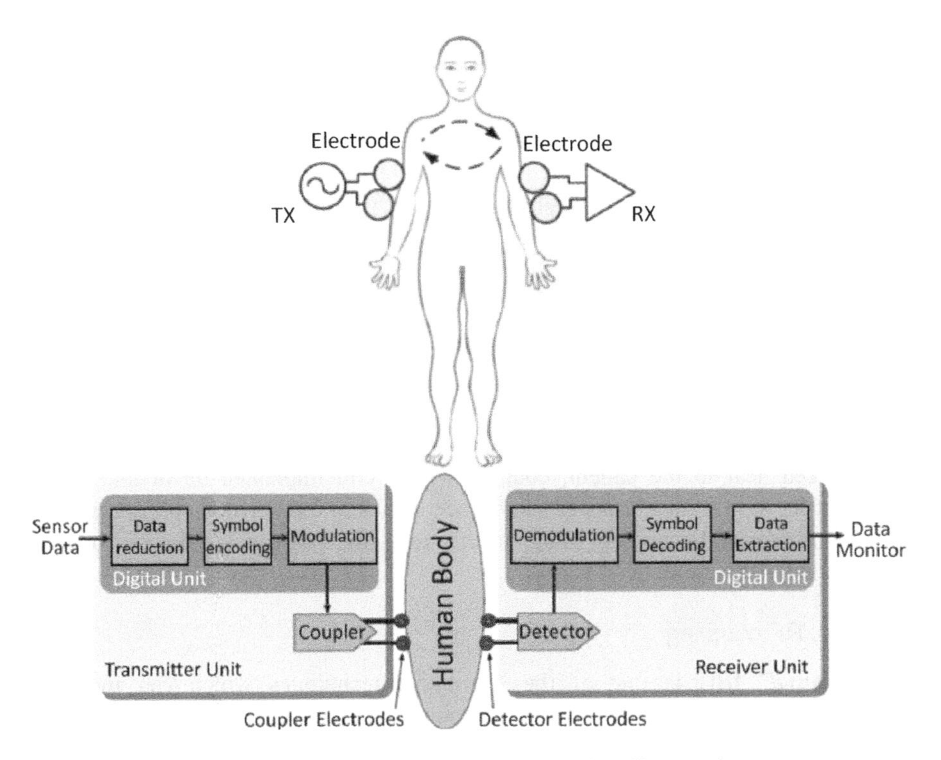

Figure 5.29. *IBC using galvanic coupling. For a color version of this figure, see www.iste.co.uk/paret/smartpatches.zip*

5.7.3.4. *IBC using capacitive coupling*

In capacitive coupling mode, Figure 5.30 gives a very simple overview of the principle and indicates the main components and application to the body with a transmitter Tx, which is composed of an encoder for the modulation of the data, and a coupling unit attached to the body. One of the coupler's two electrodes is connected *capacitively* to the body, while the other is connected to the ground in the environs. In summary, the signal's outbound path runs through the human body, and the return path closes the loop through the surrounding environment.

This way of operation allows for easy interconnection of devices which are in use on the same surface of the skin or of those which are in proximity to the patient, without needing to be in direct contact with the skin, but which could also be used for communication with and between implanted devices. Of course, the strength and quality of the received signal are affected by the position of the transmitter in relation to the receiver, the number of electrodes connected to the body, the size of the receiver's ground plate and the environment. In short, everything has an impact on everything else.

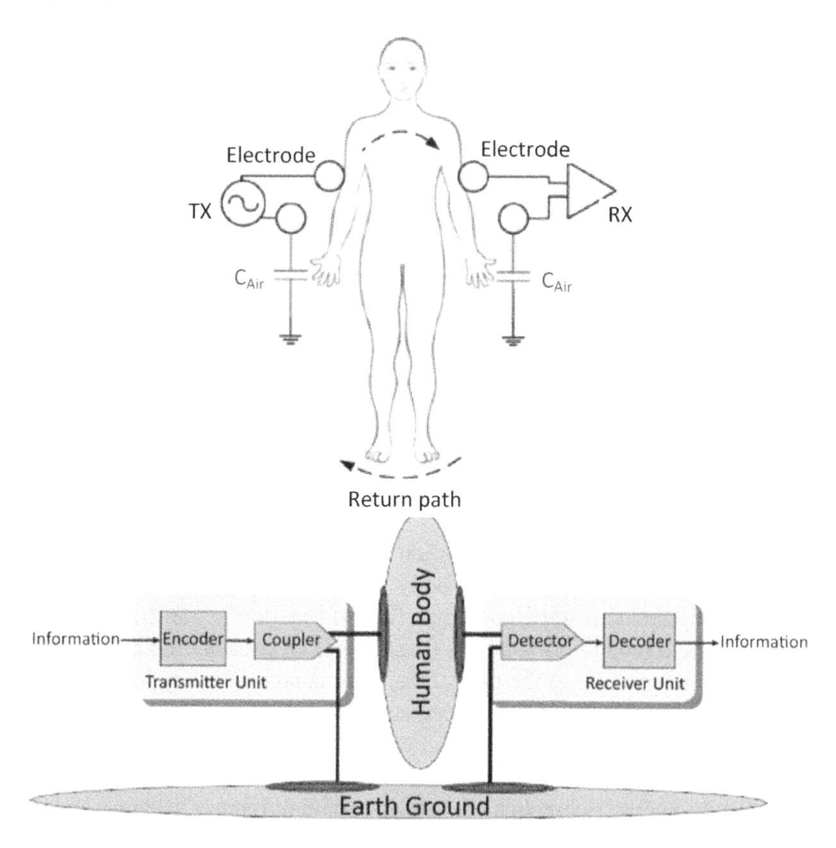

Figure 5.30. *IBC using capacitive coupling. For a color version of this figure, see www.iste.co.uk/paret/smartpatches.zip*

Capacitive IBC is capable of delivering high datarates and less path loss in comparison to the galvanic method, in particular over longer communication distances in the body. In addition, it should be noted that the conformities with the security standards relating to the admissible value of current density and the values of the electric/magnetic fields generated can be established by circuit-based simulations using the finite element method (FEM). These results indicate that the currents and fields in capacitive IBC are orders of magnitude less than the specified safe limits.

With the application of flexible patches, which are unintrusive to wear, capacitive coupling can be established between a graphene-based patch directly integrated with/into smart apparel (e.g. a cotton T-shirt) and the skin/body, which provides the medium for communication between patches and elements, and also Internet connection and communication.

5.7.3.4.1. History of capacitive IBC

The earliest works to gain widespread recognition in the techniques and technologies of IBC were written by researchers in the media laboratory at MIT, dating from around 1994. These works showed that capacitive coupling to the human body and its environment, and to certain parts of the near field, could be exploited so that the human body acts as a means of data transmission. On the receiver side, there is an element recording a signal (a potential difference) between its two electrodes. One is connected to the body and the other to the ground, and the system then demodulates the signal transmitted and received.

Thomas G. Zimmerman's seminal publication

The first system actually developed and described was devised by Zimmerman (1996):

> A PAN is a wireless communication system which allows electronic devices on and near the human body to exchange digital information by near-field electrostatic coupling. The information is transmitted by electric field modulation and the (capacitive) electrostatic coupling of picoampere currents in the body. The body conducts a very low-level current (e.g. 50 pA) to the receivers mounted to the body. The environment ("and the floor in the room") provides a return path for the transmitted signal. A low carrier frequency (e.g. 330 kHz) is used so that no energy is propagated, thus minimizing remote listening and interference with other nearby PANs. The digital information is transferred by transmission using *On-Off Keying*, with quadrature detection to reduce errant interferences and increase the sensitivity of the receiver.

5.7.3.4.2. Principle of operation of IBC

Figure 5.31 gives a full view of the principle of operation of a model based on an electrical near field produced by a PAN/BAN transmitter in proximity to the body, with capacitive coupling (above the left arm in the figure – note the small arrows indicating the directions of the fields in the figure).

The PAN/BAN assembly is composed of three elements: firstly the medium, the body; secondly, a transmitter; and finally, a receiver. Both the transmitter and the receiver have a pair of electrodes (plates).

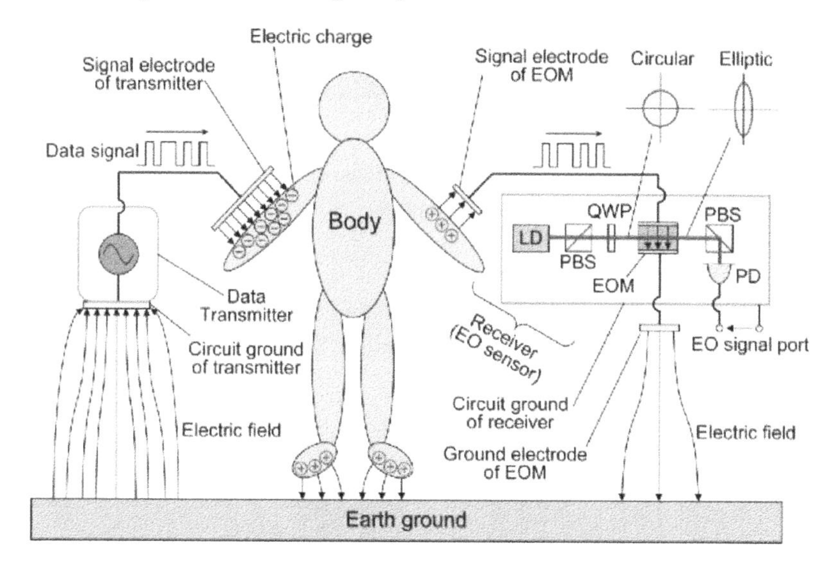

Figure 5.31. *Operational principle of a near-field IBC model. For a color version of this figure, see www.iste.co.uk/paret/smartpatches.zip*

Simply put

This book mainly describes IBC with capacitive coupling. A simplified electrical diagram of this is given in Figure 5.32.

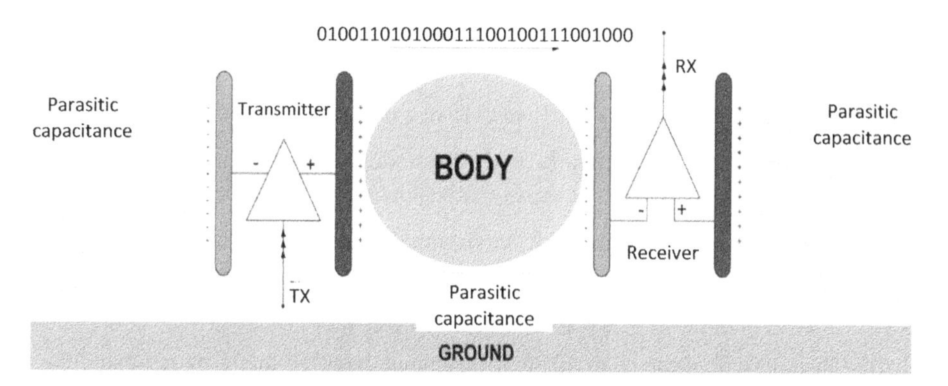

Figure 5.32. *IBC with capacitive coupling – simplified electrical diagram. For a color version of this figure, see www.iste.co.uk/paret/smartpatches.zip*

In this example:

– capacitive mode uses all capacities present between the different elements and the ground through the lines (shown in green) of the electric field;

– the signal runs from a transmitter (on the left) to a receiver (on the right);

– in principle, the communication channel is symmetrical, so data can be communicated bidirectionally;

– the shorter the distances between the transmitter and the receiver, and the grounding path, the less the signal will attenuate.

NOTE.–

– If the communication signal has an operating frequency lower than or around 10 MHz, the user's body cannot (or cannot well) act as an antenna, and the signal is not radiated.

– Direct contact between the antenna and the IBC module and the user's skin is not strictly necessary (for example, they may wear gloves, shoes, other articles of clothing, etc.). In this case, datarates between 100 kbit/s and 10 Mbit/s can be achieved.

In detail

To allow readers to fully understand, and thereafter be able to solve problems in the design and realization of a product (textiles, clothing, etc.), it is necessary to perform an in-depth study of the model of electrical communication employed in capacitive-coupling IBC, identifying the numerous paths that may be taken by the electrical fields present in the system.

For greater clarity in our explanations, Figure 5.33 illustrates the whole IBC electrical model of a transmitter T communicating with a receiver R, and all the different paths of the electrical fields E_x existing between all the organs with different potentials.

– A transmitter T and a receiver R each have a pair of plates (electrodes).

– On the one hand, we will mark "body electrodes" – with the index "b" – as the electrodes of the transmitter Tb or receiver Rb situated closest to the body.

– On the other hand, we will mark "environment electrodes" – with the index "e" – as the electrodes of the transmitter Te or receiver Re situated closest to the environment, and therefore further away from the body.

NOTE.– In principle, there is no inherent difference between them; by reversing the body electrode "b" and the environment electrode "e" for the transmitter or the receiver, we create a 180° phase shift, respectively, in the transmitted or received signals.

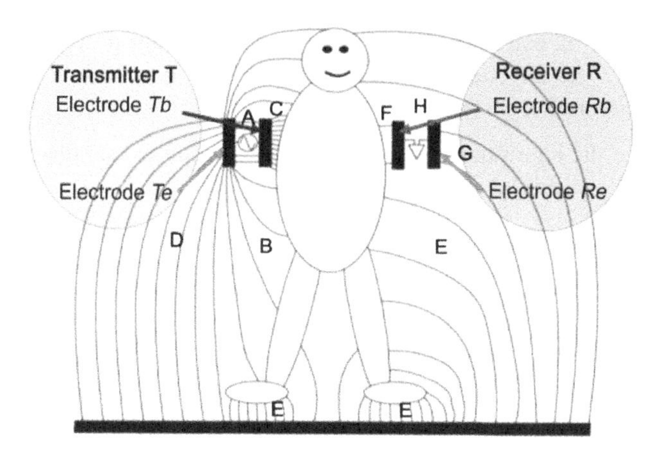

Figure 5.33. *Distribution of electrical fields. For a color version of this figure, see www.iste.co.uk/paret/smartpatches.zip*

In addition, we suppose that all objects situated around the PAN/BAN devices are electrically neutral (no static or alternating potential), or at least do not have the same exact potential (frequency, amplitude and phase) as the BAN transmitter. Let us take a look at how they work:

– alternating voltage is created between the two electrodes Te and Tb of the transmitter T, which are electrically isolated;

– the oscillatory frequency of this alternating voltage is chosen to be around a MHz (with a wavelength of over 300 meters);

– these electrodes are centimetric in scale, so they radiate no appreciable energy;

– at the transmitter, an electrical field C is formed between the transmitter body electrode Tb (placed near to the body) and the human body itself;

– that field C, produced by the BAN transmitter, penetrates the body:

- the body is modeled as a perfect conductor, electrically isolated from the potential of the surrounding environment, and acts as a large plate for a capacitor, being charged and discharged at the same rate as the transmitter,

- the charge of the body gives the organism a different potential from its environment, leading in part to the flow of an electrical field E between the body and the environment (through the feet, which are in direct contact with the ground),

- the potential of the body refers to a part B of the field C which has passed through the body to the transmitter environment electrode Te,

- the receiver detects an electrical field F exiting the body,

- this field F is extremely weak, because a significant portion of the initial electrical field C is diverted by the electrical field E established around the transmitter's "ground" electrode;

– the path of the return transmission established by the second plate of each unit by the floor (the ground) harvests the part D of the field C;

– the receiver environment electrode Re can better "see" the environment (it has a lower impedance) than it can see the body; consequently, the body electrode Rb can better see the body than it can the environment;

– the potential of the body engenders an electrical field F at the receiver body electrode Rb;

– as a result of the electronic circuits present in the receiver R (differential amplifier), the alternating potential of the receiver body electrode Rb tends to maintain the receiver environment electrode Re at the same potential, producing an electrical field G for/toward the environment. In practice, the receiver cannot precisely maintain the two at the same potential, so a small electrical field H is created between the body and the receiver environment electrode Re. This asymmetry between Rb and Re allows the receiver to detect the potential difference between the body and the environment, which means that a weak current can be measured by the receiver circuit.

EXAMPLE.– Near-field communication can function at low frequencies and with low transmission power. As a prototype PAN transmitter, Zimmerman proposed operating:

– at a frequency of 330 kHz;

– with an alternating signal at the terminals of the transmitter's plates of around 30 Vpp;

– with energy consumption for transmission of the order of 1.5 mW to charge the capacity of the electrode;

– with a datarate of 2.4 kbit/s;

– with OOK carrier modulation;

– with quadrature detection to reduce interference and increase the receiver's sensitivity.

For many years, Zimmerman's demonstrator has served as the basis, and the reference, for the principle of the PAN system for capacitive transmission via the human body, and the applicative potential of this technology lies in the identification systems and biomedical systems.

5.7.3.4.3. A few additional theoretical points

The capacitive couplings present in the IBC system allow electrical charges to circulate through the body (so in principle, we have the circulation of an electrical current i = dq/dt), generating an electromagnetic field on the surface of the skin:

– the buildup of electrical charges on a surface (electrode, plate, etc.) creates an electrical field E;

– with capacitive coupling, these charges move/circulate/are displaced on or under the skin. Over time, electrical currents (= dq/dt) come to circulate in the skin and create a magnetic field H;

– a variable electrical field gives rise to a magnetic field, and vice versa;

– the human body then becomes a conductor and a transmitter, at the frequency of variation of movement of the charges (for example, from 60 kHz to 10 MHz);

– another element/unit in the IBC system thus constituted on the body harvests the radiated energy, and can send a response using the same method.

Using Maxwell's equations, it is possible to predict the behavior of the electromagnetic (EM) fields:

$$\begin{aligned}
&\nabla \cdot \mathrm{D} = \rho_v \\
&\nabla \cdot \mathrm{B} = 0 \\
&\nabla \times \mathrm{E} = -\frac{\partial \mathrm{B}}{\partial t} \\
&\nabla \times \mathrm{H} = -\frac{\partial \mathrm{D}}{\partial t} + \mathrm{J}
\end{aligned} \qquad [5.1]$$

– Maxwell's equations:

$$\begin{aligned}
&\oint\!\!\!\oint_s \mathrm{D} \cdot \mathrm{dS} = Q_{enc} = \text{Quantity of charges within the surface} \\
&\oint\!\!\!\oint_S \mathrm{B} \cdot \mathrm{dS} = 0 \\
&\oint_L \mathrm{E} \cdot \mathrm{dL} = -\iint_s \frac{\partial \mathrm{B}}{\partial t} \cdot \mathrm{dS} \\
&\oint_L \mathrm{H} \cdot \mathrm{dL} = I_{enc} + \iint_s \frac{\partial \mathrm{D}}{\partial t} \cdot \mathrm{dS} \\
&\qquad \left[I_{enc} = \iint_s \mathrm{J} \cdot \mathrm{dS} \right]
\end{aligned} \qquad [5.2]$$

– Relations between D, E, B, H and the current density J:

$$D=\varepsilon E \quad B=\mu H \quad J=\sigma E \tag{5.3}$$

Considering that the human body behaves like an assembly/collector or individual sources/charges, with a total charge Q, using the left-hand side of Maxwell's second equation, we can solve for Q_{enc}, which gives the field E as a function of the distance R:

$$E = \frac{Qenc}{4\pi\varepsilon_0 R^2}$$

This equation indicates that the intensity of the electrical field is proportional to the total charge, given by Q = CV, and inversely proportional to the square of the distance from the source. In addition, the voltage V is directly linked to the electrical field E by the equation:

$$E = -\nabla V \quad E = -\text{grad } V$$

Thus, the equation of the voltage as a function of distance is:

$$V = \frac{Q_{enc}}{4\pi\varepsilon_0 R}$$

How to harvest the electrical field E received

As we have just seen, capacitive IBC uses the electrical field to transmit information.

– The intensity of the field E is directly proportional to the quantity of electrical charges Q.

– The quantity of charge varies with the voltage applied to the electrode.

– Any conductor placed in an electrical field will tend to equalize the potential of the field around it.

Figure 5.34 shows what happens when we place two conductors (conductive plates) in the field. A potential difference is created between the coupling plate electrode and its common reference. The usable signal can be found by subtracting between the two potentials. This is the principle used to harvest voltage.

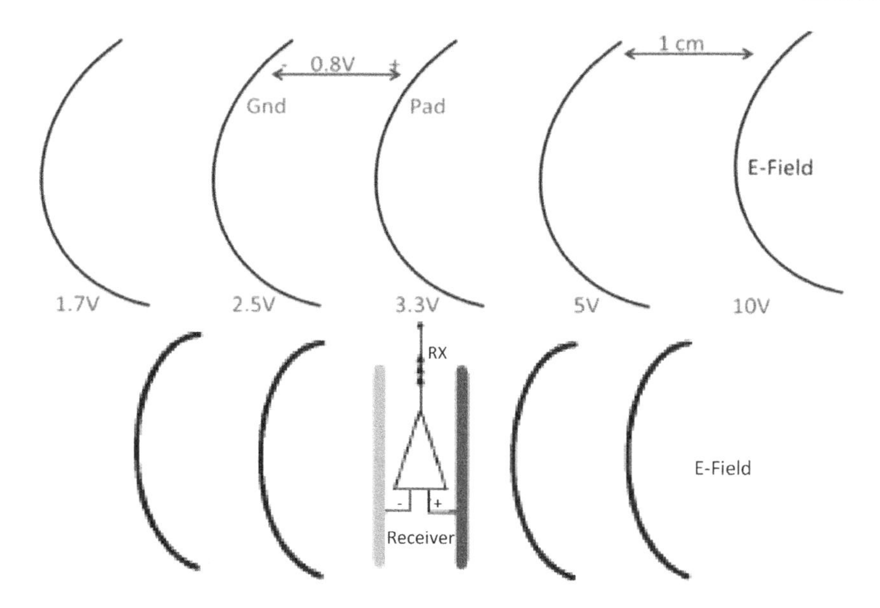

Figure 5.34. *Principle used to harvest voltage due to an electrical field. For a color version of this figure, see www.iste.co.uk/paret/smartpatches.zip*

How to harvest the field H

If we place a coil/inductor L in the magnetic field H, it will induce a current when harvesting the energy contained in the field. When, with that coil, we form a tuned circuit LC, it is possible to capture that energy in a capacitor (see Figure 5.35).

5.8. Capacitive IBC system

5.8.1. *Communication between two network elements*

The idea of Personal Area Networks has helped illustrate that electronic devices near to a human body can form a network and exchange data by means of capacitive coupling. This is the situation that we examine throughout this chapter in the domain of patches and smart apparel integrating multiple bio-sensors, where the idea of remote monitoring is omnipresent. In addition, capacitive IBC technology uses the human body as the transmission medium, making miniaturized transceivers and making communications more robust against interference, and shielded from attacks compromising the privacy of the transmitted data.

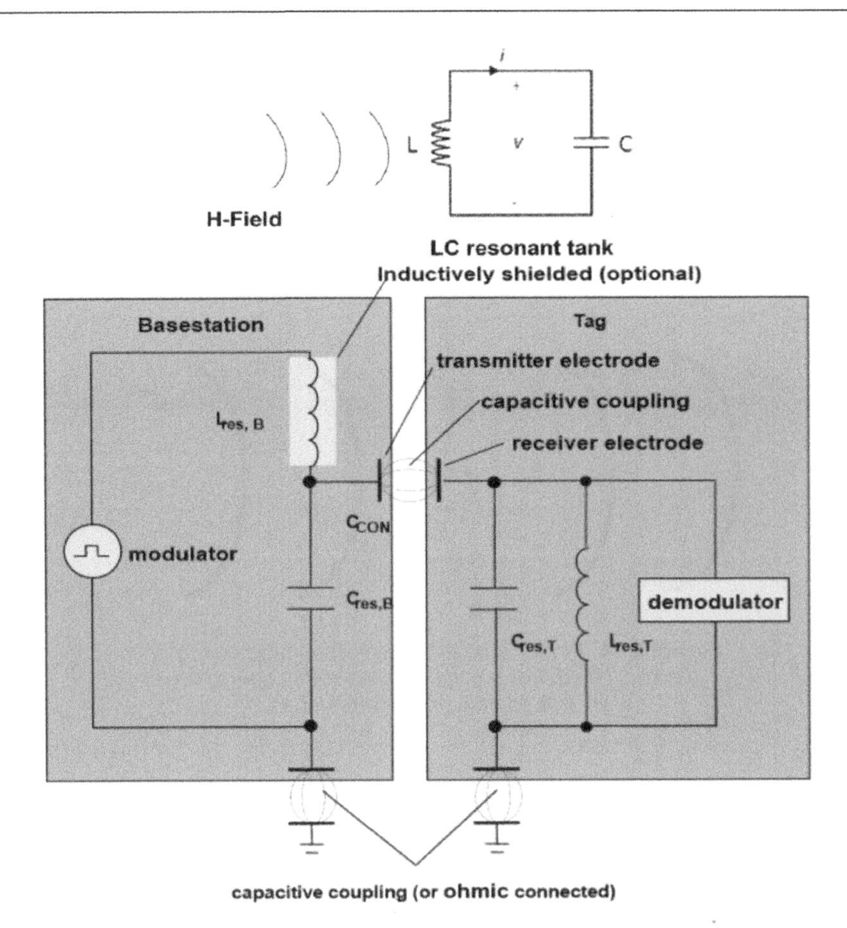

Figure 5.35. *Principle used to harvest voltage due to a magnetic field. For a color version of this figure, see www.iste.co.uk/paret/smartpatches.zip*

IBC is based on the use of two different things: physics, and its concrete implementation.

5.8.1.1. *Physical principles*

As we have just seen:

– close capacitive coupling (or contact) between two objects – a "hub" and a "peripheral" device;

– the electromagnetic field created around the human body when we touch an electrically polarized object.

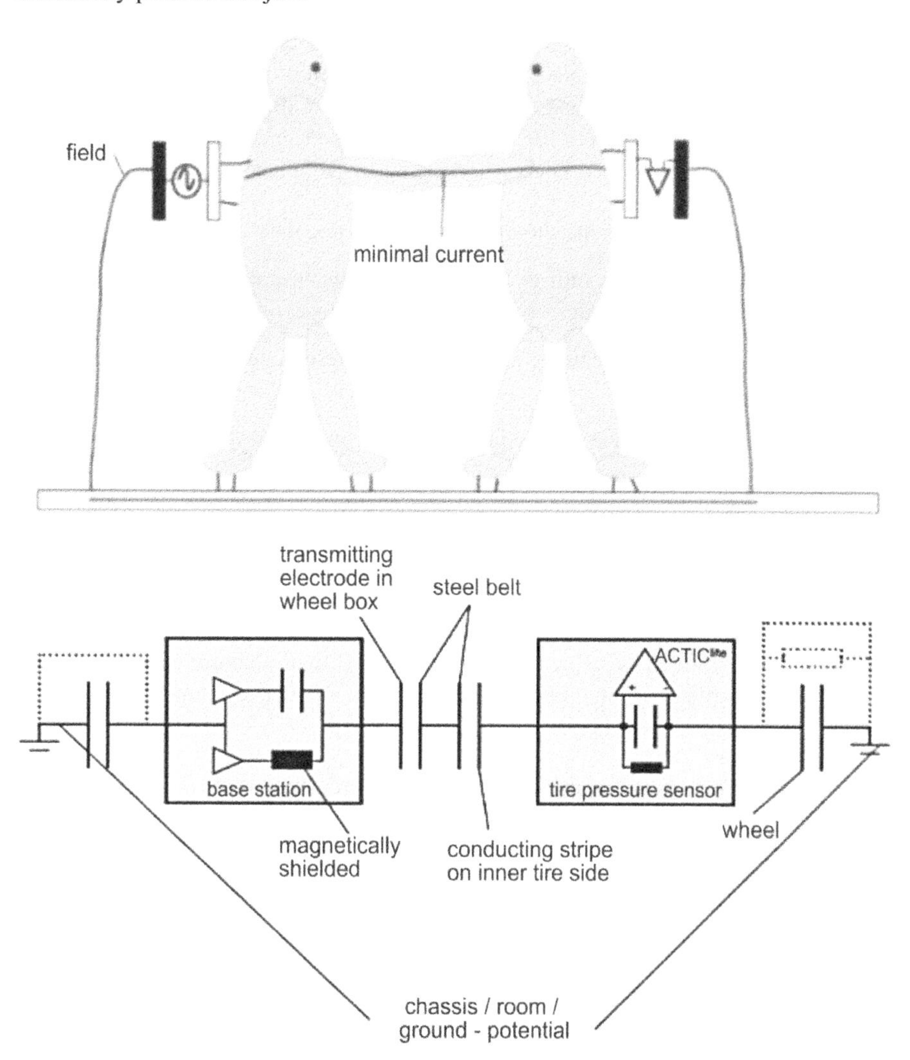

Figure 5.36. *Communication between two network elements. For a color version of this figure, see www.iste.co.uk/paret/smartpatches.zip*

5.8.1.2. *Hardware*

Numerous elements are necessary for the operation of an IBC system in capacitive mode.

5.8.1.2.1. The basic module: the patch

The "basic" electronic module – the patch with its biosensor – worn by the individual includes a biosensor; a specific bioprocessor electronic circuit, already discussed at length, which is also very similar to a chip-card circuit, because in numerous applications, it must contain the user's digital rights; an NFC transmitter/receiver; a capacitive IBC transmitter/receiver; a power source such as a battery or remote power supply, energy harvesting by RF, piezoelectric sensor.

The following considerations should be taken into account:

– Does the hub/patch transmit all the time, seeking to pair with other patches? Alternatively, does it transmit at its own applicational pace, or on request?

– Can we create multichannel solutions to differentiate the patches or the functions within a patch?

– The principle of IBC does not require a high throughput.

– IBC generally uses one-directional data communication, from IBC-compatible peripheral devices to another module or to the IBC hub.

– In capacitive mode, for the system to function, the module does not need to be in constant direct contact with the body. Indeed, such contact may occur only intermittently (for example, the sleeve of a garment is not necessarily always in contact with the skin). It is sufficient for it to be less than a centimeter away from the skin, for example integrated into a patch, a watch, a jewel, a garment, a belt, etc., in order for capacitive communication to be established.

5.8.1.2.2. The body as a "communication medium"

Generally, the body, hand or finger of the user/individual serves as a "communication medium" for the exchange of information between a base (a patch) and a peripheral device (another patch or a different device). The connection (the circulation of current) takes place through the "skin" (epidermis, dermis, hypodermis, fat, etc.).

5.8.1.2.3. The IBC peripheral device

The IBC peripheral device may be any IBC-compatible object (another IBC module, another patch, a computer keyboard, a car door handle, an electronic payment terminal, etc.).

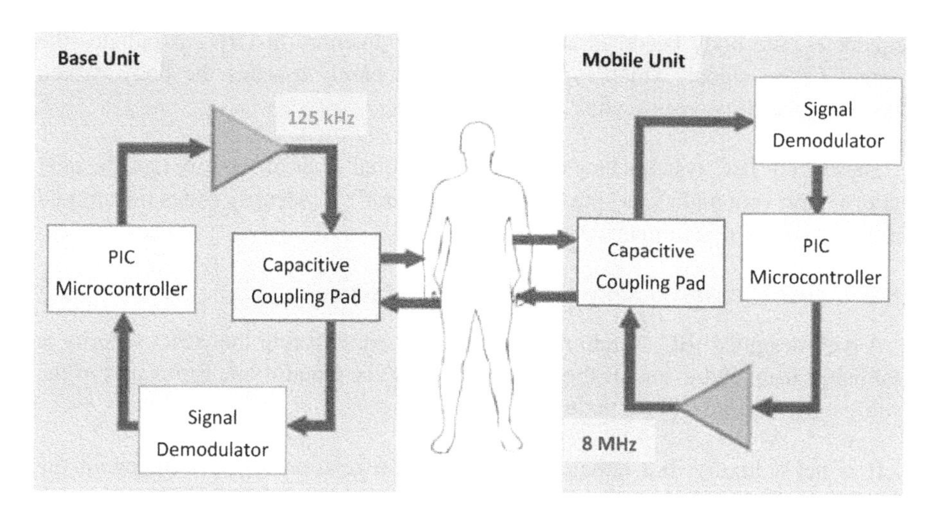

Figure 5.37. *Communication by means of contact between IBC elements. For a color version of this figure, see www.iste.co.uk/paret/smartpatches.zip*

5.8.2. *IBC and bodily non-radiation*

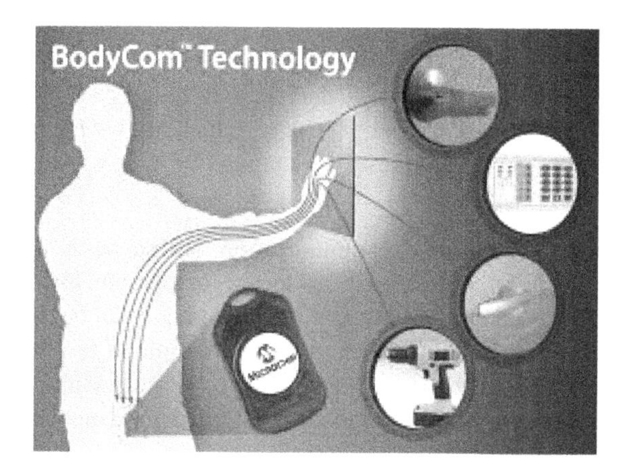

Figure 5.38. *Examples of industrial communications through contact between IBC elements (source: Gemplus). For a color version of this figure, see www.iste.co.uk/paret/smartpatches.zip*

Traditional wireless communication systems using the body as a medium generally/often operate by radiofrequency, using UHF carrier frequencies between 400 and 900 MHz, and 2.45 GHz and offer the possibility of instituting a function such as remote control with the body (see Figure 5.38). However, at these frequencies, the body becomes and serves as an antenna. In UHF, it radiates the communication signal, which is very weak, and barely touches the body, which serves as a sort of waveguide for "creeping" waves.

These UHF IBC systems thus require an additional layer of security (see the next section) to prevent malicious spies from "sniffing out" the security codes transmitted during the course of communication.

5.8.2.1. *Just in case: security measures to be taken*

A well-designed IBC system needs to be secured, ensuring that strict security is established from end to end of the chain, including the cloud if one forms part of the chain – otherwise it is all for nothing.

It is not a luxury, but rather, an obligation, in relation to the function of the assembly and to privacy, both today and in years to come, because there are too many risks of piracy, hacking, phishing, etc., of data.

In this context, where the choice of frequency could lead to inappropriate levels of bodily exposure to radiation, the following sections indicate the vast range of solutions that may, conventionally, be implemented.

5.8.2.1.1. Pairing

When required by the security of the application (e.g. with military smart apparel, PPE, etc.), the principle of easy pairing, which allows for simple and quick connection between patches, for example, requires two wireless technologies to be put in place:

– The first option for wireless communication, which has a deliberately very short operating range, is to establish a local connection for the purpose of transferring short, essential streams of data. This technique can be used to create and initialize a second means of communication. For example, a few hundred bits (the transfer request data) are quickly sent to the patch (only the essential data, password, etc.), allowing the carrier to be paired rapidly with the communication channel (IBC, NFC, BT, UWB).

– The second option for wireless communication, over which the applicative data are exchanged, requires high performance specifications (better operating ranges, from a few meters to a few hundred meters, and higher datarates).

5.8.2.1.2. Authentication

Sticking with the case of systems in which security is particularly important, if we touch any IBC-compatible object (a computer keyboard, a door handle, a medicine drawer, an electronic payment terminal, etc.), a simple capacitive touch-detection mechanism can be activated, and a sequence of question-and-answers and an authentication phase spontaneously take place between the base module/patch worn by the individual and that which is present in the touched object, each time the user's presence is detected.

5.8.2.1.3. Security

Once the phases of pairing and authentication have been completed, the security data are sent by the chosen RF technology (UHF, BLE, UWB, etc.). Every day, in order to personalize the patches of the application, the user can/must identify themselves by providing their fingerprints. This considerably reduces the risk of theft.

5.8.2.1.4. In concrete terms

The following are a number of ways to overcome these limitations, in order to use the user's body as a data transmission medium:

– capacitively coupling the user's body to the base unit and mobile units (patches);

– due to capacitive coupling (by means of an impedance of the type $Z = 1/C\omega$), the signals are more attenuated at low frequencies than they are at high frequencies, requiring the transmitted signals to have a higher amplitude in the case of lower frequencies;

– due to the high permittivity of the human body at low frequencies, the transmissions may take place by near-field communication, using a communication frequency between 60 kHz and 10 MHz, and a format based on *Amplitude Shift Keying* (ASK) carrier modulation;

– the portable "unit" (the patch), located on the body, is powered either by power cells/batteries or by energy harvesting, where energy consumption is a matter of high priority;

– the choice of the frequency used and the power levels involved will therefore be the result of a trade-off between the patch's energy consumption and the availability of integrated circuits with low consumption on the market.

Industrially speaking, the establishment of an IBC system of this type must always respect the following order of priorities:

– very low energy consumption by the patch;

– a rapid response and a fairly high datarate;

– a stable and robust communication system, with error-detection mechanisms;

– a limited field of action (a few centimeters) to allow for identification;

– low cost and low complexity.

5.8.3. *Fundamental concept of IBC*

One way to illustrate the communication mechanism of IBC is to say that the transmitter is capacitively coupled, through the human body, to the receiver, that at the frequencies at which the system operates, the RF signal circulating in the body produces little or no radiation[13], and that a return path therefore needs to be produced for the signal to flow from the receiver to the transmitter.

This return path is provided by the air (which is a dielectric), and the ground, which refers to the conductors and dielectrics present in the environment, nearby to the IBC devices, as illustrated in Figure 5.35.

The electrodes (plates) of the transmitter and the receiver of the IBC modules may be modeled as capacitor plates, with the whole of the environment being present between them. Remember the conventional battery of formulas relative to electric capacities and field:

$$C = \varepsilon_0 \, \varepsilon_r \, S/e$$

where:

– C is in Farad;

– permittivity of air $\varepsilon_0 = 8.85 \times 10^{-12}$;

– relative permittivity ε_r (see Figure 5.39);

– surface area S is in m^2;

– thickness e is in m.

$E = V/l$; E in volts per meter; V = potential in volts; l = distance in meters.

The surfaces of metallic plates increase the capacity value. Dielectrics may allow a greater quantity of charges Q to accumulate on the plates for a given level of voltage V, and also increase the capacity ($C = Q/V$).

---

13 In principle, in general, an IBC transmitter, module, patch, etc., operating by RF connections through the body will cause some disturbance to the electrical potential of the environment and the receivers. In addition, if there is a radiated field, it is known that for an ideal dipole, it is reduced with the cube of the distance.

Conductive elements in the environment include metal furniture, wires, plumbing, reinforcing bars, metal building skeletal elements, stationary, ventilation shafts, bodies of water and the ground.

Dielectrics	$\varepsilon_r =$
Air	1
Wood	3-4
Glassy dry geological materials	4-10
Plastic	2-10
Rubber	3-7
Water	78

Figure 5.39. *Values of the main dielectrics*

The materials in the environment must be electrically isolated from the body so that no coupling takes place between the body and the return path, which would short out the current loop, with the effects becoming evident when we study an electrical model of the communication channel.

5.9. Modeling of an IBC system

In this book, the human body is characterized as a transmission medium for electrical currents, as a result of the dielectric properties and electrical models developed of human tissues.

In this context, models of an IBC system using the human body as a wireless transmission channel without any other form of connection can be classified into three major categories:

– modeling of human body transmission channels with the relevant signal measuring methods;

– the diagram of the optimum types of modulations;

– frame synchronization to achieve a desired datarate with a given BER[14], and the practical implementation of an IBC transceiver.

NOTE.– The following sections deal only with techniques of IBC by capacitive coupling, which offers better performances than galvanic coupling when using

14 BER: bit error rate.

frequencies higher than 60 kHz for the desired datarates to handle WBAN applications.

5.9.1. *Model of an IBC system and channel analysis*

The analysis of an IBC system shows that there is a model of a communication channel for digital transmission by capacitive coupling. Indeed, capacitive coupling uses a signal electrode for transmission, one for reception of the signal, and two floating earth electrodes, on both sides – the transmitter Tx and the receiver Rx.

– The signal induced by the Tx generates an electrical potential with the ground electrode connected to the floor.

– The Rx recognizes the signal transmitted by detecting changes in the electrical field.

The properties of the transmission channel are, of course, affected by the different postures in which the human body is placed, and different environments, which cause variations in conductance between the ground and the return paths.

5.9.2. *Modeling of human-body communication channels*

In IBC, models of communication channels using the human body are generally differentiated on the basis of the signal transmission methods:

– galvanic coupling, in which a signal is controlled by the flow of applied current, with the human body serving as a waveguide;

– capacitive coupling, in which a signal is transmitted by creating an electrical field, induced by the signaling electrode Tb.

5.9.2.1. *Quality of a communication channel in IBC*

For (W)BANs in IBC, the standard IEEE 802.15.6 presents and details the quality of a "human-body transmission channel" on the basis of the sizes of the ground plates of Tx and Rx, and the distance between Tx and Rx, as being linked to the measurement of an amplitude response to a particular impulse (a cardinal sine function – sinc), where:

– the duration must be sufficiently short to cover the maximum desired frequency for the application of the channel response;

– the transmitted impulse signal and the received signal passing through the human body are measured at the same time as a trigger signal, in order to measure the different transmission delays (and to cancel out the delays due to the cables, the measurements of the impulse signal which are generally carried out by means of optical measurements – see Figure 5.40).

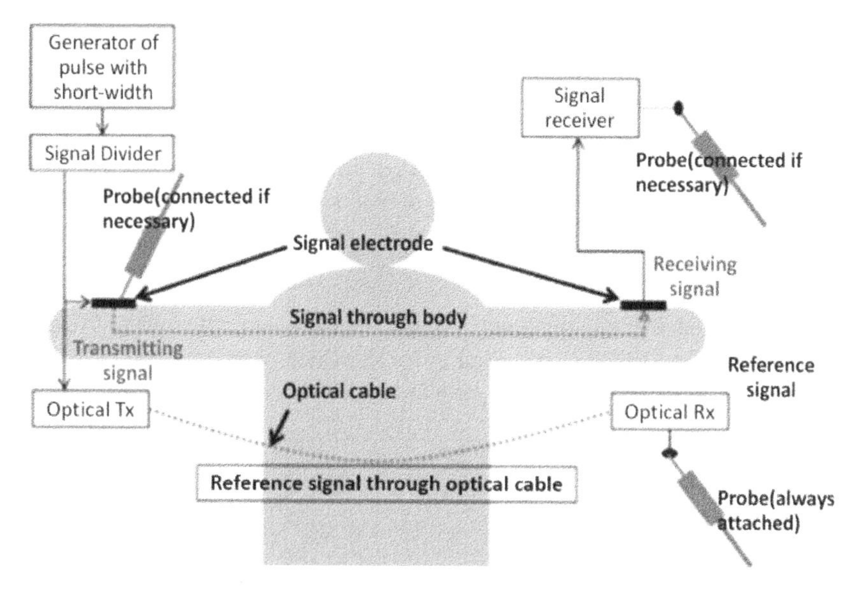

Figure 5.40. *Model of measurement of the communication channel. For a color version of this figure, see www.iste.co.uk/paret/smartpatches.zip*

The properties of the channel are based on certain results of analyses of the measured data, such as:

– The *transmission path loss* as a function of the frequency and transmission range;

EXAMPLE.– To measure only the transmission path loss, the transmitter Tx transmits a signal containing a single frequency to be measured, and receiver Rx measures the amplitude or power of the received signal.

NOTE.–

– The path of the signal through human tissue is also interpreted as an electrical circuit with electric field equations.

– In order to more easily isolate the transmission path losses between the module Rx and the equipment, instead of using an (asymmetrical) passive probe, which could skew the measurement, the measuring probe is connected to the measuring equipment either by a balun (balanced-unbalanced interface) or, more usually, with a differential probe which leaves the measured assembly totally unencumbered.

– The *time-delay parameters of an impulse response* of a channel, such as the effective value of the spread of the time delay (its root mean square) and the coherence bandwidth.

Another point in relation to the quality of the transmission channel is also important. Although, in IBC, there are various signal modulation techniques (Frequency Shift Keying, FSK; Phase Shift Keying, PSK, etc.), let us focus for a moment on digital transmission, because an IBC transceiver using this technology can more easily be implemented without a digital-to-analog converter (DAC) or an analog-to-digital converter (ADC), or RF blocks, which reduces the apparatus' bulk and energy consumption.

One of the fundamental structures in digital transmission is Frequency-selective digital transmission (FSDT) (see Figure 5.41), which was adopted in IBC for transmission Tx in the IEEE 802.15.6 standard on WBANs.

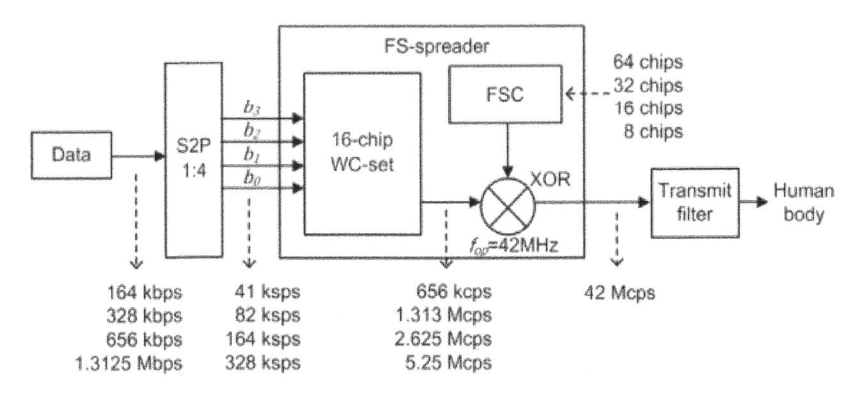

Figure 5.41. *Examples of a block diagram for an FSDT transmitter*

In an IBC BAN, the transmission FSDT technique offers:

– the use of an algorithm selecting electrode pairs to find the best reception channel between two IBC patches;

– the use of the algorithm LineSync, synchronizing the data and compensating for the data polarity during long sections of data transmission between transmitter and receiver, taking care of high-throughput data transmission, which prevents the signal attenuation effect;

– the use of methods and techniques to increase datarates, improve spectral efficiency, reduce the complexity of calculating the Hamming distance (HD), and improve the bit error rate (BER). In addition, new preamble structures and frame detection methods have been proposed to improve synchronization performances for digital transmission.

Because IBC is based on the mechanism of electrical field coupling between a transceiver and the human body, maximizing the coupling effect and minimizing the ambient interference in Figure 5.42, the methodology for implementing the IBC transmitters and receivers has a significant influence on the systems' energy yield.

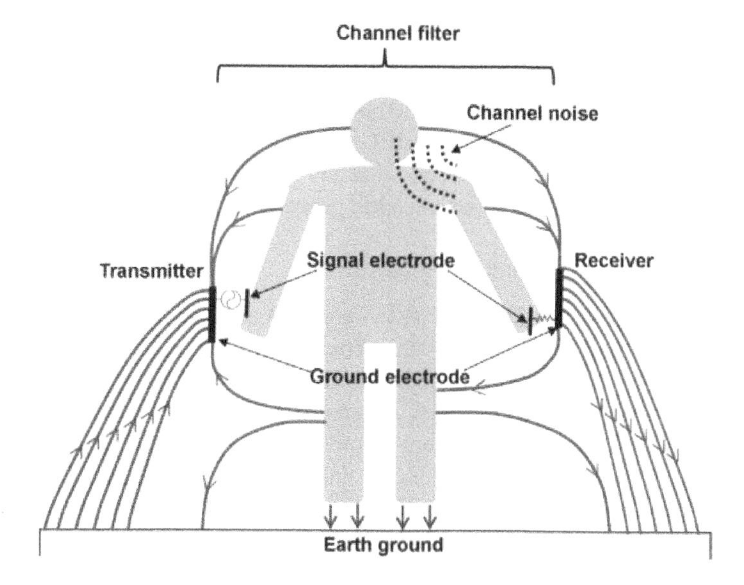

Figure 5.42. *Communication channel filter. For a color version of this figure, see www.iste.co.uk/paret/smartpatches.zip*

5.9.2.2. *Noise*

Noise is one of the key factors in improving the quality performances of a transmission channel.

If the channel model is considered as the set of paths taken by the signal from the Tx to the Rx, it may be simplified to a channel filter composed of effective paths through the air and through the body. Knowing that the channel path loss increases as the size of the ground plates shrinks and the distance between Tx and Rx increases, the interference signals from the electromagnetic waves generated by various electronic devices in the surrounding environment act as a "channel noise" parameter. In general, the channel noise exhibits a spectrum with a Gaussian distribution. In IBC, over the years, multiple implementation techniques have been put forward as a way of structurally improving the signal-to-noise ratio of the assembly, including by FSK modulation, direct digital transmission using a Manchester bit code, and digital transmission using the Walsh error-correction code.

5.9.3. *Electrical model of the medium*

Let us now examine the main electrical characteristics of the medium – the human body – with regard to its application as a communication channel.

5.9.3.1. *Characteristics of the human body*

IBC transmitters and receivers use the human body as a communication channel. Thus, in order to design and assess an IBC system, it is necessary to examine the characteristics of that communication channel. There is an abundant body of literature on this domain, and on the electrical properties of the human body.

5.9.3.1.1. Electrical Impedance Tomography

Electrical Impedance Tomography (EIT) provides a certain image of the internal organs and structure of the body, by measuring the resistance of human tissues. In order to do this, an EIT system involves multiple electrodes placed on/around the body (for example, 24 electrodes). A constant current is applied to one of the electrodes (a transmitter) and the system records the potentials received on the other electrodes (receivers). These potentials are linked to the impedance values of the tissues between the transmitter and receivers. By sequencing the electrodes, we can collect sets of impedance measurements, and then process those measurements to reconstruct the impedance matrices of the tissues and formulate a low-resolution image of the organs and tissues of the body.

Parameters measured	ρ in Ω/meter	Conditions
Human arm	2.4	Longitudinal
	6.75	Transverse
Lung tissue	1.6-51	Interstitial conditions/intracellular conditions
Blood	1.5	
Fat	12.75	
Wet bovine bone	166	

Figure 5.43. *Approximate resistivity values of mammal tissues*

Although the resistivity values measured for any human tissue may vary, depending on numerous factors, it can be broadly stated that, roughly speaking, the resistivity of the human body is around 10 ohms per meter, or less.

5.9.3.1.2. Approximation of the human body as a perfect conductor

Using the classic relation $R = \rho(l/s)$, where R is the resistance in ohm Ω, ρ is the resistivity in ohm/meter, "l" is the length of the conductor in meters and "s" is the surface area of the conductor in square meters, we can estimate the internal resistance R of the human body (beneath the skin). For example, for an adult male, 1.8 meters in height and of an average diameter of 0.3 meters, with a mean resistivity ρ of 10 ohms/meter, his resistance is around 250 Ω.

If we do not take account of the IBC peripheral electrodes, the skin, the air and the clothing, which, collectively, have significant impedance of the order of mega-ohms or even giga-ohms, and in view of the very low internal resistance of the human body (see above), it is generally considered a perfect conductor ($R \sim 0\ \Omega$).

5.9.3.1.3. Body-to-environment electrical capacity

The body-to-environment capacity (measured in farad), which we shall represent hereafter by the symbol "BE":

– can be considered a capacitor formed of two plates, separated by an insulative material, firstly the body, which is a large three-dimensional plate, and secondly the environment, which is a huge plate, in addition to the air and other insulative materials forming the dielectric;

– has a value which depends on the geometry of the body (its size and shape), its spatial relationship with the environment and the composition of the environment.

The value of this body-to-environment capacity is conventionally determined by empirically measuring the resonance frequency of a coil whose value is $L = 33.5\ \mu H$, parallel to the body. The resonance frequency is first measured to take account of parasitic increases in the body's capacity, by effectively decreasing the distance between the plates. The skin, which is composed of dead cells and water-filled cells contained within a fatty membrane, is in fact an insulator and a dielectric. The insulating skin covers the tissues with low resistance and the conductive blood.

The organism's capacity can also be calculated analytically. A simple method to do this is to approximate the human body as a large sphere. Given that, with electricity, it can be shown that an isolated, charged sphere produces a uniform radial electrical field and the value of this capacity is $C = 4\pi R\varepsilon_0$ – an equation in which R is its radius and ε_0 is the permittivity of the medium, considering that a person 1.85 m in height and 95 kg is equivalent to a sphere one meter in radius, we obtain a value of the capacity as ~110 pF for an electrically isolated person.

– To carry out this measurement, the ground mass electrode is affixed to the subject's forehead using a 1-in^2 piece of copper conductive adhesive tape.

– When the person grips/touches a piece of metal that is electrically grounded, the body's measured capacity increases to 845 pF.

– If the radius of the sphere equivalent to the person is less than one meter, this is compensated by the fact that the body is not insulated; the placement of zero potential is much closer than infinity, and is made up of all the conductors which account for the potential of the surrounding ground. To calculate it exactly, using Laplace's equation, we would need to calculate the potential around the body, and apply Gauss' law around a surface enveloping the body in order to calculate the charge on the body.

5.9.4. *Electrical model of the channel*

Figure 5.44 presents a simplified electrical model of the application of an IBC transmitter and receiver, in the form of an equivalent circuit, in the wake of the description of the electrical fields. Certain indicative values are also given.

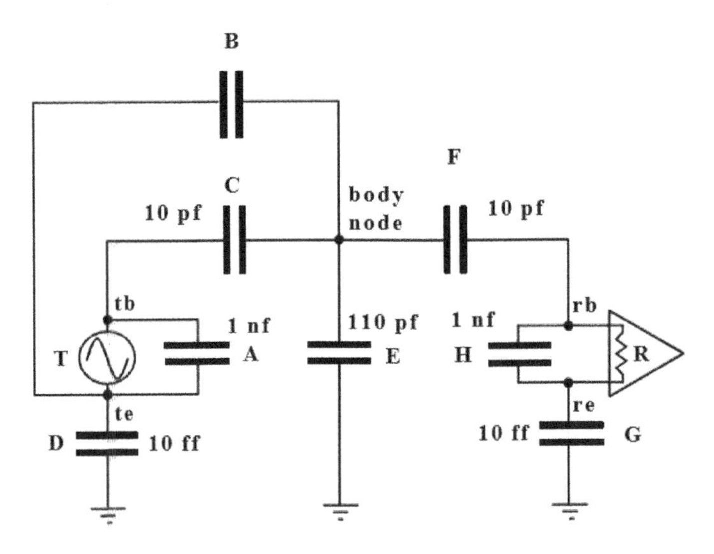

Figure 5.44. *Electrical model of the channel*

– As indicated previously, the transmitter "T" is capacitively coupled to the receiver "R" through the modeled body, appearing to be a perfectly conductive node with zero resistance, as mentioned previously (thus, it is represented by a dot in the diagram).

– Capacitive coupling of the transmitter to the body takes place through the capacity C = 10 pF on the diagram.

– The ground provides the return signal path.

– For example, a current is capacitively coupled, from a transmitter worn on the left wrist of a subject to a receiver on the right wrist. The subject is wearing sports shoes, in specified conditions of measurement of the electrode's capacity. The measured motional current is attenuated –12 dB when the subject is barefoot and –28 dB when an earthing wire is connected to the subject's forehead.

5.9.4.1. *On the side of the transmitter*

– The *capacity A* represents the capacity present between the electrodes Tb and Te.

– The *capacity B* represents the capacity due to capacitive coupling between the transmitter environment electrode Te and the body.

– The *capacity C* represents the capacity due to capacitive coupling between the transmitter body electrode Tb and the body itself.

– The *capacity D* represents the capacity due to capacitive coupling between the transmitter environment electrode Te and the environment.

– The *capacity E* represents the capacity due to capacitive coupling between the body and the environment.

NOTE.– The electrical model indicates that the value of the capacity E (110 pF) between the body and the environment is harmful to the IBC system's communication performance, because, to a large extent (due to its comparatively high value), it short-circuits to the ground (a dividing bridge between C and E with a ratio of 10) the potential that the transmitter T is attempting to impose on the body.

5.9.4.2. *On the side of the receiver*

– The *capacity F* represents the capacity due to capacitive coupling between the body and the receiver body electrode Rb.

– The *capacity G* represents the capacity due to capacitive coupling between the receiver environment electrode Re and the ground in the environment.

NOTE.–

– The electrical model suggests that the feet are the best place to put IBC modules, the body and environment electrodes respectively having strongest coupling for the body and the environment. This is particularly true for the environment electrode, which is the "weak link" (with the highest impedance) in the IBC system. Placing an environment electrode with a large surface area as close as possible to the physical mass (which is generally the shoes) maximizes the amplitude of the communication signal (the shoes are also an ideal place to harvest a portion of the energy dissipated or recovered while walking; see the earlier discussion on energy harvesting).

– These remarks also indicate that the performances of IBC devices would suffer when, for example, a user took hold of a metal water pipe.

5.9.4.3. *Disequilibrium in the symmetry of the communication channel*

Figure 5.45 shows the equivalent diagram of an IBC transmitter IBC, an IBC receiver and the network of impedances representing the capacitive coupling between the electrodes. The transmitter (oscillator) and the receiver (differential amplifier) are two devices on this network. Each of them has a local ground, but they are electrically isolated from one another, and therefore do not have any electrical mass in common.

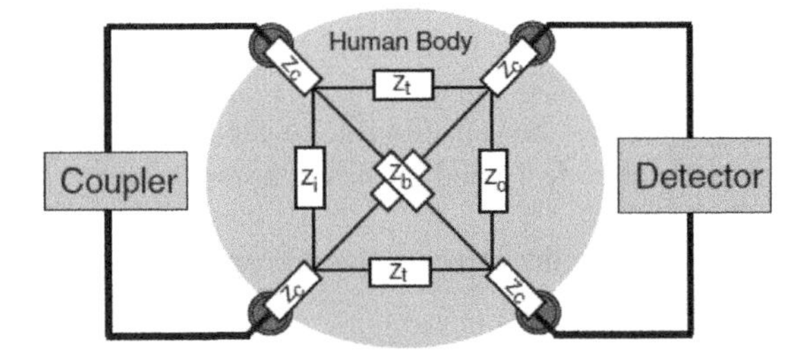

Figure 5.45. *Simplified electrical model of the human body connected to the electrodes Tx and Rx. For a color version of this figure, see www.iste.co.uk/paret/smartpatches.zip*

In Figure 5.46, the four impedances A, B, C and D present between the four electrodes (pairs of transmitter electrodes and pairs of receiver electrodes) are significant in aiding our understanding of the electrical communication.

– The intra-electrode impedance between the transmitter electrodes presents a charge to the transmitter, which is ignored, because the transmitter's oscillator is treated as an ideal voltage source.

– The intra-electrode impedance between the receiver electrodes is dominated by the current amplifier, which aims to keep the receiver electrodes at the same level of potential, leading to a slight impedance between the plates of the receiver, and this too is ignored.

The drawing of the impedances in this figure is rearranged to show how the communication of the IBC device actually operates as a Wheatstone bridge. In this diagram, the four impedances are represented as purely reactive (capacities), although they may contain real (resistive) components.

The fundamental principle behind the communication of an IBC device is to break the symmetry by which the transmitter electrodes "see" the receiver electrodes. Indeed, any imbalance in the relation A/B = C/D in the Wheatstone bridge will give rise to a potential difference, and therefore a current, through the receiver.

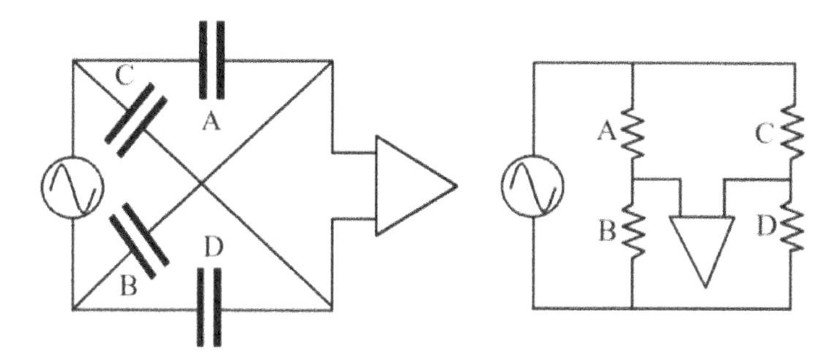

Figure 5.46. *Equivalent Wheatstone bridge of the electrical model*

Given that these ratios must be exactly equal in order to cancel out the current in the circuit formed on the median of the bridge, and that, in principle, IBC devices positioned on the body will be continuously moving, provided the receiver is sufficiently sensitive to detect the imbalance, there will always be a channel for electrical communication.

5.10. Simulations

Very commonly, in preliminary studies for IBC devices, digital simulations are carried out. Often, they are performed using the Finite Element Method.

Next, these simulations are compared to in vivo measurements. For this purpose, generally, a measuring device applies an alternating current of 1 mA peak in the frequency range 10 kHz-1 MHz to the human body, performing data transfers at up to 64 kbit/s with BPSK modulation (see the example results given in Figure 5.47).

Parameters	Frequency in Hz with 12-bit	Sampling rate in kbit/s
Blood pressure	60	1.44
1-point EKG	250	6
12-point EKG	250	72
1-channel EEG	200	4.8
192-channel EEG	200	921.6
Pulse oximetry (SpO_2)	300	7.2
Body temperature	0.1	2.4

Figure 5.47. *Methods for measuring parameters*

5.11. Examples of smart apparel solutions using IBC

When smart apparel is worn for personal or professional purposes, using the user's body as a BAN communication channel is an interesting approach to send data from the biosensors on one part of the body to a different part of the body, or indeed to another person, particularly when it is not possible to use external conductive material to transmit the data. This may be highly useful for e-textile projects. Also, to illustrate the lengthy discussion in this section, we present below two examples of real-world technical solutions:

– the first, very simple in its principle, is patches for one-directional communication (i.e. a patch with a transmitter, communicating with another patch that has a receiver);

– the second, which is more industrial and complex, is evolved patches with two-way communications, with more advanced protocols to manage potential collisions, secured with a coupled transmitter and receiver, to create an IBC mini-network within a garment.

To conclude, we shall look at how to interrogate and remotely power all patches in a network, which is no easy task.

5.11.1. *Example 1: for beginners*

The example of an "experimental" solution with capacitive-coupling IBC presented below is that which was proposed and published at MIT in 1995, in Zimmerman's original paper, *Personal Area Networks (PAN): Near-Field Intra-Body Communication*, so although not particularly recent, revisited many times over, and still holding true today. The objective then was to present, very simply, circuitry to demonstrate the principle of operation of a data *communication channel with capacitive* coupling *involving a number of participants*.

NOTE.– Although this example suggests wearing the transmitter near to the body, the project presented here does not allow for this solution, industrially speaking. When the transmitter is placed too near to the body, the transmitted and received signals are disturbed, and the transmission itself may also suffer disturbance. Shielding the transmitter may help it to function better. A more recent version of the same example solves the problem.

It is clear that this type of experimental setup, very simple in design, may serve as a closing example in an argument or thesis, or may be used for Proof of Concept (POC) of feasibility, but it is a long way from being an industrial and commercial product. In the latter case, it is necessary to go through research programs, patent filings, solve numerous technical problems on an industrial level, and problems relating to reproducibility (see conclusion) which are far beyond the scope of this book, but which the coauthors (and professionals in these applications) have solved for themselves after months of work for the companies with which they are affiliated. We shall now provide a technical description.

5.11.1.1. *Transmitter*

In order to transmit the bitstreams of digital data (= square wave signal) from the biocontroller of the IBC patch, it is necessary to:

– create a sine signal at a frequency of around 330 kHz, to serve as a carrier for the communication of digital data (using an oscillating LC circuit; see the note below);

– apply OOK modulation to that carrier by the flow of digital data;

– obtain a voltage level for the signal which is appropriate for the intended application;

– apply the signal of that modulated wave to the electrode Te of the patch;

– apply that electrode directly to, or near to, the skin (including in an article of clothing) to achieve capacitive coupling between the IBC system and the skin.

NOTE.–

– The LC oscillating circuit is not simple to make. It is composed of a small coil L (~220 μH), arranged serially with a capacitor C (~1 nF), one of whose extremities is connected to the mass and whose mid-point is directly applied to the electrode Te. The whole setup is tuned to the value of the desired carrier (resonance frequency of the tuned LC circuit) (see Figure 5.48). This configuration generates a sine wave whose amplitude depends on the quality coefficient $Q = L\omega/R$ of the oscillating circuit and on the incident signal from the bioprocessor.

– In the case of the choice of IBC operating not at 330 kHz but at 13.56 MHz, the value of the inductance is reduced in the vicinity of 5 μH (which is achievable in some spires of the printed circuit), and the capacity is also reduced to a value of around 30 pF (directly realizable in the integrated circuit) (see the Conclusion).

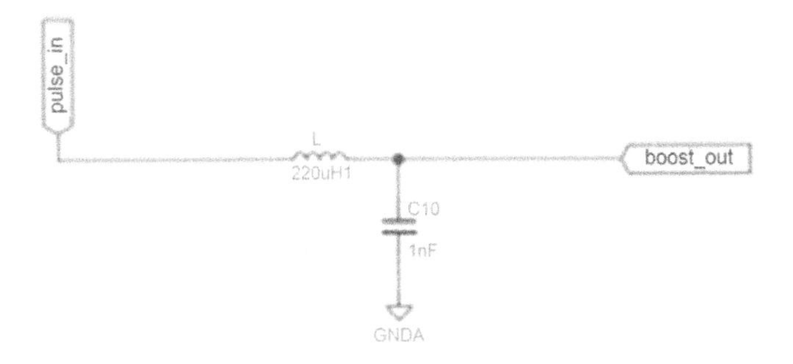

Figure 5.48. *Oscillating LC circuit of the transmitter at 330 kHz. For a color version of this figure, see www.iste.co.uk/paret/smartpatches.zip*

EXAMPLE.–

– If the square signal stream at 330 kHz exiting the biocontroller has an amplitude of the order of ~3.7 Vpp, with an LC circuit whose Q is of the order of 7.5–8, at the mid point of the serial LC circuit, we obtain the output voltage of approximately ~35.6 Vpp (see Figure 5.49a).

– This gives us 1.5 mW of power consumed during transmission.

– It is necessary to electrically isolate the electrodes to prevent resistive coupling, because we only want capacitive coupling to take place.

– Depending on the components used, direct contact with the human skin may draw too much current out of the receiver, and interfere with its function.

Obviously, the square wave present on the output pin of the bioprocessor is slightly disturbed (shown in yellow) by the presence of the LC oscillating circuit (Figure 5.49a). This effect may be cancelled out entirely by inserting a buffer stage, which would also enable us to increase the amplitude of the signal applied to the electrode (the plate) (see Figure 5.49b).

In this case, it is important not to exceed a certain voltage. Too high a voltage would be dangerous to use (48 Vp should be regarded as an absolute maximum for use on the body)!

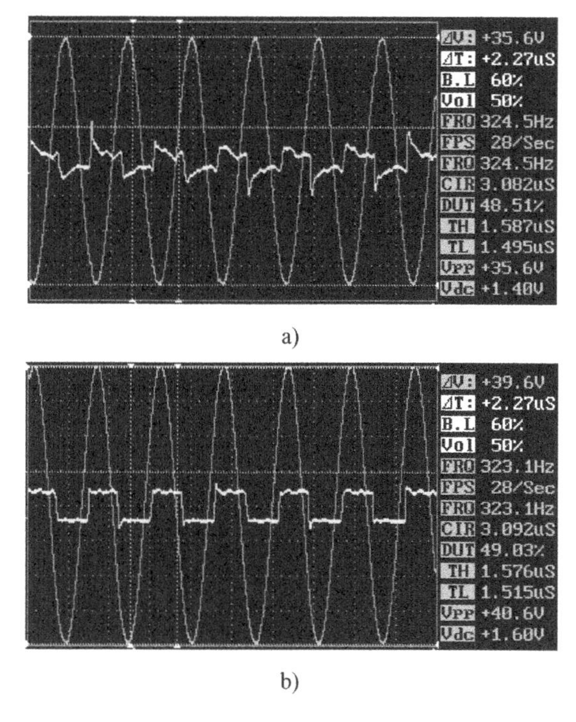

Figure 5.49. *Output signals for an oscillating circuit and a microcontroller (a) and with a buffer zone (b). For a color version of this figure, see www.iste.co.uk/paret/smartpatches.zip*

5.11.1.2. *Data coding*

– Modulation of the carrier wave, at the frequency of ~330 kHz, by digital data is performed by means of OOK, in which a "logic 1" activates the wave and a "zero" cuts it off.

– The bit duration is chosen as 200 μs, which translates to a bitrate of 5 kbits/s.

– To identify the start of the data transmission, generally a simple 8-bit preamble (known as a run code) is created and send in the form of a bitstream 10101010. Only when this preamble is detected will the decoder accept the subsequent bits as input data.

– A simple serial communication protocol, similar to RS 232 encoding with the addition of a parity bit and a stop bit, can be used to transmit data. Other types of bit coding can, of course, be used to transmit the data clock, etc.

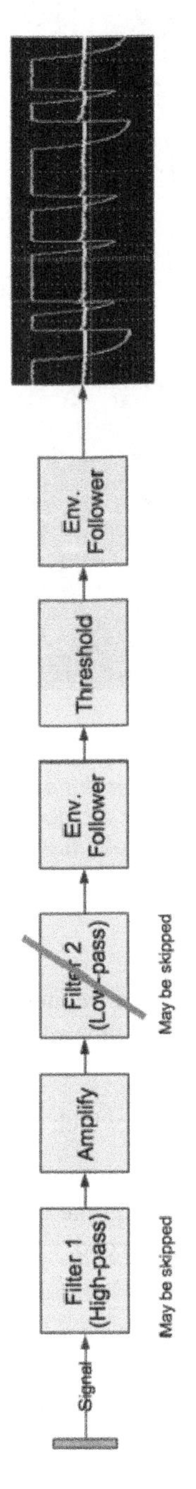

Figure 5.50A. *Example of the stages in the receiver chain. For a color version of this figure, see www.iste.co.uk/paret/smartpatches.zip*

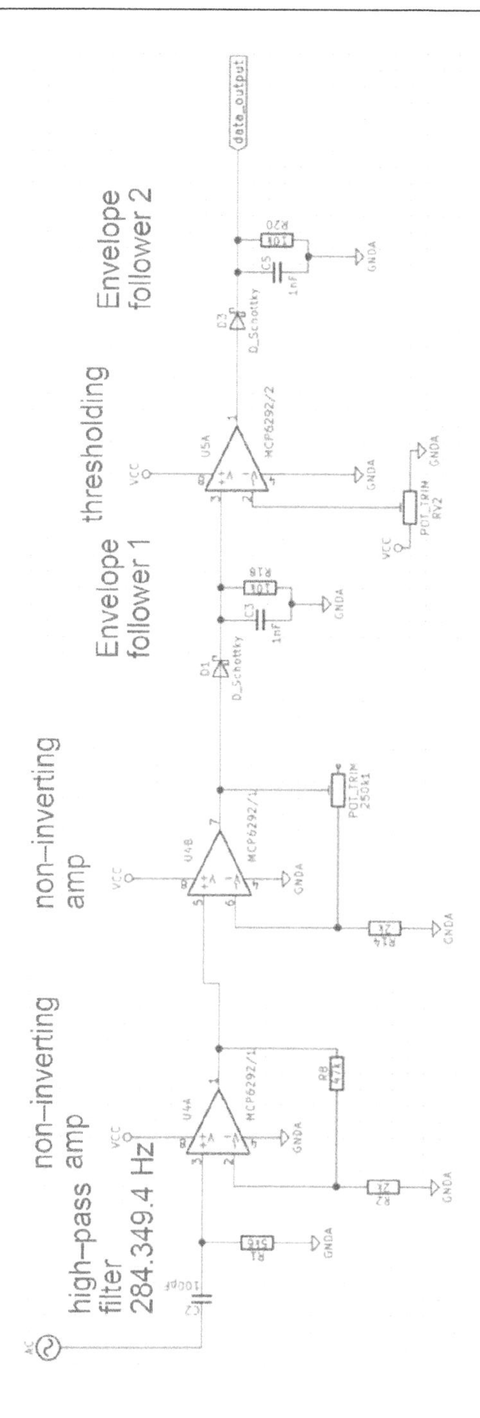

Figure 5.50B. *Concrete example of the receiver chain (cont.). For a color version of this figure, see www.iste.co.uk/paret/smartpatches.zip*

5.11.1.3. *Receiver*

The receiver must be able to receive and record the signal. For this, it is necessary to amplify, filter and (after thresholding) clean up the signal. The receiver circuit, for example, uses two envelope tracking stages to smooth the signal and avoid breakages in the bitstreams (see Figures 5.50A and 5.50B). In addition, once again, it is necessary here to isolate the receiver electrode to avoid resistive coupling, because in this setup, we are only looking for capacitive coupling.

5.11.1.4. Signal decoding

Obviously, to decode the signal, we simply reverse the encoding process.

5.11.2. *Example 2: for the initiated*

Since the publication of Zimmerman's original paper, a vast number of studies have been conducted the world over on the development of IBC devices and hundreds of studied have been carried out on transmission channels: how is it possible to reduce transmitted power, and what are the best types of bit coding depending on the desired datarates? Of course, all of this is in keeping with the standards and regulations in force. The tune to which the researchers dance is well known!

There are varieties of IBC systems in which some are more oriented toward medical applications and some more toward identification purposes in industry or professional settings. These systems differ mainly in terms of the operating frequency ranges chosen, the signal modulation methods, the bitrates achieved and the values of the electrical fields emitted.

To guide you in your choices for your own applications, the following conclusions of these works will help you choose the solutions which are best suited to the problems at hand, depending on the types of technical performances of biosensors and biocontrollers you wish to implant in your garment to make it smart.

5.11.2.1. *Experiments with datarates, modulation and consumption*

– In the United States, MIT conducted a detailed investigation of the physical limitations of IBC. The in-depth analysis of the transmission channel showed that on the body, amplifier noise and diaphony with other IBC devices influence the received signal. It was concluded that, for the final design of hardware, it would be best for the modulation to use FSK, which is capable of delivering a digital throughput of 9.6 kbit/s.

– Also in the United States, the University of Washington developed and described a system, based on Zimmerman's prototype, with which the research team achieved a datarate up to 38.4 kbit/s using carrier frequencies of 180 kHz and 140 kHz, with FSK modulation and a signal with amplitude 22 V. They then compared the BER and the levels of signals received with different distances between the body and the electrodes, and different positions of the electrodes on the body.

– Lindsey et al. studied the volume reduction of implantable elements (case for in vivo biometric systems), and developed a method which uses the ionic properties of the human body for signal transmission. The best results obtained for these applications were with a current of 3 mA and a carrier frequency of 37 kHz, using frequency modulation.

– In Japan, researchers have developed an EKG signal monitoring system with very low power consumption. The signal is transmitted galvanically between an EKG detector and a receiver with an amplitude of 20 μA. The device uses a carrier frequency of 70 kHz and Pulse Width Modulation (PWM). The system's total energy consumption is around 8 μW.

– At NTT (Human Interface Laboratories, Japan), Fukumoto et al. developed a portable system based on a variant of capacitive IBC, using analog frequency modulation with carrier frequencies between 50 and 90 kHz, and power consumption of 1.75 mW.

– Fuji and Ito (Japan) studied the transmission characteristics of the human body in the IBC system, and described the FDTD (Finite-Difference Time-Domain) simulation of the distribution of the electrical field around the digital model of a human arm, approximated by a parallelepiped measuring 5 cm × 5 cm × 45 cm, with relative permittivity $\varepsilon_r = 81$ and conductivity o = 0.62 s/m (dielectric characteristics of muscle). An IBC transmitter generating a signal at a frequency of 10 MHz and amplitude 3 V was used as the source of electrical field. The results obtained by simulation were compared with the measured results using the equivalent phantom arm in biological tissue. Findings showed that the signal was distributed in the form of a surface electromagnetic wave (creeping wave) along the surface of the skin. The simulation results were confirmed on realistic models of adults (both men and women), showing the spatial distribution of the electrical field around the digital model of the arm on which the IBC system was in place.

5.11.2.2. *In summary*

Figure 5.51 gives an overview of the solutions presented.

IBC capacitive coupling	Carrier		Signal amplitude	Signal encoding	Datarate
	Frequency	Modulation	Volt	Type	Bit/s
Fukumoto	90 kHz	FM	21		0.1 k
Zimmerman	330 kHz	OOK	30	RS232	2.4 k
Fujii/Ito	10 MHz	OOK	3		10 M
NTT/Docomo	10 MHz	OOK	25		10 M
Reynolds	70 kHz	FSK	10		9.6 k
Partridge	160 kHz	FSK	22		38.4 k
Hachisuka	10.7 MHz	FSK	1		9.6 k

Figure 5.51. *Summary of solutions presented here*

5.11.2.3. *Implementation of a professional IBC transceiver*

Given that the communication channel between the transmitter Tx and the receiver Rx using the human body is greatly different from a conventional hardwired or wireless communication channel, special implementation techniques are needed for the design of communications within the human body. Although each solution for an IBC transceiver listed in the previous sections has its own specific performances and has advantages and disadvantages in comparison to others, by way of example, we have selected two of them to illustrate in the following sections.

5.11.2.3.1. First solution

In this first solution, the IBC transceiver operating by adaptive FSK constantly scans in the frequency band from 30 to 120 MHz, searching for the frequencies at which the best performances can be achieved, the cleanest frequency range with low interference from ambient noise; it then uses the selected channel to transmit the data with FSK. This frequency shift technique is introduced in order to attenuate the variable loss, as a function of time or of space, over a given transmission channel.

To concretely implement frequency shifting with FSK modulation of the baseband signal, several RF blocks are needed. These RF blocks include an FSK modulator, a mixer and an LNA (see Figure 5.52). These of course lead to additional energy consumption.

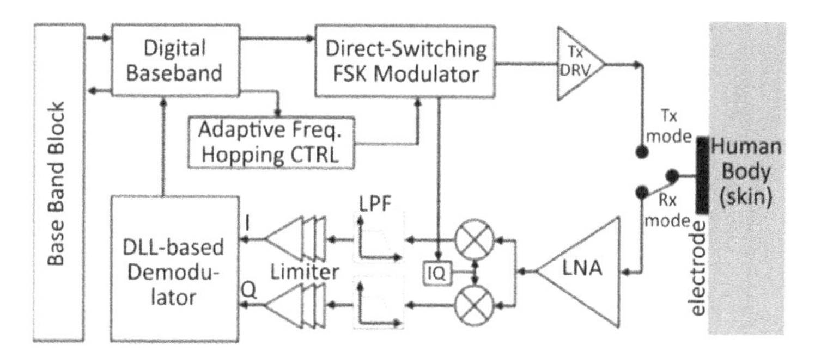

Figure 5.52. *Example of an IBC transceiver operating by adaptive FSK*

5.11.2.3.2. Second solution

In this second solution, the IBC transceiver uses digital transmission. In digital transmission, each element of the transceiver (including the AFE), the HF head operates in the baseband block. Instead of using an FSK modulator in the transmitter Tx and a demodulator in the receiver Rx, the IBC transmitter uses an FS-spreader and an FS-despreader to obtain processing gain on the received signal.

The use of digital transmission offers numerous benefits in comparison to other systems in terms of simplicity of implementation. Unlike the design using RF blocks for FSK modulation for a frequency shift, digital transmission does not need an RF block, which simplifies the electronic structures of the Tx and Rx, and also reduces power consumption. An example of the architecture of a transceiver is given in Figures 5.53A and 5.53B.

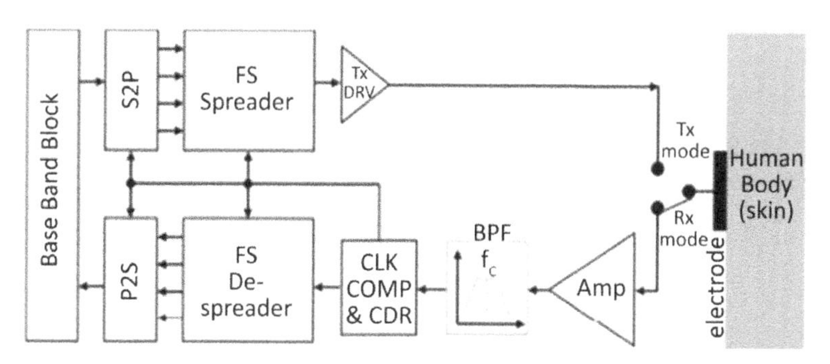

Figure 5.53A. *Principle of an IBC transceiver functioning in digital transmission*

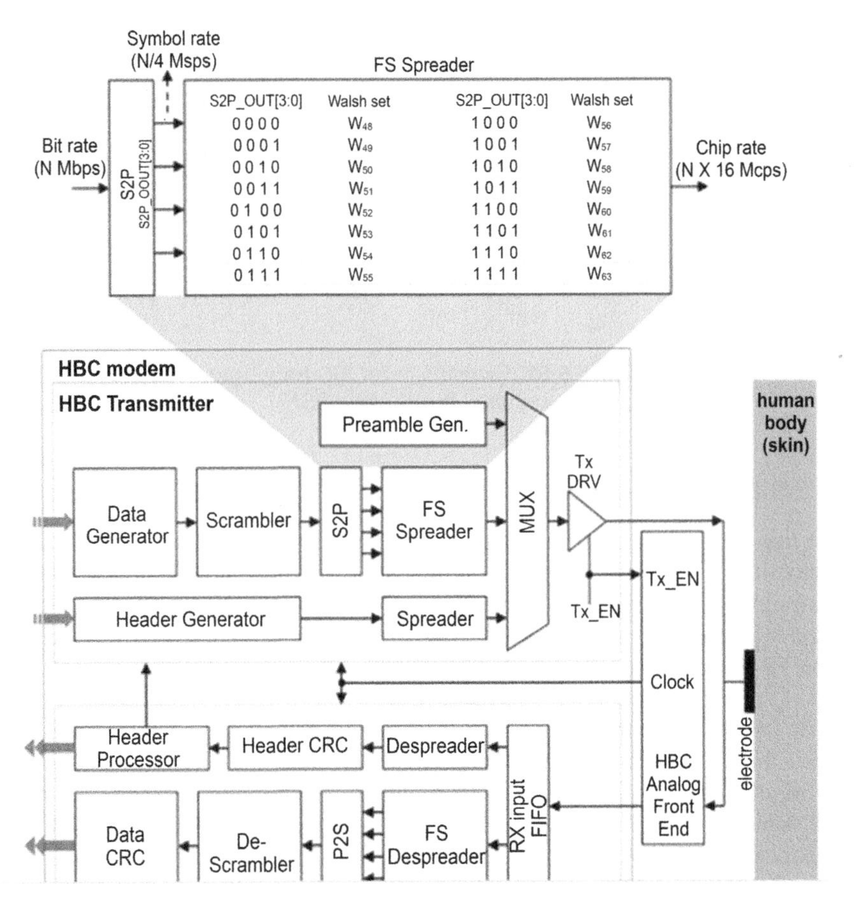

Figure 5.53B. *Diagram of an IBC transceiver functioning in digital transmission (cont.)*

This concludes this part of the book. We can now shift our focus to more concrete concepts.

Conclusion

Concrete Implementation of a Solution

By way of introduction to the final part of this book, let us briefly return to our starting point, and look again at our original goal in relation to applications for smart apparel and patches for health, wellness, etc.

We embarked on this long journey to illustrate one of the best possible ways to successfully implement clothing equipped with (multiple) biosensors for consumer electronics or for professional applications, allowing for the use of small, lightweight patches, implementable in flexible garments, operating without batteries, and capable of being installed in a network. Of course, at the time of writing (in mid-2021), the technologies are not yet all in place on an industrial level, but it expected that they will be in the very near future:

– Graphene technology will have become an industrial standard, and the graphene-based multiple biosensors that exist even today in the form of prototypes will be available industrially.

– As there will be a clearer picture of the quantities required for the applications, the manufacturers of biocontroller integrated circuits will have enriched their ranges of components with inclusive options (management of NFC, IBC, etc.) and also reduced the power consumption and costs of their products.

– Energy harvesting systems will be in place and high-value supercapacitors, rapidly rechargeable and not bulky, will be available.

– Small, washable patches will be available.

It will no longer be a dream, but a reality, and this book aims to present a clear picture of how it was possible, technically and financially, to bring this reality about.

C.1. Concrete realization of patches for smart apparel

From the very start of this book, we have presented our remarks from the point of view of concrete industrial implementation of technologies. Thus, this is the moment you have been waiting for! Of course, the examples given here are just that: examples, which everyone will need to adapt to their own applications; everyone will need to rerun the calculations at the time they wish to start up, but we have given some ideas of the orders of magnitude, which can be used from the very outset of a project to prevent certain disappointments. We hope this last section makes for enjoyable reading.

C.1.1. *Integration into clothing*

The following sections present examples of canvases of simple smart (and connected) apparel, in which the biosensor/bioprocessor and patch are included. Obviously, to construct such an example and remain near to the real world, it is necessary to make a number of hypotheses, estimated timescales, development costs, scheduling for implementation and estimations of the manufacturing price of prototypes and of the end product, which we shall present in the coming paragraphs. Obviously, you will still need to refine a given scenario to suit your own needs, in terms of costs, timescales and updates, and/or assess other scenarios if need be.

C.1.2. *Breakdown of electronic technologies*

From the start, let us give an idea of the functional content of an example of smart apparel, equipped with an OBC and IBC biosensor patch, operating without a battery, as a function of its breakdown into electronic blocks (see Figure C.1).

The communicating smart apparel which we have chosen to illustrate our example is a fairly complete PPE-type garment, so the scenario can actually be simplified to apply to readers' own projects. From a functional point of view, it includes the main elements: firstly, one or more garments (in the style of a parka), including the patches, and secondly, an external paired base (such as a smartphone).

The base is the part situated on the users' site, and the smart apparel (PPE) includes certain rights of access, allowing the different personnel on site to perform various functions with the base. Of course, the two elements of the PPE system (the base and the PPE garment itself) have various parts, each of which include electronic components.

In spite of the almost biblical aspect described below, the characteristics of the system are as follows.

Breakdowns of the blocks and functions	Block contents	Technology used	Integrated technology
Physical or biological data sensors			
A/D conversion			
Digital processing of original data	Filtering functions: Fourier, wavelet, refinement, etc.		
Algorithms/data fusion	Obtaining the pertinent data		
Data management	µC		
Local power supply	Energy harvesting	Piezo and NFC	
OBC and IBC connectivity	NFC	Near-field	
	IBC	Local	
	Deported *via* BLE	Mobile	
User applicational functions	Local		
	Deported	Mobile	

Figure C.1. *Breakdown of technologies present in smart apparel. For a color version of this figure, see www.iste.co.uk/paret/smartpatches.zip*

C.1.2.1. *Base*

NOTE.– The base is not described in detail in this book.

In addition to RF connection by BLE, which is necessary to ensure communication between the patches in the smart apparel and the base, in order to communicate with the outside world and provide the "I" side (Internet) of IoT, the base has multiple versions/options of RF connections, which must be possible on the basis of the degrees of the final applications decided upon with customers: a Wi-Fi connection to communicate with the user's box, a LPWAN SIGFOX and/or LoRa connection to communicate over a long distance (multiple kilometers) with an Internet gateway, or a GPRS/4G connection with the option of a 4G modem.

Figure C.2 gives an overview of the base.

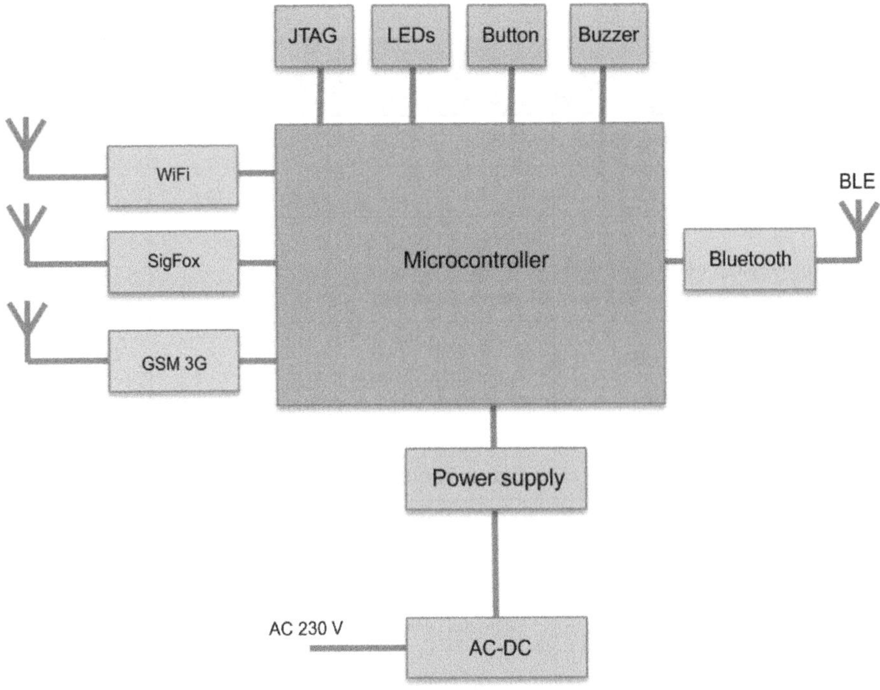

Figure C.2. *Overview of the base. For a color version of this figure, see www.iste.co.uk/paret/smartpatches.zip*

C.1.2.2. *Patch*

Two scenarios can be envisaged for the application of patch technologies.

C.1.2.2.1. Technology to apply the patch directly to the individual

Either the patch is stuck directly to the individual's skin, by means of a dressing, for example. Thus, all of the technology in the patch must satisfy the numerous criteria set out in the regulations in force, in particular, on health (see Chapter 1).

Flexible printed circuits for certain purposes, and rigid circuits for others

Let us begin our discussion with the printed circuits which accommodate the electronics of the patch. In this area, electronics engineers have gone from one surprise to the next, along with smart-apparel manufacturers. How is it possible to achieve an understanding between these two domains which are so different? On the one hand, we have electronics engineers, for whom integrated circuits and other components, mounted on printed circuits which were made first from Bakelite, then later from epoxy glass, and nowadays, thin, "flexible" polyester films. On the other hand, we have textile producers, for whom those elements are often too hard, too rigid, not flexible enough, and inapt for clothing in relation to the intended applications. In addition, these flexible films must truly be able to be folded without breaking or deteriorating, notably during washing-machine cycles; not be noisy when folded (for example, in the past, certain RFID tags were disallowed from use in the collars of top-of-the-range shirts because they caused a small amount of noise when the wearers turned their heads); not scratch, or cause skin irritation; etc.

There has already been major progress with graphene-based patches with polymer-based supports. The same is true of connections and connectors.

Electrodes

Of course, this is an important point, because quality of contact and information recording is of prime importance. Normally, the electrodes must be placed in as close contact as possible with the skin, including taking account of hair and sweat. When they are integral part of a patch (through the analyte), or of a garment (using capacitive coupling), they are included in the fabric, and therefore the information is subject to additional noise. Thus, by principle of merit, it requires more in-depth signal processing. In addition, stable contact must be maintained between the wearer and the sensor, with sufficient pressure being applied. The sensor must be comfortable to wear, stretchy enough to adapt to the users' body morphology, with as few stitches as possible and composed of skin-compatible fibers.

C.1.2.2.2. Technology to integrate the patch into a garment

Alternatively, the patch can be inserted into an article of clothing. In this case, it may be stitched, glued, woven, etc., into the thickness of the fabric or onto its surface. In addition, its placement depends on the desired application, the parameters to be measured and whether it is acceptable for the sensor to have intermittent contact with the skin.

Figure C.3 shows the example, in the manufacture of a garment, of the insertion of a patch, stitched into the thickness of the shoulder pad, where the sensors are not in direct contact with the skin. In this case, IBC takes place by means of a network (or not) capacitively coupled through the fabric of the garment, and then through the skin.

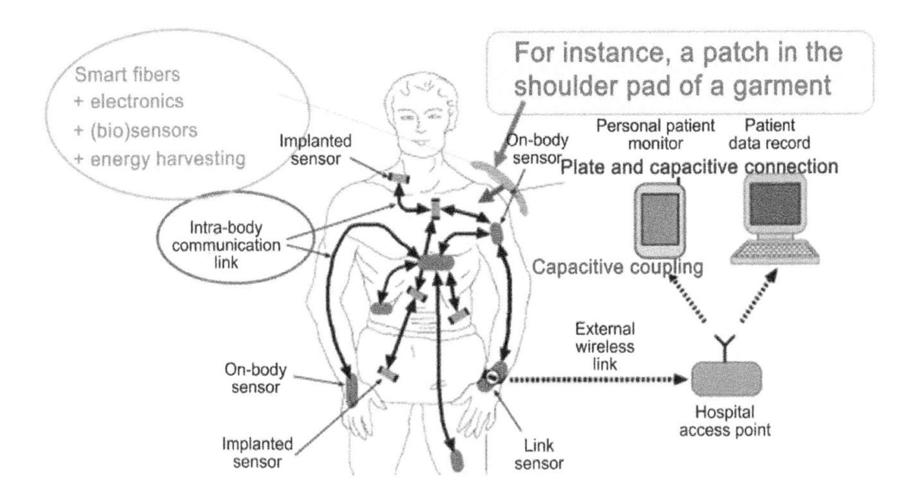

Figure C.3. *Capacitive IBC: applications included in a smart garment. For a color version of this figure, see www.iste.co.uk/paret/smartpatches.zip*

C.2. Example of the manufacture of a patch

Consider the example of a patch (on a graphene-based medium) placed on the surface of the body and mounted either directly onto the skin or indirectly into a garment over the skin surface. The purpose of the patch will be to constantly measure parameters and signals which are physical, biological and chemical in nature (example: evolution of temperature levels, pH, biomarkers, etc.) at a given frequency.

C.2.1. *Overall overview of the patch*

Figure C.4 presents an overview of the patch.

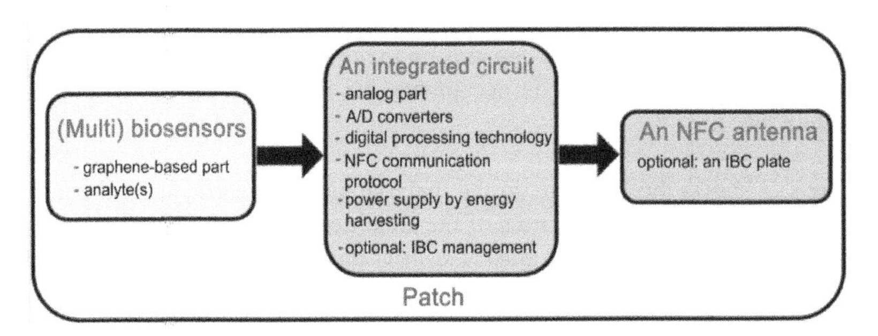

Figure C.4. *Diagrammatic overview of the patch. For a color version of this figure, see www.iste.co.uk/paret/smartpatches.zip*

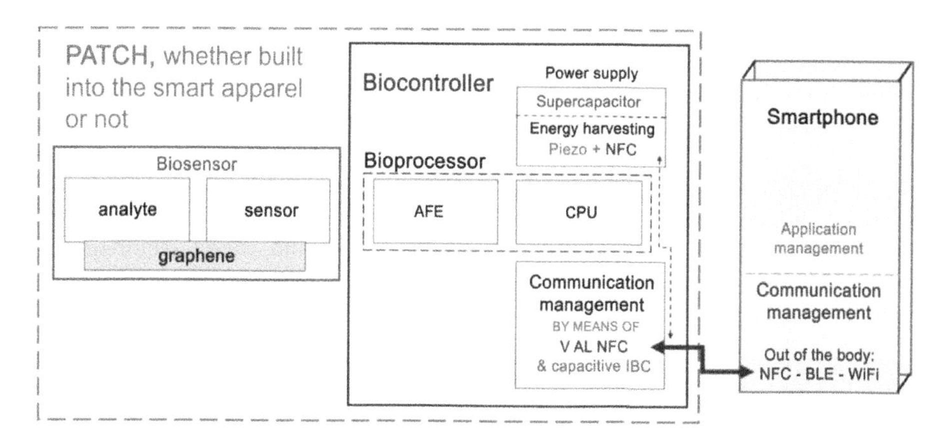

Figure C.5. *Diagrammatic overview of a patch for smart apparel. For a color version of this figure, see www.iste.co.uk/paret/smartpatches.zip*

The functional block diagram of the electronics in a patch that can be implemented on/in a smart garment is given in Figure C.5. The whole constituted by the patch and its environment comprises five distinct elements:

– a *biosensor* (analyte + sensor) designed around a graphene support (see Chapter 2);

– a *biocontroller* which includes (see Chapter 3):

- an AFE system, its analog part, and its A/D converters,

- a CPU which comprises a RAM, a ROM (E2PROM),

- the management of OBC protocols by means of NFC (whose capacity of tuning of NCF of around 60 pF is etched on the silicon of the bioprocessor's integrated circuit) and IBC,

- management of the energy source and recharging by HF, through NFC communications, or by means of mechanical vibrations with integrated piezoelectric sensors on the chip of the biocontroller;

– a discrete *supercapacitor* made of graphene, acting as an energy reserve for the *energy harvesting* system (see Chapter 4);

– the *NFC antenna* (see Chapter 5);

– a dedicated reader or a *smartphone communicating* by *NFC*.

C.2.1.1. *Physical realization of the smart apparel patch*

Figure C.6 summarizes the technical choices, the reasons for which we have investigated this at great length in this book.

This offers readers a glimpse of the technological way in which a patch can be concretely implemented. The patch in question may, if required, be inserted directly into a smart garment, and comply with the numerous and varied requirements expressed in the earlier chapters: it needs to be flexible, non-intrusive, batteryless, washable, tough, etc.

Functions	Components	Proposed technologies		
Sensor	External, discrete	Graphene-based FET	Graphene	
Biosensor		Graphene-based FET + analytes	Analytes dedicated to the applications	
AFE	Integrated mono-circuit	Converters	Low consumption	
Microcontroller		CPU ARM		
Power supply to the patch		Energy harvesting	RF → NFC	Boost, buck, MPTT
			Vibratory energy → Piezoelectric	
		Reservoir	Supercapacitor	External
Out-of-Body Communications (OBC)		NFC Forum ISO 15693	Backscattering, ALM	Distance 5–8 cm
		Small loop antenna	High-Q graphene	Relay to smartphone
Intra-Body Communications (IBC)		Capacitive		Either alone or in a BAN
		Small plate antenna	Metal film	

Figure C.6. *Breakdown of the electronic technologies present in the patch. For a color version of this figure, see www.iste.co.uk/paret/smartpatches.zip*

Figure C.7 shows the breakdown of the possible types of applications/uses for each technology.

Patch	Uses			Technologies				
Types	Lifespan	Application	Implementation	Biosensor	Bioprocessor (AFE + CPU)			Specificity
Biosensors	Days			Material	Security and GDPR	Power supply	Communication	Washable
Mono	1	Medical sport	Stuck to the skin	Graphene + Specific analytes	µC with a secure element	Energy harvesting with NFC and piezoelectric components	OBC by NFC	N/A
	3–5	Medical sport						30 to 50 times
	15	Pseudo-medical						
Multi	A few days	Pseudo-medical						
	30	Wellness, sport	Integrated into a garment					
Mono	365	Pseudo-medical, Wellness, sport					OBC NFC IBC capacitive	
Multi and network								

Figure C.7. *Breakdown of technologies present in patches. For a color version of this figure, see www.iste.co.uk/paret/smartpatches.zip*

C.2.1.2. *Physical implementation of a patch*

Figure C.8 illustrates a possible example of the physical production of a satisfactory patch, either for an isolated patch without IBC, to be affixed to the skin, or a solution with a patch with IBC for integration into a smart garment.

NOTE.– Often, the physical make-ups of these devices are subject to numerous patents, either current or in the process of certification. Thus, the construction presented in Figure C.8 and described below must be considered a generic (and patented) example of a possible construction.

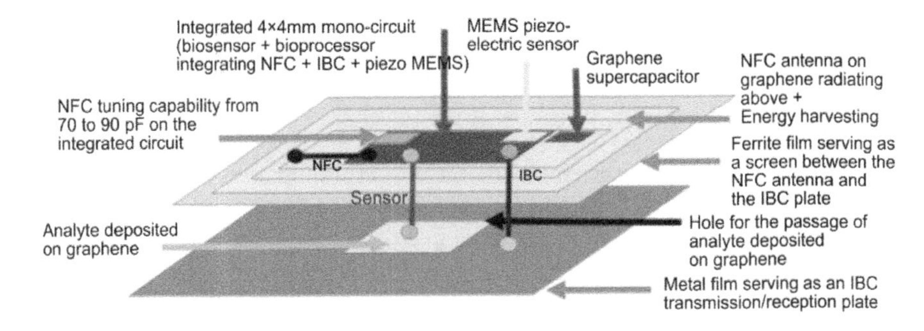

Figure C.8. *Breakdown of the layers of the patch (thickness of the stack of layers of the patch < 600 µm, including NFC and IBC). For a color version of this figure, see www.iste.co.uk/paret/smartpatches.zip*

C.2.1.2.1. The different layers of the patch and their roles

We shall now describe the multi-layered structure of the patch with IBC, one layer at a time:

– The central yellow layer is built on a thin film of graphene, for a number of reasons:

- At the center of this layer is the integrated bioprocessor monocircuit including all the electronics for the analog and digital processing of the elements biosensor + AFE + CPU + NFC communication management (and the tuning capacity of around 50 pF of the NFC antenna) + management of the IBC system + MEMs piezo sensor + management of energy harvesting via piezoelectrics and NFC + security/crypto/secure element.

- It directly accommodates the engraving on graphene of the NFC antenna (serving as a relay with a smartphone). Because of the patch's small dimension (inductance of the antenna of the order of 5 µH), and the desired communication range, the electrical characteristics of that antenna need this type of highly conductive material to ensure its quality coefficient Q. The whole of the surface of the conductors in the NFC antenna also serves as the surface for the plate Te of the IBC antenna operating at 13.56 MHz.

- This layer also accommodates the engraving on graphene of the "supercapacitor" (around 700 mF) constituting the energy reserve of the energy harvesting system, ensuring the patch remains operational between recharges (every two or three days.

– The beige layer, for its part, is made up of a thin film of plastoferrite with very low magnetic reluctance, designed to guide and reinforce the magnetic field H created, radiated upwards by the NFC antenna in order to improve the communication range and the signal-to-noise ratio, and also, to simultaneously form a magnetic screen below the patch, where the IBC capacitive emission plate is located.

– Finally, the blue layer, which serves as the electrode plate Tb with capacitive transmission for IBC, is formed of a thin sheet of polyester, covered in a metal film with, at its center, a part reserved for the deposition of a layer of graphene, on which the analyte will be laid, to form the structure of the desired biosensor detector.

– The local connections in the patch are formed by a jet of graphene ink between 50 and 100 μm thick.

NOTE.–

– Energy harvesting for the "piezoelectric and NFC" system. The energy harvesting by which the supercapacitor, made of graphene technology, is charged and recharged, uses two simultaneous techniques: NFC and piezoelectricity:

- during each NFC consultation/interrogation of the patch with a smartphone, the supercapacitor is recharged simultaneously, and can function autonomously for 2–3 days;

- in parallel, every time the person moves (and thus their clothing also moves), the piezoelectric sensor provides further charging to the same supercapacitor and thus provides part of the patch's permanent power supply.

– Quality factor of the patch's NFC antenna. A patch with an NFC solution, which operates at close range and at a high frequency of 13.56 MHz requires the use of a "loop" antenna with several spires, whose dimensions must be compatible with the small dimensions of the patch (approximately the size of a €2 coin). As pointed out previously, this leads to a low inductance value L of the antenna (of the order of a μH), meaning that the value of the serial resistance R of its coil must also be low in order to satisfy a L/R ratio giving a quality coefficient of the antenna $Q = L\omega/R$ conforming to ISO 15 693, used in the NFC Forum V specification.

This constraint is solved by producing a graphene antenna of small dimensions (copper is not suitable for the purpose), whose mechanical robustness also means that it will be stretchy, washable and withstand wash cycles to which the smart apparel is subjected.

C.2.2. *Concrete industrial example of smart apparel*

Before describing an industrial solution of smart apparel, in order to define the framework of our discussion, let us briefly provide a grounding (see Figure C.9): which biological parameters do we wish to measure? How do we want to measure them? With what degree of accuracy? Where on the body do we need to take these readings? What will the cost be? Etc.

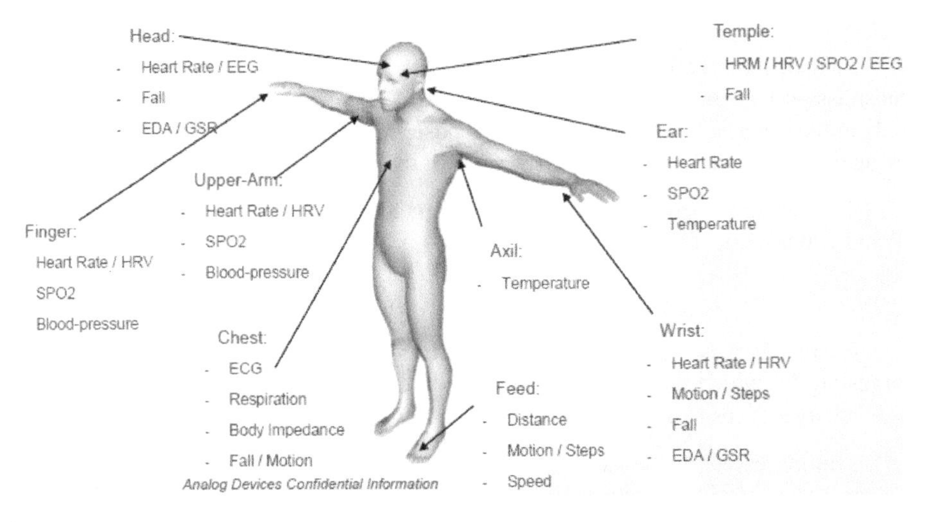

Figure C.9. *Potential locations for probes. For a color version of this figure, see www.iste.co.uk/paret/smartpatches.zip*

Having, at length, reflected upon and answered these questions and drawn their own conclusions, in a manner similar to the company BioSerenity (Paret and Crégo 2018), the French company Chronolife produced a professional, connected, smart garment (not yet available to the consumer electronics market), oriented toward the market of health monitoring and predictive analysis, collecting and transferring, to its secure, authorized health servers, data taken from physiological measurements (electrocardiograms, abdominal breathing, thoracic breathing, body temperature, pulmonary impedance and overview of physical activity).

This smart-apparel solution is built on a proprietary algorithm capable of data fusion from a continuous stream of complex data from multiple sources; basing its analysis on pattern deformation or overall changes in state, just as a human doctor would; raising alerts to predict changes in a person's state of health; operating on processes with low power consumption, with local processing using little energy and without requiring a connection to a server, which ensures a high level of security.

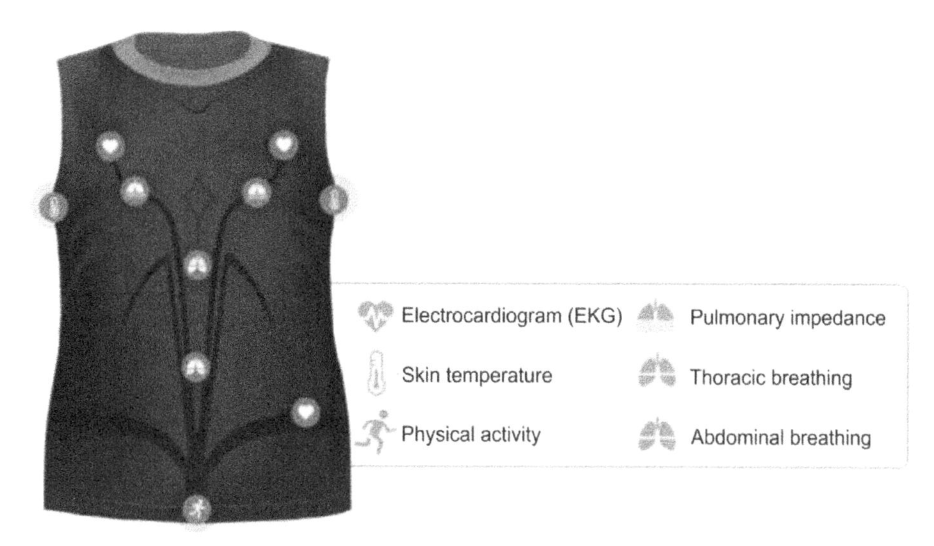

Figure C.10. *Example of application: Chronolife gilet. For a color version of this figure, see www.iste.co.uk/paret/smartpatches.zip*

This smart garment, designed to resemble a "normal" garment, is machine washable, without the need to remove the electronics (the sensors) or the battery, which are totally integrated into the fabric. The data harvested are then analyzed on the patient's smartphone application, which includes the proprietary algorithm technology.

Beyond personal use, this smart garment allows pharmaceutical groups to test and validate their solutions in clinical trials and therapeutic effectiveness, notably thanks to telesurveillance technology classed and certified 2a in healthcare. It allows insurance companies to develop and manage services to support elderly and vulnerable people in their own homes, with better risk prevention, the aim being to prevent hospital (re)admissions and the associated healthcare costs. The smart garment can also be used by organizations in the world of leisure and sport to support their programs to train and monitor high-level athletes.

It is worth noting that this solution relies more on the content, the software technology of its system, than on the container – the hardware or fabric upon which the software is hosted. The advantages of this solution and these choices echo those which we have outlined at length throughout this book:

– a neuromorphic algorithm to record physiological data;

– local computational tasks using low-capacity processors and multi-parameter predictive analysis performed on the user's smartphone for integration and continuous analysis;

– data storage locally without an energy recharge (~20 hours);

– connection to cloud servers by a smartphone or Wi-Fi;

– no cabling: the fabric is conductive to prevent unwanted friction;

– greater comfort for users; the devices are comfortable and pleasant to use;

– conforms to medical requirements, improved observance of treatments by patients and therefore relief of the pressure on hospitals, constant collection and analysis of data, which is of interest to pharmaceutical groups wishing to verify the observance of treatment undergoing testing, and prove that it is effective.

In the short term, the primary objective is to rapidly drive down the production costs of smart apparel (which was around €250 per unit in 2020 for professional equipment, not yet available to the general public, and an additional subscription or around €20 per month) so that their price can be covered by reimbursement mechanisms linked to medical devices (see Chapter 1).

C.3. Industrial patch technologies

In this book, we wish to set forth the main points of an example of concrete industrial and commercial production of a smart garment based on the production of a few thousand units each year (for example, some 5000 smart garments each year, produced over the course of 4 years, totaling 20,000 units). To bring such a product to fruition, there are two main paths that we could go down:

– The development phases including: the design of the electronic parts with a view of industrialization and serial production; integration of firmware of the application and the onboard software and adaptations depending on the design of the patch; design of the "mechanics" with a view to producing working prototypes.

– The industrial phase of the project, including certification tests, conformity, etc. (CEM, "CE", etc.), industrialization, serial production fabrications of the product for the intended quantities, and all of this often takes months.

Following this technical overview, let us now move on to estimating the costs of these solutions.

C.3.1. *Cost aspects: CAPEX and OPEX*

In the wake of the foregoing chapters evoking the points of architecture of solutions, technical principles, components, etc., the next section deals with a point which is generally somewhat problematic: the costs of the electronic part of a patch solution for a communicating smart garment. We will need to address this aspect some day, and it is best to do so fully informed.

Knowing that, in general, numerous economic and financial points need to have been examined – in particular, the viability of the project – meaning the financial and marketing aspects, i.e. the establishment of the provisional budget to be allocated to the study of the project, let us now turn our attention to a host of new and very pedestrian aspects: CAPEX and OPEX. These two notions are fundamentally important in the means of industrializing and commercializing a patch for a connected smart garment.

– *CAPEX* (capital expenditure), refer to fixed investments, i.e. expenses which have a positive impact in the long term. For example, CAPEX include all costs for the development of a connected smart parka (non-consumable), which can be useful later on.

– *OPEX* (operational expenditure), are the running costs to operate a product, a business, or a system. Examples include the annual cost of cloud services, paid to Amazon for services provided by its cloud.

C.3.1.1. *CAPEX*

In relation to connected smart apparel, CAPEX are divided mainly into two areas:

– The cost of the hardware, otherwise known as the BOM: Bill of Material.

– The cost of the software specific to the smart apparel itself (management of communication protocols, etc.) which includes its connection to the cloud, so as to provide a spinal column to the applicative software.

We shall start with the BOM for the electronic part.

C.3.1.1.1. BOM (Bill of Material) of smart apparel

The BOM (cost of material, components, etc. excluding taxes) of the electronic part of the smart apparel itself is only the beginning of the story: the point at which, unfortunately, far too many people stop.

C.3.1.1.2. Development of the electronics and the environments

It is necessary to invest time and spend money on the design, development and test procedures for the electronic cards making up the products. In parallel, there are various phases of development of the embedded software, including the low-level software layers for resource management (for the NFC, IBC, etc.), the on-board application itself with its links to the mobile phones, tablets, its look, and the creation of the environment for tests and validation, which will also serve during the production process.

C.3.1.1.3. Planning

Let us now look at the time aspect. This often takes the form of a detailed estimated schedule of the development phases of the project and an estimation of the preserial development phase, with a PERT chart (Program Evaluation and Review Technique).

C.3.1.1.4. Development costs

To perform all the numerous segments of the PERT mentioned above for each part of the smart apparel, this takes time which, multiplied by an hourly rate, costs money, of course excluding the costs of warranty, after-sales serves and transport depending on the production sites (Europe, the Far East) and delivery in Europe, the size of the production batches (multiple batches or a so-called mono-shot) and shipping, delivery, etc. In short, all of this gives us an initial idea of the "entry ticket" due to the BOM + research + industrialization costs.

C.3.1.1.5. Accreditations and certifications

This is all very well, but in order to have the right to legally commercialize the product, obviously, we must have it accredited and certify all the stacks of protocols (Bluetooth, NFC, SIGFOX, etc.), which requires knowledge, energy, time, and therefore, money, which must be added to the CAPEX. By way of example, below are listed a few types of accreditations which we must be able to obtain: CE precertifications and certifications, compliance with the security regulations IEC/EN 62 061, compliance/certification with the RF regulations issued by the ETSI and FCC, security certification with France's ANSSI (*Agence nationale de la sécurité des systèmes d'information*), operational/standard compliance, etc.

Above all, it is crucial to be honest about a system's performances, so as not to be prosecuted by the DGCCRF (*Direction générale de la concurrence, de la consommation et de la répression des fraudes*: Directorate General for Competitition, Consumption and Fraud Prevention) for unfair competition and misleading advertising. Put simply, there is a very long path that we must walk for every new model of a patch and/or garment that is launched into the industry, and contrary to popular belief, the cost involved is quite significant.

C.3.1.1.6. Serial cost price of a patch

Knowing the projected quantity of patches/garments that we wish to manufacture, having reached this stage, we have sufficient information to make an initial estimate of the gross serial manufacture price of the patch. To this must, of course, be added the cost of NRE (Non-Recurring Engineering) for each item (one-time cost of research, development, design and testing, averaged per patch). We then obtain the estimate of the factory cost price (e.g. for the patch, see Figure C.11).

Cost price of a patch			**Price exc. tax**
BOM of components			€32
Assembly (~30 %)			€10
Gross salary and costs/year (social security contributions, etc.)	€40 k/year average × 1.4	€56 k/year	
Number of hours worked per year	35 hrs × 45 weeks	1575 hours/year	
Hourly price	€56 k/1575 hrs	35,50 €/h	
Research and development costs	1000 hrs × €35.50/hr	€35.5 k	
Intended production run of garments	5000 per year over 4 years	20,000	
Number of patches per garment	3		
NRE per patch	35,500/60,000		€0.60
TOTAL			**€42.60**

Figure C.11. *Initial estimate of the serial manufacture cost price of a patch*

C.3.1.1.7. Sale price of a patch

On the market, it is usual for the ratio of the sale price excluding tax to the factory cost price excluding tax to be around 1.7–1.8. This price difference generally includes: a portion dedicated to funding research into future products, the cost of after-sales services, the cost of communication, advertising, a payout for the company's shareholder, and above all, the remuneration/margin demanded by the players in the distribution network.

Consider the example of a sale price of the product to the end customer (professional excluding tax or consumer electronics including tax, with 20% VAT), around:

42.6 exc. tax × 1.8 = 76.70 exc. tax × 1.2 = ~€92 exc. tax per patch or €115 inc. tax

Readers are, of course, invited to run their own calculations with their own data.

C.3.1.1.8. Software development of smart apparel applications via the Internet

Once the hardware part of the product has been set in stone, we come to the very heart of the matter, because now, we need to turn our attention to the development and costs of the software parts belonging to Cloud Computing, which will be situated and hosted in the cloud (data management, brokering, security, etc.) and how these aspects will interface with the end user. These costs may, in part, be included in the NRE.

C.3.1.1.9. Cost of the platform and the application in the cloud

For the CAPEX, the first time that a company embarks on such an adventure of a patch for smart apparel, starting from scratch, to produce the functional totality of the applicative software platform (management of all protocols: HTTP, MQTP, client interface, user-friendliness of sites, etc.), we need to account for an engineering cost of person/year of salary + employer social security contributions + premises + etc. = in short, a total of around €xx k. This software part often costs as much as the hardware. Thus, it needs to be added to the final sale price, excluding tax. Then, for any new project, based on the groundwork laid previously, we simply need to make adaptations to fit each specific applicative project. It is up to you to balance up all these costs and charge your end client accordingly, and formulate a plan as to how to achieve a return on investment.

C.3.1.2. *OPEX*

How high will operating expenditure run? Once again, we need to subdivide the smart-apparel solutions in order to gain an idea of the OPEX for a product, a business, or a system, and in particular, of whom these costs will be borne by:

– What are the connection costs, with and without Internet connection?

– Is the operator unique and/or proprietary to the network (Orange, Bouygues, SigFox, LoRA, etc.)?

– What is the cost of rental of the LPWAN used?

– Will the smart apparel be talkative (which requires a great deal of bandwidth), or less so (requiring less bandwidth)?

– Are the smart garments near or far from the cloud host?

– What are the impacts of these parameters on the costs of each of these communications?

– Once we know the monthly or yearly cost of communications by the smart apparel, multiplied by millions for products released on an industrial scale, is that sum to be paid on a monthly, yearly, one-off (etc.) basis?

– If the network extends overseas, which roaming costs are applied by the network provider to reintegrate the data locally?

– Is the operator remunerated for feeding back the information and transmitting it to the users? In what form does this data come in? Raw data? Preprocessed data? Where do they come from?

– There are many more questions of a similar ilk!

In fact, this is where all the work starts.

Cost of the applicative platform in the cloud

Once the pseudo-operators or operators have been paid, you need to create your software applications for the end clients and those surrounding the broker, which are both situated in the cloud, and of course, finance them partially under consumables (OPEX). Thus, it is up to you to obtain the relevant information from your preferred service providers.

C.3.2. *In conclusion*

If the marketing department of your company believes that the intended smart apparel application (e.g. a total of 20,000 units produced) may be of interest to 5 different clients, you need to be sure that for each of these target clients, the amount invested as CAPEX for the company (event before the first item is sold) represents interest, before commencing any activities.

To conclude, let us reiterate the famous phrase quoted much earlier, at the start of the book: “The ‘sellable’ sale price of the product must be compatible with the ‘buyable’ price for the end customer to whom we wish to sell it”.

Now, all you need to do is choose your camp!

Epilogue

We have now come to the end of this book, the goal of which was to present a broad overview of graphene-based biosensors, with a view to applications in patches in the context of smart apparel and other smart connected devices (securable), the regulatory, normative, economic (etc.) frameworks applied to them, their applications, their technologies, and to open the door to an understanding of the complete chain in smart apparel which can be connected via the Internet, from the ideas behind their designs to the industrial realizations.

Throughout this book, we have sought to provide as clear an indication as possible about all the lengthy and various steps that we need to know about before venturing out into the vast jungle of connected smart apparel. It is true that certain commercial propositions may appear to make this journey simpler, and may lead people to believe that it is easy to implement such applications, but it is important to be completely aware of what happens behind the scenes, in order to guard against future surprises.

Thus, our main aim was to train the reader on how to set up permanent gateways between the smart apparel, hardware and software, between the electronic components and the computer system, between financial and societal values, etc.

We hope we have partly fulfilled the task we set out to perform in writing this book. If you have any questions, comments or (constructive) remarks, please do not hesitate to contact the authors on the below e-mail addresses. We will always be glad to hear from you. Constructive exchanges can only be enriching for all involved!

Dominique Paret dp-consulting@orange.fr

Pierre Crégo pierre.crego@mercury-technologies.fr

Pauline Solère pauline.solere@yahoo.com

Glossary

	Term
BAN	Body Area Network
BASN	Body Area Sensor Network
BCH	Bose, Ray-Chaudhuri, Hocquenghem Code
BSN	Body Sensor Network
CCA	Clear Channel Assessment
CRC	Cyclic Redundancy Check
CSMA/CA	Carrier Sense Multiple Access with Collision Avoidance
DTDMA	Reservation-Based Dynamic TDMA Protocol
EFC	Electric Field Communication
FCS	Frame Check Sequence
FM-DTDMA	Frequency-Based DTDMA
FM-UWB	Frequency Modulation Ultra-Wideband
H-MAC	Heartbeat Driven MAC Protocol
HARQ	Hybrid Automatic Repeat Request

HBC	Human Body Communications
HCS	Header Check Sequence
IR-UWB	Impulse Radio Ultra-Wideband
ISM	Industrial, Scientific and Medical
LPL	Low-Power Listening
MAC	Media Access Control
MICS	Medical Implant Communications Service
NB	Narrow Band
PB-TDMA	Preamble-based TDMA
PHR	Physical Layer Header
PHY	Physical or Physical Layer
PLCP	Physical Layer Convergence Protocol
PPDU	Physical Layer Protocol Data Unit
PSDU	Physical Layer Service Data Unit
SC	Scheduled-Contention
SFD	Start-of-Frame Delimiter
SHR	Synchronization Header
TDMA	Time Division Multiple Access
UWB	Ultra Wideband
WBAN	Wireless Body Area Network
WBASN	Wireless Body Area Sensor Network

Authors

Dominique Paret

Holder of an electronics engineering degree from Breguet-ESIEE, and of a DEA in physics from the Faculté des sciences, Paris, Dominique was, for many years, Emerging Business and Innovation & Systems – Advanced Technical Support at NXP. His work looks at both consumer-electronics and professional engineering technologies: RFID, contactless, NFC, geolocation, IoT, Geoloc, Zigbee, BT, BLE, UWB, UNB, IEEE 804-xxx, and automotive technology. He sits on numerous committees at AFNOR, the ISO, the CEN, the ETSI, and the BNA, etc.

In parallel to this work, Dominique has taught as a final-year professor at 15 engineering schools, both in his native France and elsewhere in the world. He is the founder and CEO of the businesses dp-Consulting – providing consultation and technical advice – and IBC Research. He has also authored over 40 (very large) technical manuals in French (Dunod, ISTE), English (ISTE/John Wiley), Spanish (Paraninfo), Korean (ACCORN) and Chinese (PHIE).

Pierre Crégo

Pierre holds a degree in electrical and microelectronic engineering from ESIGELEC. He has worked for STMicroelectronics, Alcatel (Nokia), AT&T and Gemalto. Pierre is an expert in bank payment security, contactless technologies (readers and applications), e- and m-Commerce applications, multi-application cards and electronic signatures. At present, he is the CEO of Mercury Technologies, and has also created and approved a number of payment establishments.

With Dominique Paret, Pierre has co-authored the book *Wearables, Smart Textiles & Smart Apparel* (published in French and English by ISTE (Paret and Crégo 2018)), and is the co-founder of the expert group RGPD Associates. In addition, Pierre is the accredited Lead Auditor for GDPR Audits and the establishment of Data Protection Officer (DPO) structures.

Pauline Solère

Currently approaching the end of her engineering studies at ESIEE Paris, and highly interested in e-health technologies, Pauline has received training in biotechnology and e-health on the end-to-end chain of health data acquisition, from sensor to exploitation, value creation from the data and, for example, the implementation of solutions relating to the mechanisms of the cardiorespiratory system. She is also interested in the fields of Big Data, programming and cybersecurity.

In parallel to her studies, Pauline has worked with Pierre and Dominique in fundamental studies and applied research on graphene.

References

Arvind, D.K. and Bates, A. (2008). The speckled golfer. In *Third International ICST Conference on Body Area Networks*. Belgium.

Barakah, D.M. (2012). A survey of challenges and applications of wireless body area network (WBAN) and role of a virtual doctor server in existing architecture. In *Third International Conference on Intelligent Systems, Modelling and Simulation*. Kota Kinabalu, Malaysia.

Baskaran, R. (2012). An overview of applications, standards and challenges in futuristic wireless body area networks. *International Journal of Computer Science Issues*, 9(2).

Bluetooth Special Interest Group (2010). Bluetooth specification version 4.0. Adopted Bluetooth Core Specification.

Chartier, D. (2007). IEEE launches new working group for Body Area Network tech: A new short-range, low power standard could allow future gadgets to... [Online]. Available at: https://arstechnica.com/uncategorized/2007/12/ieee-launches-new-working-group-for-body-area-network-tech/.

Chatterjee, G. and Somkuwar, A. (2018). Design analysis of wireless sensors in BAN for stress monitoring of fighter pilots. In *16th IEEE International Conference on Networks*. Canada.

Chávez-Santiago, R., Khaleghi, A., Balasingham, I., Ramstad, A. (2009). Architecture of an ultra-wideband wireless body area network for medical applications. In *2nd International Symposium on Applied Sciences in Biomedical and Communication Technologies*. Slovakia.

Chen, M. and Gonzalez-Valenzuela, S. (2011). Mobility support for health monitoring at home using wearable sensors. *IEEE Transactions in Information Technology in Bio-medecine*, 15(4).

Chin, C.-A. (2021). Advances and challenges of wireless body area networks for healthcare applications. In *International Conference on Computing, Networking and Communications*. United States.

Chung, W.-Y., Lee, Y.-D., Jung, S.-J. (2011). A wireless sensor network compatible wearable u-healthcare monitoring system using integrated ECG, accelerometer and SpO2. In *30th Annual International Conference of the IEEE Engineering in Medicine and Biology Society*. Canada.

Dam, V.T. and Langendoen, K. (2003). An adaptive energy-efficient mac protocol for WSNs. In *Proceedings of the 1st ACM Conference on Embedded Networked Sensor Systems (Sen-Sys)*. United States.

Ghasemzadeh, H., Jafari, R., Prabhakaran, B. (2010). A body sensor network with electromyogram and inertial sensors: Multimodal interpretation of muscular activities. *IEEE Transactions on Information Technology in Biomedicine*, 14(2).

Ghasemzadeh, H., Loseu, V., Guenterberg, E., Jafari, R. (2011). Sport training using body sensor networks: A statistical approach to measure wrist rotation for golf swing. In *Fourth International Conference on Body Area Networks*. Germany.

Gonzalez-Valenzuela, S., Leung, V., Chen, M. (2021). *Body Area Networks (BAN)*. Springer, Berlin.

Harada, T., Uchino, H., Mori, T., Sato, T. (2003). Portable orientation estimation device based on accelerometers, magnetometers and gyroscope sensors for sensor network. In *IEEE International Conference on Multisensor Fusion and Integration for Intelligent Systems*. Japan.

Heinzelman, W.R., Chandrakasan, A., Balakrishnan, H. (2000). Energy-efficient communication protocol for wireless microsensor networks. In *Proceedings of the 33rd Annual Hawaii International Conference*. Hawaii.

Hou, J., Chang, B., Cho, D.-K., Gerla, M. (2009). Minimizing 802.11 interference on zigbee medical sensors. In *Fourth International Conference on Body Area Networks.*

Hoyt, R. (2008). SPARNET – Spartan data network for real-time physiological status monitoring. Report, US Army Telemedicine Partnership Series 2008: "Personal Health Monitoring", ADA490258.

IEEE Computer Society (2007). IEEE standard for a smart transducer interface for sensors and actuators – Common functions, communication protocols, and transducer electronic data sheet (teds) formats. IEEE Standards.

IEEE Computer Society (2012a). IEEE Standard for Local and metropolitan area networks – Part 15.6: Wireless Body Area Network. IEEE Standards.

IEEE Computer Society (2012b). IEEE Standard for Local and metropolitan area networks – Part 15.4: Low-Rate Wireless Personal Area Networks (LR-WPANs). Amendment 1: MAC sublayer. IEEE Standards.

Jovanov, E., Milenkovic, A., Otto, C., de Groen, P. (2003). A wireless body area network of intelligent motion sensors for computer assisted physical rehabilitation. *Journal of Neuro-Engineering and Rehabilitation*, 2(6).

Khan, N.P. and Boncelet, C. (2007). PMAC: Energy efficient medium access control protocol for wireless sensor networks. In *Proceedings of IEEE Military Communications Conference*. United States.

Latré, B., Braem, B., Moerman, I., Blondia, C., Demeester, P. (2011). A survey on wireless body area networks. *Wireless Networks*, 17(1).

Lee, J., Cho, S., Lee, J., Lee, K., Yang, H.-K. (2007). Wearable accelerometer system for measuring the temporal parameters of gait. In *29th Annual International Conference of the IEEE EMBS*. France.

Lu, G., Krishnamachari, B., Raghavendra, C. (2004). An adaptive energy efficient and low-latency MAC for data gathering in sensor networks. In *Proceedings of the 4th International Workshop: Algorithms Wireless, Mobile, Ad Hoc Sens. Netw.*

Martin, T., Jovanov, E., Raskovic, D. (2000). Issues in wearable computing for medical monitoring applications: A case study of a wearable ECG monitoring device. In *International Symposium on Wearable Computer.*

Miaoxin, L. and Mingjie, Z. (2012). An overview of physical layers on wireless body area network. In *International Conference on Anti-Counterfeiting, Security and Identification*. Taiwan.

Ng, K.-G., Wong, S.-T., Lim, S.-M., Goh, Z. (2010). Evaluation of the cadi thermo-sensor wireless skin-contact thermometer against ear and axillary temperatures in children. *Journal of Pediatric Nursing*, 25(3).

Otto, C. and Milenkovic, A. (2006). System architecture of a wireless body area sensor network for ubiquitous health monitoring. *Journal of Mobile Multimedia*, 1(4).

Paret, D. (2008). *RFID en ultra et super hautes fréquences UHF-SHF*. Dunod, Paris.

Paret, D. (2012). *NFC – Principes et application de la communication en champ proche*. Dunod, Paris.

Paret, D. (2016a). *Design Constraints for NFC Devices*. ISTE Ltd, London, and John Wiley & Sons, New York.

Paret, D. (2016b). *Antennas Designs for NFC Devices*. ISTE Ltd, London, and John Wiley & Sons, New York.

Paret, D. and Crégo, P. (2018). *Wearables, Smart Textiles & Smart Apparel*. ISTE Ltd, London, and John Wiley & Sons, New York.

Paret, D. and Huon, J.-P. (2017). *Secure Connected Objects*. ISTE Ltd, London, and John Wiley & Sons, New York.

Piotrowski, K., Sojka, A., Langendoerfer, P. (2010). Body area network for first responders – A case study. In *Fifth International Conference on Body Area Networks*. Greece.

Polastre, J., Hill, J., Culler, D. (2004). Versatile low power media access for wireless sensor networks. In *Proceedings of the 2nd ACM SenSys Conference*. United States.

Poon, C.C.Y. and Zhang, Y.T. (2005). Cuff-less and noninvasive measurements of arterial blood pressure by pulse transit time. In *27th Annual International Conference of the Engineering in Medicine and Biology Society*. China.

Prabh, K.S., Royo, F., Tennina, S., Olivares, T. (2012). BANMAC: An opportunistic MAC protocol for reliable communications in body area networks. In *Distributed Computing in Sensor Systems: 2012 IEEE 8th International Conference*. China.

Rajendran, V., Garcia-Luna-Aveces, J.J., Obraczka, K. (2005). Energy-efficient, application-aware medium access for sensor networks. In *Proceedings of IEEE Mobile Adhoc and Sensor Systems Conference*. United States.

Schurgers, C., Tsiatsis, V., Srivastava, M.B. (2004). STEM: Topology management for energy efficient sensor networks. In *Proceedings of Aerospace Conference*. United States.

Simon, P. and Acker, D. (2008). La place de la télémédecine dans l'organisation des soins. Report, Ministère de la Santé et des Sports, Direction de l'hospitalisation et de l'organisation des soins.

Smeaton, A., Diamond, D., Kelly, P., Moran, K., Lau, K., Morris, D., Moyna, N., O'Connor, N., Zhang, K. (2008). Aggregating multiple body sensors for analysis in sports. In *International Workshop on Wearable Micro and Nanosystems for Personalised Health-pHealth*. Norway.

Thales (2012). Body area networks in defence applications. In *7th International Conference on Body Area Networks*. Norway.

Ullah, S. and Latré, B. (2012). A comprehensive survey of wireless body area networks. *Journal of Medical Systems*, 36(3).

Ullah, S., Shen, B., Riazul Islam, S.M., Khan, P., Saleem, S., Sup Kwak, K. (2009). A study of MAC protocols for WBANs. *Sensors*, 10(1), 128–145.

Wang, C., Wang, Q., Shi, S. (2012). A distributed wireless body area network for medical supervision. In *IEEE International Instrumentation and Measurement Technology Conference 2012*. Austria.

Will, H., Hillebrandt, T., Kyas, M. (2012). Wireless sensor networks in emergency scenarios: The feuerwhere deployment. In *1st ACM International Workshop on Sensor-Enhanced Safety and Security in Public Spaces*. United States.

Williams, G., Doughty, K., Bradley, D.A. (1998). A systems approach to achieving CarerNet – An integrated and intelligent telecare system. *IEEE Transactions in Information Technology in Biomedecine*, 2(1).

Younis, O. and Fahmy, S. (2004). Heed: A hybrid, energy-efficient, distributed clustering approach for ad hoc sensor networks. *IEEE Trans. Mob. Comp.*

Zhang, Z., Jung, T.-P., Makeig, S., Rao, B.D. (2012). Compressed sensing of EEG for wireless telemonitoring with low energy consumption and inexpensive hardware. *IEEE Transactions on Biomedical Engineering*, 60(1).

Zhao, Y.J., Davidson, A., Bain, J., Li, S.Q., Wang, Q., Lin, Q. (2005). A mems viscometric glucose monitoring device. In *13th International Conference on Solid-State Sensors, Actuators and Microsystems*. South Korea.

Zigbee (2012). ZigBee wireless sensor applications for health, wellness and fitness. Report, Zigbee Alliance.

Zimmerman, T.G. (1996). Personal area networks: Near-field intrabody communication. *IBM Systems Journal*, 35(3), 609–617.

Zimmerman, T.G. (2021). Personal Area Networks (PAN): A technology demonstration. Report, IBM Research.

Index

E, F

G, H

I, L

M, N

O, P

R, S

T, V, W

Other titles from

in

Science, Society and New Technologies

2023

ELAMÉ Esoh
The Sustainable City in Africa Facing the Challenge of Liquid Sanitation
(Territory Development Set – Volume 2)

JURCZENKO Emmanuel
Climate Investing: New Strategies and Implementation Challenges

2022

AIT HADDOU Hassan, TOUBANOS Dimitri, VILLIEN Philippe
Ecological Transition in Education and Research

CARDON Alain
Information Organization of The Universe and Living Things: Generation of Space, Quantum and Molecular Elements, Coactive Generation of Living Organisms and Multiagent Model
(Digital Science Set – Volume 3)

CAULI Marie, FAVIER Laurence, JEANNAS Jean-Yves
Digital Dictionary

DAVERNE-BAILLY Carole, WITTORSKI Richard
Research Methodology in Education and Training: Postures, Practices and Forms
(Education Set – Volume 12)

ELAMÉ Esoh
Sustainable Intercultural Urbanism at the Service of the African City of Tomorrow
(Territory Development Set – Volume 1)

FLEURET Sébastien
A Back and Forth Between Tourism and Health: From Medical Tourism to Global Health
(Tourism and Mobility Systems Set – Volume 5)

KAMPELIS Nikos, KOLOKOTSA Denia
Smart Zero-energy Buildings and Communities for Smart Grids
(Engineering, Energy and Architecture Set – Volume 9)

2021

BARDIOT Clarisse
Performing Arts and Digital Humanities: From Traces to Data
(Traces Set – Volume 5)

BENSRHAIR Abdelaziz, BAPIN Thierry
From AI to Autonomous and Connected Vehicles: Advanced Driver-Assistance Systems (ADAS)
(Digital Science Set – Volume 2)

DOUAY Nicolas, MINJA Michael
Urban Planning for Transitions

GALINON-MÉLÉNEC Béatrice
The Trace Odyssey 1: A Journey Beyond Appearances
(Traces Set – Volume 4)

HENRY Antoine
Platform and Collective Intelligence: Digital Ecosystem of Organizations

LE LAY Stéphane, SAVIGNAC Emmanuelle, LÉNEL Pierre, FRANCES Jean
The Gamification of Society
(Research, Innovative Theories and Methods in SSH Set – Volume 2)

RADI Bouchaïb, EL HAMI Abdelkhalak
Optimizations and Programming: Linear, Non-linear, Dynamic, Stochastic and Applications with Matlab
(Digital Science Set – Volume 1)

2020

BARNOUIN Jacques
The World's Construction Mechanism: Trajectories, Imbalances and the Future of Societies
(Interdisciplinarity between Biological Sciences and Social Sciences Set – Volume 4)

ÇAĞLAR Nur, CURULLI Irene G., SIPAHIOĞLU Işıl Ruhi, MAVROMATIDIS Lazaros
Thresholds in Architectural Education (Engineering, Energy and Architecture Set – Volume 7)

DUBOIS Michel J.F.
Humans in the Making: In the Beginning was Technique
(Social Interdisciplinarity Set – Volume 4)

ETCHEVERRIA Olivier
The Restaurant, A Geographical Approach: From Invention to Gourmet Tourist Destinations
(Tourism and Mobility Systems Set – Volume 3)

GREFE GWENAËLLE, PEYRAT-GUILLARD DOMINIQUE
Shapes of Tourism Employment: HRM in the Worlds of Hotels and Air Transport (Tourism and Mobility Systems Set – Volume 4)

JEANNERET Yves
The Trace Factory
(Traces Set – Volume 3)

KATSAFADOS Petros, MAVROMATIDIS Elias, SPYROU Christos
Numerical Weather Prediction and Data Assimilation (Engineering, Energy and Architecture Set – Volume 6)

KOLOKOTSA Denia, KAMPELIS Nikos
Smart Buildings, Smart Communities and Demand Response (Engineering, Energy and Architecture Set – Volume 8)

MARTI Caroline
Cultural Mediations of Brands: Unadvertization and Quest for Authority (Communication Approaches to Commercial Mediation Set – Volume 1)

MAVROMATIDIS Lazaros E.
Climatic Heterotopias as Spaces of Inclusion: Sew Up the Urban Fabric (Research in Architectural Education Set – Volume 1)

MOURATIDOU Eleni
Re-presentation Policies of the Fashion Industry: Discourse, Apparatus and Power (Communication Approaches to Commercial Mediation Set – Volume 2)

SCHMITT Daniel, THÉBAULT Marine, BURCZYKOWSKI Ludovic
Image Beyond the Screen: Projection Mapping

VIOLIER Philippe, with the collaboration of TAUNAY Benjamin
The Tourist Places of the World
(Tourism and Mobility Systems Set – Volume 2)

2019

BRIANÇON Muriel
The Meaning of Otherness in Education: Stakes, Forms, Process, Thoughts and Transfers
(Education Set – Volume 3)

DESCHAMPS Jacqueline
Mediation: A Concept for Information and Communication Sciences
(Concepts to Conceive 21st Century Society Set – Volume 1)

DOUSSET Laurent, PARK Sejin, GUILLE-ESCURET Georges
Kinship, Ecology and History: Renewal of Conjunctures
(Interdisciplinarity between Biological Sciences and Social Sciences Set – Volume 3)

DUPONT Olivier
Power
(Concepts to Conceive 21st Century Society Set – Volume 2)

FERRARATO Coline
Prospective Philosophy of Software: A Simondonian Study

GUAAYBESS Tourya
The Media in Arab Countries: From Development Theories to Cooperation Policies

HAGÈGE Hélène
Education for Responsibility
(Education Set – Volume 4)

LARDELLIER Pascal
The Ritual Institution of Society
(Traces Set – Volume 2)

LARROCHE Valérie
The Dispositif
(Concepts to Conceive 21st Century Society Set – Volume 3)

LATERRASSE Jean
Transport and Town Planning: The City in Search of Sustainable Development

LENOIR Virgil Cristian
Ethically Structured Processes
(Innovation and Responsibility Set – Volume 4)

LOPEZ Fanny, PELLEGRINO Margot, COUTARD Olivier
Local Energy Autonomy: Spaces, Scales, Politics
(Urban Engineering Set – Volume 1)

METZGER Jean-Paul
Discourse: A Concept for Information and Communication Sciences
(Concepts to Conceive 21st Century Society Set – Volume 4)

MICHA Irini, VAIOU Dina
Alternative Takes to the City
(Engineering, Energy and Architecture Set – Volume 5)

PÉLISSIER Chrysta
Learner Support in Online Learning Environments

PIETTE Albert
Theoretical Anthropology or How to Observe a Human Being
(Research, Innovative Theories and Methods in SSH Set – Volume 1)

PIRIOU Jérôme
The Tourist Region: A Co-Construction of Tourism Stakeholders
(Tourism and Mobility Systems Set – Volume 1)

PUMAIN Denise
Geographical Modeling: Cities and Territories
(Modeling Methodologies in Social Sciences Set – Volume 2)

WALDECK Roger
Methods and Interdisciplinarity
(Modeling Methodologies in Social Sciences Set – Volume 1)

2018

BARTHES Angela, CHAMPOLLION Pierre, ALPE Yves
Evolutions of the Complex Relationship Between Education and Territories
(Education Set – Volume 1)

BÉRANGER Jérôme
The Algorithmic Code of Ethics: Ethics at the Bedside of the Digital Revolution
(Technological Prospects and Social Applications Set – Volume 2)

DUGUÉ Bernard
Time, Emergences and Communications
(Engineering, Energy and Architecture Set – Volume 4)

GEORGANTOPOULOU Christina G., GEORGANTOPOULOS George A.
Fluid Mechanics in Channel, Pipe and Aerodynamic Design Geometries 1
(Engineering, Energy and Architecture Set – Volume 2)

GEORGANTOPOULOU Christina G., GEORGANTOPOULOS George A.
Fluid Mechanics in Channel, Pipe and Aerodynamic Design Geometries 2
(Engineering, Energy and Architecture Set – Volume 3)

GUILLE-ESCURET Georges
Social Structures and Natural Systems: Is a Scientific Assemblage Workable?
(Social Interdisciplinarity Set – Volume 2)

LARINI Michel, BARTHES Angela
Quantitative and Statistical Data in Education: From Data Collection to Data Processing
(Education Set – Volume 2)

LELEU-MERVIEL Sylvie
Informational Tracking
(Traces Set – Volume 1)

SALGUES Bruno
Society 5.0: Industry of the Future, Technologies, Methods and Tools
(Technological Prospects and Social Applications Set – Volume 1)

TRESTINI Marc
Modeling of Next Generation Digital Learning Environments: Complex Systems Theory

2017

ANICHINI Giulia, CARRARO Flavia, GESLIN Philippe,
GUILLE-ESCURET Georges
Technicity vs Scientificity – Complementarities and Rivalries
(Interdisciplinarity between Biological Sciences and Social Sciences Set – Volume 2)

DUGUÉ Bernard
Information and the World Stage – From Philosophy to Science, the World of Forms and Communications
(Engineering, Energy and Architecture Set – Volume 1)

GESLIN Philippe
Inside Anthropotechnology – User and Culture Centered Experience
(Social Interdisciplinarity Set – Volume 1)

GORIA Stéphane
Methods and Tools for Creative Competitive Intelligence

KEMBELLEC Gérald, BROUDOUS EVELYNE
Reading and Writing Knowledge in Scientific Communities: Digital Humanities and Knowledge Construction

MAESSCHALCK Marc
Reflexive Governance for Research and Innovative Knowledge
(Responsible Research and Innovation Set - Volume 6)

PARK Sejin, GUILLE-ESCURET Georges
Sociobiology vs Socioecology: Consequences of an Unraveling Debate
(Interdisciplinarity between Biological Sciences and Social Sciences Set – Volume 1)

PELLÉ Sophie
Business, Innovation and Responsibility
(Responsible Research and Innovation Set – Volume 7)

2016

BRONNER Gérald
Belief and Misbelief Asymmetry on the Internet

EL FALLAH SEGHROUCHNI Amal, ISHIKAWA Fuyuki, HÉRAULT Laurent, TOKUDA Hideyuki
Enablers for Smart Cities

GIANNI Robert
Responsibility and Freedom
(Responsible Research and Innovation Set – Volume 2)

GRUNWALD Armin
The Hermeneutic Side of Responsible Research and Innovation
(Responsible Research and Innovation Set – Volume 5)

LAGRAÑA Fernando
E-mail and Behavioral Changes: Uses and Misuses of Electronic Communications

LENOIR Virgil Cristian
Ethical Efficiency: Responsibility and Contingency
(Responsible Research and Innovation Set – Volume 1)

MAESSCHALCK Marc
Reflexive Governance for Research and Innovative Knowledge
(Responsible Research and Innovation Set – Volume 6)

PELLÉ Sophie, REBER Bernard
From Ethical Review to Responsible Research and Innovation
(Responsible Research and Innovation Set – Volume 3)

REBER Bernard
Precautionary Principle, Pluralism and Deliberation: Sciences and Ethics
(Responsible Research and Innovation Set – Volume 4)

VENTRE Daniel
Information Warfare – 2nd edition

Printed and bound by CPI Group (UK) Ltd, Croydon, CR0 4YY

12/09/2023

08113668-0001